Springer Proceedings in Physics 47

Springer Proceedings in Physics

Managing Editor: H. K. V. Lotsch

Volumes 1 – 29 are listed on the back inside cover

Number Theory and Physics

Proceedings of the Winter School, Les Houches, France, March 7–16, 1989

Editors: J.-M. Luck, P. Moussa, and M. Waldschmidt

With 45 Figures

Springer-Verlag Berlin Heidelberg New York
London Paris Tokyo Hong Kong

Dr. Jean-Marc Luck
Dr. Pierre Moussa
Service de Physique Théorique, C.E.N. Saclay, F-91191 Gif-sur-Yvette Cedex, France

Professor Dr. Michel Waldschmidt
Université de Paris 6, Institut Henri-Poincaré, Problèmes Diophantiens,
F-75231 Paris Cedex 05, France

ISBN-13:987-3-642-75407-4 e-ISBN-13:978-3-642-75405-0
DOI: 10.1007/978-3-642-75405-0

Softcover reprint of the hardcover 1st edition 1990

2154/3150-543210 – Printed on acid-free paper

Preface

Les Houches

Number theory, or arithmetic, sometimes referred to as the queen of mathematics, is often considered as the purest branch of mathematics. It also has the false reputation of being without any application to other areas of knowledge. Nevertheless, throughout their history, physical and natural sciences have experienced numerous unexpected relationships to number theory. The book entitled *Number Theory in Science and Communication*, by M.R. Schroeder (Springer Series in Information Sciences, Vol. 7, 1984) provides plenty of examples of cross-fertilization between number theory and a large variety of scientific topics.

The most recent developments of theoretical physics have involved more and more questions related to number theory, and in an increasingly direct way. This new trend is especially visible in two broad families of physical problems. The first class, dynamical systems and quasiperiodicity, includes classical and quantum chaos, the stability of orbits in dynamical systems, K.A.M. theory, and problems with "small denominators", as well as the study of incommensurate structures, aperiodic tilings, and quasicrystals. The second class, which includes the string theory of fundamental interactions, completely integrable models, and conformally invariant two-dimensional field theories, seems to involve modular forms and p-adic numbers in a remarkable way.

Several theoretical physicists, having experienced this evolution, have found it worthwhile to bring their individual and occasional relations with mathematicians working in number theory into a more official and profitable framework. This idea has met with a very positive echo from the mathematicians' side. Thus a first version of the scientific committee given below gathered in an informal way in order to organise an interdisciplinary meeting on "Number Theory and Physics". Our aim was twofold. Physicists were given, mostly through the programme of lectures, the opportunity to become acquainted with the basic methods and results of number theory that are the most used in their branch. Mathematicians discovered in most of the seminars novel domains of the physical sciences where methods and results related to arithmetic have been useful in recent years.

The meeting was held at the Centre de Physique, Les Houches (Haute-Savoie, France), March 7 – 16, 1989. Most of the 58 participants, both mathematicians and physicists, attended the totality of the scientific programme, in spite of the charm of the mountainous setting, the fine weather and the neighbouring ski slopes.

Although the timetable was already rather dense, with 32 lectures and 33 seminars, many informal, but passionate, discussions and meetings often lasted late into the night.

The Scientific Committee that organised *Number Theory and Physics* was composed of eight French scientists: J. Bellissard (Theoretical Physics, Marseille), C. Godrèche (Solid State Physics, CEN Saclay), C. Itzykson (Theoretical Physics, CEN Saclay), J.-M. Luck (Theoretical Physics, CEN Saclay), M. Mendès France (Mathematics, Bordeaux), P. Moussa (Theoretical Physics, CEN Saclay), E. Reyssat (Mathematics, Caen), and M. Waldschmidt (Mathematics, IHP, Paris). The organisers were assisted by an International Advisory Committee, composed of M. Berry (Physics, Bristol, UK), P. Cvitanović (Physics, Copenhagen, Denmark), M. Dekking (Mathematics, Delft, The Netherlands), and G. Turchetti (Physics, Bologna, Italy).

The following institutions are most gratefully acknowledged for their generous specific financial support of the meeting: the Département Mathématiques et Physique de Base of the Centre National de la Recherche Scientifique; the Institut de Recherche Fondamentale of the Commissariat à l'Energie Atomique; the Direction des Recherches, Etudes et Techniques de la Délégation Générale pour l'Armement (under contract no. 88/1474); the French Ministère de l'Education Nationale; the French Ministère des Affaires Etrangères; and the Commission of the European Communities. The regretted absence of support from NATO finally turned out to allow a more flexible organisation.

The variety and abundance of the material presented during lectures and seminars led to the decision to publish two separate volumes of proceedings. This first volume is devoted to the proceedings of the seminars. These talks, limited on purpose to 40 minutes, offered to most participants, both mathematicians and physicists, the possibility of illustrating, through their own research activity, the variety of specific physical contexts which involve number-theoretic questions in a natural though sometimes unexpected way. The 32 contributions to this volume have been grouped into five parts for the sake of clarity: (I) conformally invariant field theories, integrability, quantum groups; (II) quasicrystals and related geometrical structures; (III) spectral problems, automata, and substitutions; (IV) dynamical and stochastic systems; (V) further arithmetical problems, and their relationship to physics.

An extensive programme of 32 lectures gave to all participants a rather complete view of the different areas of number theory that, according to us, play a role in physics. The first intensive set of 24 lectures made physicists more familiar with basic concepts and methods in the branches of number theory that have recently turned out to be useful in physics. The contributors were F. Beukers, J.B. Bost, P. Cartier, G. Christol, H. Cohen, R. Gergondey, E. Reyssat, H.M. Stark, and D. Zagier. The subjects covered included p-adic numbers, Galois theory and number fields, elliptic functions and modular forms. A second series of lectures displayed some theoretical physics problems where number theory plays a prominent role, namely the quasiperiodicity theory in dynamical systems and quasicrystals. The contributors were J. Bellissard, P. Cvitanović, M. Duneau, A. Katz, M. Senechal and J.C. Yoccoz. We plan to publish a written version of these lectures in a separate volume.

Although "Number Theory and Physics" may have seemed to be both too ambitious and too exotic a theme for a 10-day workshop, this meeting has been quite fruitful, according to the participants, in spite of its brevity, and allows one to expect an even richer variety of subjects of future collaborations between mathematicians and physicists, as well as other meetings around similar themes.

Saclay and Paris
June 1989

J.-M. Luck
P. Moussa
M. Waldschmidt

Word of Appreciation

On behalf of those who have attended this Workshop I wish to thank Jean-Marc Luck and the other organisers of this meeting. It is always a great pleasure for me to revisit Les Houches. I greatly enjoy the intense concentration on the subject which is characteristic of meetings here, combined with the opportunity to enjoy the beautiful surroundings. I have found this to be an enormously stimulating meeting. In this group of mathematicians and physicists from different disciplines we have had to face the challenge of understanding other languages and unfamiliar sets of priorities, and we have had to translate our own enthusiasms into language that people from other disciplines can understand. This has all been made easier by the smooth running of the Workshop by the organisers.

David Thouless

Contents

The talks by Professor M. Berry and Professor P. Freund are not reported in this volume because most of the material presented has already appeared in the following:

M.V. Berry, J. Goldberg: Renormalisation of curlicues. Nonlinearity *1*, 1–26 (1988)

M.V. Berry: Semiclassical formula for the number variance of the Riemann zeros. Nonlinearity *1*, 399–407 (1988)

P.G.O. Freund: "The Arithmetic of Strings", in *New Theories in Physics*, ed. by Z. Ajduk, S. Pokorski, A. Trautman (World Scientific, Singapore 1989) pp. 369–382

Part I

Conformally Invariant Field Theories, Integrability, Quantum Groups

Z/NZ Conformal Field Theories

P. Degiovanni

LPTENS, 24 rue Lhomond, F-75005 Paris, France
Unité propre de Recherche du Centre National de la Recherche Scientifique, associée à l'Ecole Normale Supérieure et à l'Université de Paris-Sud

Abstract : In this paper, we give a brief introduction to Conformal Field Theory (CFT) following the presentation of G. Segal. We explain how to reconstruct part of a CFT from its fusion rules. The possible choices of S matrices are indexed by some automorphisms of the fusion algebra. We illustrate this procedure by computing the modular properties of the possible genus-one characters when the fusion algebra is the representation algebra of a finite group. We also classify the modular invariant partition functions of these theories. We recover as special cases the $A_N^{(1)}$ WZW theories and the rational gaussian model.

Introduction

It is now clear that the last few years have witnessed a dramatic convergence of interests and methods between mathematics and theoretical physics. String theory and more recently, Conformal Field Theory perfectly illustrate this evolution.

CFT is believed to be of basic interest both in particle physics and statistical mechanics. According to Friedan [1], the classical states of string theory (i.e. the propagation of a string in a fixed background space-time) are in one-to-one correspondence with Conformal Field Theories in two dimensions; and any statistical system with a second order phase transition is scale, translationally and rotationally invariant at its critical point which is therefore described by a Conformal Field Theory.

From a mathematical point of view, the importance of CFT in two dimensions arises from the enormous symmetry of these theories. It is so huge that the hope of a possible classification has emerged. The main idea is to classify the states of a CFT using representations of the symmetry algebra of the theory. This has pointed out a connection between CFT and the representation theory of infinite dimensional algebras among which are the Virasoro algebra and the Kac-Moody algebras and their supersymmetric generalisations. Different formulations of a CFT led to various identities between partition functions and connected this field to number theory [1]. More recent works concerning the study of CFT's on arbitrary Riemann surfaces relate CFT to the study of Riemann surfaces and modular geometry [2][3][4][5]. As was shown by J. Birman [6], the modular group of an arbitrary Riemann surface is related to a braid group and CFT's naturally provide us with representations of the braid group, quantum groups and integrable models and, following E.Witten [7], of knot theory and topological field theory.

[1] The reader will remember the celebrated Dyson-Mac Donald identities

Springer Proceedings in Physics, Vol. 47 Number Theory and Physics
Editors: J.M. Luck · P. Moussa · M. Waldschmidt

This paper is an attempt to give a very brief introduction to Conformal Field Theory for mathematicians. We do not pretend to be either exhaustive or rigorous but rather to indicate the ideas and methods used in this field. In section 1, we review some basic aspects of conformal field theory. We follow the presentation of Segal [8]: a CFT associates a state of a certain Hilbert space with each Riemann surface with boundaries. We describe how this space arises from the representation theory of the chiral algebra. The case of Rational Conformal Field Theories (RCFT), where only a finite number of representations appears, provides the simplest situation. In this case, it has been possible to study modular covariance and duality in full generality [4]. We give a brief summary of these results in section 1.2. A possible method of classification of RCFT's has emerged from this study. Sections 2 and 3 are devoted to a very elementary illustration of these ideas.

Here, we start from the fusion rules of the theory and try to reconstruct as much as possible of it from these data. The unexpected relationship between the fusion rules of a RCFT and the modular properties of genus-one characters $tr(q^{L_0})$ discovered by Verlinde [9] will be our basic tool and is recalled in section 2.1. We shall use it in section 2.2 to compute the dimensions and central charge of the theory in some particular examples. Finally, section 3 contains the classification of the genus-one partition functions of these theories. This work has been motivated by our attempt to get familiar with Verlinde's work and by a recent work of C. Itzykson on $A_N^{(1)}$ level 1 theories [10]. He formulated a conjecture on the possible modular invariants which we prove in section 3: the modular invariant partition functions are indexed by divisors of N when N is odd and divisors of $N/2$ when N is even.

1 Some concepts in Conformal Field Theory

This section is devoted to a brief introduction to Conformal Field Theory for non specialists. Our presentation is inspired by Segal [8] and relies on the operator formalism [12] [11]. The interested reader can find a very clear presentation in [13]. A more practical point of view is exposed in [14] and [15]. We also describe recent results obtained by Moore and Seiberg in the case of rational conformal field theories.

1.1 Conformal Field Theory

We shall firstly recall what is generally meant by Quantum Field Theory (QFT). We start with a manifold (with a certain structure on it: metric,...) M with boundary ∂M. On this manifold exist some fields which we shall generically denote by φ; these are functions on M with values in a space X (target space). We assume that we are given a local functional $S[\varphi]$ of the fields called the action. Locality means that if we glue two different manifolds M_1 and M_2 together by their common boundaries to get M, we have $S_M[\varphi] = S_{M_1}[\varphi_1] + S_{M_2}[\varphi_2]$. The object we are interested in, is the functional

$$Z[M,f] = \int_{\varphi=f \ on \ \partial M} D[\varphi] \exp\left(-S_M[\varphi]\right).$$

A very well-known example is provided by quantum mechanics where $M = [0,T]$ and $X = \mathbf{R}^3$ (the ordinary space), S_M being the classical action for a particle moving in this space (with or without potential). Then, the functional integral defined above is the probability amplitude for the particle to go from one point to another point in time T. The set of boundary conditions of the fields on ∂M defines the Hilbert space $H_{\partial M}$ of states associated with the manifold ∂M. We are thus able to associate with M a state in $H_{\partial M}$: $|M>$ such that $Z[M,f] =< f|M>$ (scalar product) where $|f>$ is the state associated with the boundary condition f. Note that $Z[M,f]$ can be considered as the wave function associated with $|M>$. This assignment of a state to a (possibly decorated) manifold really defines the field theory.

We are now ready to define a Conformal Field Theory. The precise definition has been given by Segal [8] and here, we shall only outline his definition. A conformal field theory is defined by the assignment of a state to any compact connected Riemann surface M possibly with boundary. To be precise, the state is associated with the surface together with a metric but it is supposed to depend only on the conformal class of the metric [2]. The boundary of such a Riemann surface is a disjoint union of n circles and Segal assumes that $H_{\partial M} = H^{\otimes n}$ where H is the Hilbert space associated with a circle. In the case of a surface with no boundary, the Hilbert space is simply $\mathbf{C}$.

Then, we need "sewing axioms" which express the locality of the theory. Let us consider two Riemann surfaces with boundaries and the surface obtained by gluing the two surfaces together along one of the boundary circles. Let $|M_1>\in H^{\otimes n_1}$ and $|M_2>\in H^{\otimes n_2}$ be the states associated with the two surfaces. We suppose for clarity that the circles we consider are associated with the first spaces. Let ϕ be the canonical antilinear bijection from H onto its topological dual deduced from the Hilbert space structure. Then, we have

$$|M>= ((\phi \otimes \mathbf{1} \otimes \ldots \otimes \mathbf{1})|M_1>).|M_2>$$

where the linear form acts on the first Hilbert space H in the tensor product $H^{\otimes n_2}$. The state we find clearly belongs to $H^{\otimes(n_1+n_2-2)}$. From a surface with at least two boundary circles, we can construct a new surface by gluing along two circles. The associated state is obtained by the same kind of contraction process. These sewing axioms enable us to reconstruct the state associated with any Riemann surface from the state associated with the three holed sphere: we only have to partition the surface into three holed spheres. The state we obtain through this process does not depend on the particular partition of the sphere. By contraction of the state associated with a Riemann surface with suitable states, we can compute physical correlation functions. Inserting a given field on the surface is equivalent to cutting a little disk around the insertion point and putting appropriate boundary conditions on the boundary circle that we have created. Hence, any local field corresponds to an operator acting on H.

In the case of a two-dimensional conformal field theory, it is possible to show that the Hilbert space H is a representation (in general reducible) of two copies of the Virasoro algebra Vir. This algebra is infinite dimensional and is generated by the L_n's for $n \in \mathbf{Z}$ and c. The commutation relations are

$$\begin{aligned} [L_n, L_m] &= (n-m)L_{n+m} + \frac{c}{12}n(n^2-1)\delta_{n,-m} \\ [L_n, c] &= 0 \end{aligned}$$

[2] Up to a normalisation factor: a Weyl transformation of the metric multiplies the state by a number which can be computed [1] [11]. This phenomenon is called the conformal anomaly.

The L_n's are the Laurent coefficients of the analytic component of the energy momentum tensor $T(z)$ (the other copy arises from the antianalytic component $\bar{T}(\bar{z})$). The eigenvalues of the L_0 and $\bar{L}_0$ operator are called the conformal dimensions of the state. The representations we consider are highest weight representations because $L_0 + \bar{L}_0$ represents the energy of the system and we want it to be bounded from below. The reflexion positivity property of statistical mechanics is expressed through unitarity of the theory: H is a Hilbert space and $L_n^\dagger = L_{-n}$. We shall restrict ourselves to that case. This implies that the conformal dimensions of the states are all positive numbers. We impose the unicity of the vacuum: there exists a unique state of conformal dimensions $(0,0)$ called the vacuum and noted $|0>$. In fact, as explained in [16], with each state $|\varphi>$ is associated a field $\varphi(z)$ such that $\varphi(z)|0>=|\varphi>$. The state $|0>$ is therefore related to the identity operator.

1.2 Rational Conformal Field Theories

At present time, there is no way to study a general conformal field theory: the Hilbert space H is in some sense too complicated (infinite sum of irreducible representations of $Vir \otimes \overline{Vir}$). Of course, the greater the symmetry of the theory, the easier its study is. In some cases, there exists an extension A of the Virasoro algebra such that H is a **finite** sum of irreducible representations of $A \otimes \bar{A}$. This is the simplest situation and we call it a rational conformal field theory (RCFT). We shall write the following decomposition:

$$H = \bigoplus_{i,\bar{i}} (V_i \otimes V_{\bar{i}}) \tag{1}$$

where V_i and $V_{\bar{i}}$ are irreducible representations of A. Just as in the case of the Virasoro algebra, A is generated by the Laurent coefficients of holomorphic fields (among which $T(z)$). All the fields in A have integer dimensions (i.e. they are holomorphic fields on $\mathbf{C}^*$) [3]. The states of H arise from the one punctured sphere and the action of a Laurent coefficient of a holomorphic field can be seen as the action of a contour integral of an operator valued differential form. In some sense, we have reduced ourselves to a finite dimensional problem. To illustrate this point, we shall explain how to define the action of the extended algebra in any genus and number of punctures and then, we shall sum up the work of Moore and Seiberg which expresses duality and modular covariance through a finite number of polynomial equations satisfied by finite dimensional matrices.

The action of A generalizes to more complicated Riemann surfaces. We consider a Riemann surface M with n points P_i and a coordinate z_i defined on a disk D_i containing the point P_i. We shall call it a n-punctured surface in the following. Cutting the surface along circles of constant $|z_i|$'s is equivalent to consider a Riemann surface with n boundary circles. Let $Q(z)$ be a holomorphic field of conformal dimension $(h,0)$ and $v_i(z_i)(dz_i)^{1-h}$ a section of the $(1-h)$-th power of $T^{(1,0)}M$ over D_i. Then, we define

$$\rho(Q, M, (v_i)_i) = \sum_i \int_{\partial D_i} Q(z) v_i(z) \frac{dz}{2\pi i} \tag{2}$$

where the contour integral on C_i acts on the Hilbert space associated with ∂D_i. Thus,

[3] And $\bar{A}$ is associated with the antiholomorphic fields.

$\rho(Q, M, (v_i)_i)$ acts on $H^{\otimes n}$ [4]. In the case of a one-punctured sphere, we select a particular Laurent coefficient of $Q(z)$ in this way. In the case of a three holed sphere, one recovers the action of A on $H \otimes H$ defined by Moore and Seiberg in [4]. In the case of the energy momentum tensor, the action of the Virasoro algebra has a geometrical meaning: it represents the change of moduli of the Riemann surface with its punctures and coordinates [12].

Given an irreducible representation V_i of A, we can associate its conjugate representation $V_{\bar{i}}$ such that the chiral field associated with the highest weight state of $V_{\bar{i}}$ has a non-zero two-point function with the chiral field associated with the highest weight state of V_i. Moreover, if we define the genus-one character of V_i by

$$\chi_i(\tau) = tr_{V_i}(q^{L_0 - \frac{c}{24}}), \quad q = \exp(2\pi i \tau), \tag{3}$$

a representation and its conjugate have the same character. If V_0 denotes the representation with highest weight state $|0>$, $V_0 = V_{\bar{0}}$. The matrix $C_i{}^j = \delta_i{}^{\bar{j}}$ is called the charge conjugation matrix.

Conformal blocks.

Let Σ be the Riemann surface with n punctures and coordinates which we consider; we can write the associated vector as

$$|\Sigma> = \sum_{\substack{i_1,\ldots,i_n \\ \bar{i}_1,\ldots,\bar{i}_n};a} |\Sigma^a_{i_1,\ldots,i_n}> \otimes |\Sigma^a_{\bar{i}_1,\ldots,\bar{i}_n}>$$

where

$$|\Sigma^a_{i_1,\ldots,i_n}> \in \bigotimes_{j=1,\ldots,n} V_{i_j}$$

and the same for the antiholomorphic part. These $|\Sigma^a_{i_1,\ldots,i_n}>$ are called the conformal blocks on the Riemann surface Σ. If we contract the vector $|\Sigma>$ with an element of $H^{\otimes n}$, we compute the correlation function of the corresponding fields on Σ. Therefore, this correlation function is of the form

$$< \varphi_1, \ldots, \varphi_n >_\Sigma = \sum_a F^{(a)}(m) \bar{F}^{(a)}(\bar{m})$$

where $m, \bar{m}$ are the moduli of Σ with its punctures. For example, in the case of the four-point function on the sphere, these are the positions of the points on the sphere and the functions $F^{(a)}$ are the conformal blocks of [16].

The action of the chiral algebra A on such a state $|\Sigma^a_{i_1,\ldots,i_n}>$ is defined through contour integration of operator valued differential forms. We are thus led to certain identities, called the Ward identities which express that if one can shrink the contour to a point without encountering any singularity, the action of the operator on the state we consider should be zero. The action of the L_{-1} operator should also be associated with translations. These conditions restrict the $(|\Sigma^a_{i_1,\ldots,i_n}>)_a$ vectors to belonging to a finite dimensional vector space. In the case of the three-punctured sphere with representations i, j, k, its dimension is noted $N_{i,j,k}$. The $N_{i,j,k}$'s are called the fusion

[4] Due to the conformal anomaly, there is a subtlety in the case of the energy momentum tensor of the theory: $T(z)v_i(z)dz$ is not a one-form.

rules. Physically, they represent the number of independent couplings between the representations i, j, k which are consistent with the action of the chiral algebra. As any punctured Riemann surface can be obtained by gluing three spheres together, the fusion rules are very important data of a RCFT.

Duality and modular covariance.

If we want to find conformal blocks associated with a Riemann surface, we have to choose a decomposition of the surface into three holed spheres sewed together. This determines a ϕ^3 diagram with fixed representations on the external legs. Then, by choosing some representations in the intermediate channels and one particular coupling for each three holed sphere, we determine a conformal block. For a given ϕ^3 diagram, this gives us a basis of the space of conformal blocks associated with the surface we consider (axiom of duality). Changing the topology of the ϕ^3 diagram or gluing the different three spheres together with Dehn twists in the intermediate channels or around the external points provides us with another basis of conformal blocks. Changing the topology of the graph can be realized using braiding or fusing of the fields (see fig 1). The Dehn twists around the external points are just represented by phase factors and the Dehn twists in the intermediate channels can be computed knowing how Dehn twists are represented on the torus. The different bases of conformal blocks that we have introduced at fixed genus and number of punctures can therefore be related by matrices called the duality matrices which can be easily computed from the dimensions of the fields, the central charge of the theory, the braiding and fusing matrices and the Dehn twist matrices in genus-one. Dehn twists around non trivial loops generate the modular group of the Riemann surface with punctures and coordinates at the punctures [6]. These generators verify certain relations and this gives some equations for the different matrices we use (braiding matrices, Dehn twists matrices in genus-one,...). Moreover, starting from a given ϕ^3 graph and performing a particular sequence of braidings and fusings can give back the same ϕ^3 diagram. This gives also rise to particular identities satisfied by the braiding and fusing matrices. Famous examples of these consistency conditions are provided by the Yang-Baxter equation and the pentagonal identity of Moore and Seiberg [4].

In fact, it is possible to prove that a finite number of polynomial equations imply that all consistency conditions arising from the modular groups defining relations and closed cycles in the set of ϕ^3 diagrams with labelled external legs are verified. Henceforth, these equations ensure duality and modular covariance in any genus and with any number of punctures. This completeness theorem has been proved by Moore and Seiberg [4]. As we said before, we are reduced to a finite dimensional problem.

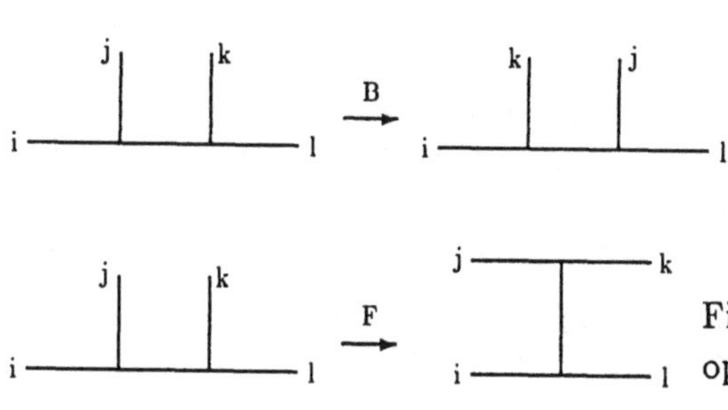

Figure 1: The braiding (B) and fusing (F) operations acting on ϕ^3 diagrams.

Example: genus-one characters.

Here, we shall give a very simple illustration of a representation of the modular group in genus-one which will be useful in the next sections of this paper. We have previously defined the genus-one characters as traces (3). In fact these characters are the conformal blocks in genus-one with zero punctures. In subsection 3.1, we shall see that the partition function of the theory on the torus is a sesquilinear combination of these characters.

As we have seen before, all fields in A have integer dimensions and therefore, from (3), we see that

$$\chi_i(\tau+1) = \exp\left(2\pi i\left(h_i - \frac{c}{24}\right)\right)\chi_i(\tau) = \sum_j T_i^{\ j}\chi_j(\tau) \tag{4}$$

where c is the central charge and h_i is the conformal dimension of the highest weight state of V_i. If τ is the modular parameter of the torus, $\tau \longrightarrow \tau+1$ represents one of the generator of the modular group in genus-one. This group is known to be isomorphic to $SL_2(\mathbf{Z})$. It is generated by s and t defined as

$$s = \begin{pmatrix} 0 & -1 \\ 1 & 0 \end{pmatrix}, \quad t = \begin{pmatrix} 1 & 1 \\ 0 & 1 \end{pmatrix}, \quad s^2 = -\mathbf{1}, \quad (st)^3 = \mathbf{1}.$$

The action of s is represented on the characters by the S-matrix

$$\chi_i(-1/\tau) = \sum_j S_i^j \chi_j(\tau) \tag{5}$$

and t is represented by T introduced just above. These matrices represent the modular group and verify

$$S^2 = C, \quad (ST)^3 = \mathbf{1} \tag{6}$$

where C is the charge conjugation matrix. The relation $S^2 = C$ can be understood by saying that geometrically, the s^2 operation reverses the orientation of the homology cycles that we have chosen on the torus. Hence, the charges turn into their conjugates [5]. Equations (6) express that we have a representation of the modular group of the torus on the genus-one, zero puncture conformal blocks space. This illustrates our previous discussion.

1.3 The reconstruction philosophy.

As explained in the introduction, our goal is to understand the classification of rational conformal field theories. A general classification picture has emerged from Moore and Seiberg's completeness theorem. One can try to classify RCFT's by solving the polynomial equations which will give us the dimensions, the S matrix, the monodromy matrices of the primary field of our theory and the central charge. But, we are not sure that there really exists a RCFT corresponding to these data. We have to solve the reconstruction problem. It consists in finding an extended chiral algebra such that the modular and monodromy properties of the characters and the primary fields are

[5] This is the reason why $SL_2(\mathbf{Z})$ appears and not $PSL_2(\mathbf{Z})$: the torus is indeed equipped with oriented cycles (see Cartier's lectures in the other volume of this school.)

precisely those we have found before. At the present time no complete answer to that problem exists.

In this paper, a similar philosophy will be followed. Namely, we shall start from a set of fusion rules. Then, using Verlinde's theorem which relates these fusion rules to the modular properties of the genus-one characters, we shall compute the S matrix in some examples. Using the defining relations of the modular group (6), it is possible to find the dimensions of the fields mod 1 and the central charge of the theory mod 8. The examples we consider are related to finite group theory. This step being completed, one can also classify modular invariant partition functions of these theories on the torus. In some cases, an explicit conformal field theory realizes these data.

2 Using Verlinde's result.

In this section, we shall briefly recall Verlinde's theorem and give a few general results useful for the reconstruction program described in section 1. Then, we will apply this to a simple case where the fusion algebra is associated with a finite group.

2.1 Verlinde's theorem and general results

E. Verlinde related the fusion rules to the modular properties of the genus-one characters. Before stating his result, we give a few properties of the $N_{i,j,k}$'s.

First of all, the projective invariance of the theory on the sphere is expressed by the following symmetry of the fusion rules:

$$N_{i,j,k} \text{ is symmetric in } i,j \text{ and } k$$

We also know some particular values of the fusion rules:

$$N_{0,i}{}^{j} = \delta_i^j$$

where $N_{i,j}{}^{k} = N_{i,j,\hat{k}}$. It is also clear that $N_{i,j,k} = N_{\hat{i},\hat{j},\hat{k}}$. We are now ready to recall Verlinde's result.

Theorem 1 (E. Verlinde) *The $N_{i,j}{}^{k}$'s are the structure constants of a commutative and associative algebra (called the fusion algebra) and S diagonalizes all the $\hat{N}_i$'s with $\hat{N}_i = (N_{i,j}{}^{k})$:*

$$\hat{N}_i = S\hat{\Lambda}_i S^{-1}$$

where $(\hat{\Lambda}_i)_{j,k} = \delta_j^k \lambda_i{}^{(j)}$ and $\lambda_i^{(j)} = S_i{}^{j}/S_0^j$.

Now, we would like to examine the reconstruction problem. Given a set of fusion rules, we can try to find all possible S matrices such that Verlinde's formula hold:

$$N_{i,j}{}^{k} = \sum_n \frac{S_i{}^{n} S_j{}^{n} S_n{}^{k*}}{S_0{}^{n}}. \qquad (7)$$

We know that the S matrix has to be a symmetric, unitary matrix such that $SC = CS = S^*$ and $S_0{}^{i} > S_0{}^{0} > 0$ for all i's. These properties follow from Verlinde's theorem and $S^2 = C$. A matrix verifying these properties and Verlinde's formula is called a possible S-matrix. In fact, we can very easily find all possible S-matrices from a given one [17].

Theorem 2 *Let S be a possible S-matrix for a given fusion algebra, then any other possible S-matrix is of the form*

$$S_i'^{j} = S_i{}^{\sigma(j)}$$

where σ defines an automorphism of the fusion algebra and verifies

$$S_{\sigma(i)}{}^{j} = S_i{}^{\sigma(j)}.$$

All such automorphisms define all possible S-matrices.

where we define:

Definition 1 *A permutation of the fields is called an automorphism of the fusion algebra if and only if it verifies*

$$N_{\sigma(i),\sigma(j),\sigma(k)} = N_{i,j,k}, \quad \sigma(0) = 0, \quad \sigma(\hat{\imath}) = \widehat{\sigma(i)}$$

Clearly, the condition $S_{\sigma(i)}{}^{j} = S_i{}^{\sigma(j)}$ ensures the symmetry of S'. This symmetry condition is absolutely crucial if we want $S^2 = C$ to hold, i.e. if we want to interpret this S' as the modular transformation matrix of some characters. We shall come back to this important point later.

It is also possible to find all possible S matrices which differ from a given one by a permutation Σ of the fields: $S_i'^{j} = S_{\Sigma(i)}{}^{\Sigma(j)}$. S' is a possible S-matrix if and only if Σ defines an automorphism of the fusion algebra.

2.2 Conformal field theories from group theory

In order to carry out this program, one has to find some "nice" fusion rules. There exists a very natural way to generate some fusion rules: we can use finite group theory [18]. Let us consider G a finite group, it has a finite number of inequivalent irreducible representations which we denote by π_i. Any tensor product can be decomposed into irreducible representations:

$$\pi_i \otimes \pi_j = \bigoplus_k N_{i,j}{}^{k}\, \pi_k$$

where $N_{i,j}{}^{k}$ is the multiplicity of the π_k representation. Thanks to the associativity and commutativity of the tensor product, the $\hat{N}_i$'s define a representation of a commutative and associative algebra. Let $\pi_{\hat{\imath}}$ be the complex conjugate representation of π_i, then it is clear that $N_{i,j}{}^{\hat{k}}$ is symmetric in i, j and k because it is the multiplicity of the trivial representation of G in the tensor product $\pi_i \otimes \pi_j \otimes \pi_k$. Finally, if π_0 is the trivial representation, then $N_{0,i}{}^{j} = \delta_i^j$. Therefore these $N_{i,j}{}^{k}$ verify all hypotheses of Verlinde's theorem. In the corresponding field theory, with each irreducible representation of G, we associate a "primary field" such that the $N_{i,j}{}^{k}$ of finite group theory give us the fusion algebra of the underlying field theory. Naturally, the trivial representation of G is associated with the identity operator and two conjugate irreducible representations of G are associated with conjugate fields. We have not yet completed the analysis of this program [6], and therefore we shall restrict ourselves to the case of finite abelian groups where things are simpler. At present time, we haven't found any non abelian finite group giving a symmetric S matrix.

We shall only give the result for the case of a cyclic group of order N where everything can be computed explicitly (for details and more about the general abelian group, see subsections 1.3 and 2.1 of [17]).

In the case $G = \mathbf{Z}/N\mathbf{Z}$, the fusion rules are given by

$$N_{i,j,k} = \delta_{i+j+k,0}, \quad i,j,k \ mod\ N.$$

The S matrix is of the form:

$$S_n{}^m = \frac{1}{\sqrt{N}} \exp(2\pi i \frac{nam}{N}). \tag{8}$$

a and N are coprime and $x \to ax$ defines, as expected, an automorphism of the fusion rules. Now, we have to solve $(ST)^3 = \mathbf{1}$ with T a diagonal unitary matrix. To state the final result, we have to distinguish between two cases.

N **even.**

In this case, a is odd. In this particular case, two possibilities can be realized:

$$\exp(2\pi i h_n) = \exp(2\pi i \frac{an^2}{2N}), \quad e^{2\pi i c/8} = \frac{1}{\sqrt{2}} S_{2N}(a)$$

$$\text{or} \ \exp(2\pi i h_n) = \exp(2\pi i \frac{an^2 + nN}{2N}), \quad e^{2\pi i c/8} = \frac{1}{\sqrt{2}} S_{2N}(a) \exp(-\frac{2\pi i N a^{-1}}{8})$$

where

$$S_N(a) = \frac{1}{\sqrt{N}} \sum_{k \in \mathbf{Z}/N\mathbf{Z}} \exp(2\pi i \frac{ak^2}{N}).$$

This Gaussian sum can be evaluated very easily. We find in this case a real central charge. We collect a few useful results about these sums in appendix A.

N **odd.**

In this case we can [7] assume that a is defined mod $2N$ and is even. We then have only one solution for each a; and, with $b = a/2$:

---

[6] Not every non-abelian finite group can be used for this purpose, we are still investigating this question. The main obstruction is the symmetry of the S matrix. We are led to a restriction on the character table of G. If χ_i is the character of π_i, there should exist a pairing between irreducible characters and conjugacy classes such that $\sqrt{|C_j|}\chi_i(C_j) = \sqrt{|C_i|}\chi_j(C_i)$ where C_i is a conjugacy class, and $|C_i|$ its order. This implies in particular that the orders of the conjugacy classes of G are squares. Using Bertrand's theorem, it is easy to show that $n!$ is not a square for n greater than 2. Therefore, by considering cycles of order n, we see that the permutation group of n elements cannot be used in this way to construct a conformal field theory for $n > 2$.

[7] This procedure avoids giving two different formulas according to a's parity.

$$\exp(2\pi i h_n) = \exp(2\pi i \frac{an^2}{2N}), \quad \exp(2\pi i \frac{c}{8}) = S_N(b).$$

This completely solves our problem.

2.2.1 Remark

We have also characterized the solutions which differ only by a permutation of the fields. In this case, what really matters is the class of a, modulo the invertible elements of the ring $\mathbf{Z}/N\mathbf{Z}$ which are squares. For example, when N is a prime number, we know that there are essentially two different theories. In fact, for $N \neq 2, 3$, the only examples we know are the $SU(N)$ level 1 theories (see below) but we don't have any physical model for the other case at present time.

3 Modular invariant partition functions of $\mathbf{Z}/N\mathbf{Z}$ theories

3.1 Modular invariance on the torus

In this section, we are interested in classifying partition functions of rational conformal field theories on the torus. Let us recall that this partition function is defined by

$$Z(\tau) = tr\,(q^{L_0 - c/24}\,\bar{q}^{\bar{L}_0 - c/24}) \tag{9}$$

where $q = \exp(2\pi i\tau)$ is the modular parameter of the torus and the trace is taken over the Hilbert space of the theory. Using the decomposition (1) of H, we can rewrite it as

$$Z(\tau) = \sum_{i,\bar{i}} \hat{N}_{i,\bar{i}}\, \chi_i(\tau)\, \overline{\chi_{\bar{i}}(\tau)}.$$

The χ_i's are the characters of the chiral symmetry algebra of the theory and $N_{i,\bar{i}}$ is the multiplicity of the $(i, \bar{i})$ field. As we explained before, this partition function is a modular invariant. This constrains the field content $(i, \bar{i})$ of the theory. To be more precise, Z, as computed in (9), is associated with a choice of two oriented cycles on the torus. We impose that not only the partition function but also the operator content of the theory are invariant. This is due to the equality of characters of conjugate representations. We are therefore led to the following equations:

$$S^\dagger \hat{N} S = \hat{N}, \quad T^\dagger \hat{N} T = \hat{N} \tag{10}$$

We also impose unicity of the vacuum: $\hat{N}_{0,0} = 1$.

We shall proceed in two steps: we first solve equation (10) with $\hat{N} \in M_N(\mathbf{C})$ and then we impose the integrality and positivity conditions on $\hat{N}$. Finally, we shall find that solutions are labelled by divisors of N when N is odd and by divisors of $N/2$ when N is even. The set of matrices satisfying (10) will be called the commutant in the following.

3.2 Classification of $\mathbf{Z}/N\mathbf{Z}$ partition functions

Here, we are interested in the case of $\mathbf{Z}/N\mathbf{Z}$ field theories. In these cases, we recall that the modular transformation matrices defined by equations (5) and (4) are given by

$$S_{\lambda,\lambda'} = \frac{1}{\sqrt{N}} \exp(2\pi i a \frac{\lambda\lambda'}{N}) \quad T_{\lambda,\lambda'} = \delta_{\lambda,\lambda'} \exp(2\pi i (\frac{a\lambda^2}{2N} - \frac{c}{24})) \tag{11}$$

where a is an integer mod 2N, a and N are coprime and a is even when N is odd. The commutant will be noted C_a in the following. We are now ready to state the main result of this paper. Then, we shall give the main ideas of the proof and discuss some physical applications. We refer the reader to the literature for technical details.

The commutant has been computed by various people in relation with the classification of modular invariant partition functions for minimal and $A_1^{(1)}$ theories [19] [20] [21]. Cappelli, Itzykson and Zuber introduced a set of matrices which gives a nice basis of C_a. Define n as N when N is odd and as $N/2$ when N is even. Let δ be a divisor of n and define $\alpha = \delta \wedge (n/\delta)$. Then, there exist $(a', b') \in \mathbf{Z}^2$ such that $a'n/\delta\alpha - b'\delta/\alpha = 1$ and we define

$$\omega(\delta) = a'n/\delta\alpha + b'\delta/\alpha \pmod{N/\alpha^2}. \tag{12}$$

We thus define the following matrices:

$$\Omega^{(\delta)}_{\lambda,\bar{\lambda}} = \begin{cases} 0 \text{ if } \lambda \text{ or } \bar{\lambda} \text{ is not zero mod } \alpha \\ \sum_{\xi \in \mathbf{Z}/\alpha\mathbf{Z}} \delta_{\bar{\lambda},\omega(\delta)\lambda+\xi N/\alpha} \end{cases} \tag{13}$$

Indeed in the case N odd, the number $\omega(\delta)$ changes by a multiple of $2N/\alpha^2$ when one changes a' and b' preserving $a'n/\delta\alpha - b'\delta/\alpha = 1$ but in the definition of the matrix $\Omega^{(\delta)}$, we only need $\omega(\delta)$ mod N/α^2. Note that these matrices are symmetric and only 0 and 1 appear as entries.

The theorem is the following:

Theorem 3 $(\Omega^{(\delta)})_{\delta|n}$ *is a basis of* C_a.

Note that all C_a's are equal to a simple vector space called C.

At this point, we have to look for linear combinations of the $\Omega^{(\delta)}$ matrices $\hat{N}$ with only positive integer entries and which verify $\hat{N}_{0,0} = 1$. The complete list of solutions is given by:

Theorem 4 *The modular invariant partition functions satisfying the integrality and positivity requirements and unicity of the vacuum are given by*

$$Z_\delta(\tau) = \sum_{(\lambda,\bar{\lambda})\in(\mathbf{Z}/N\mathbf{Z})^2} \Omega^{(\delta)}_{\lambda,\bar{\lambda}} \chi_\lambda(\tau)\overline{\chi_{\bar{\lambda}}(\tau)}, \quad \delta|n$$

Before indicating how the proof goes through, let us make a few comments on the structure of these partition functions. We rewrite Z_δ as follows:

$$
\begin{aligned}
Z_\delta &= \sum_{\substack{k\in \mathbf{Z}/\frac{N}{\alpha}\mathbf{Z} \\ \xi\in\mathbf{Z}/\alpha\mathbf{Z}}} \chi_{k\alpha}\bar{\chi}_{k\alpha\omega+\frac{N}{\alpha}\chi} \\
&= \sum_{k\in\mathbf{Z}/\frac{N}{\alpha^2}\mathbf{Z}} \sum_{(\xi_1,\xi_2)\in(\mathbf{Z}/\alpha\mathbf{Z})^2} \chi_{k\alpha+\xi_1\frac{N}{\alpha}}\bar{\chi}_{k\omega\alpha+\xi_2\frac{N}{\alpha}}
\end{aligned}
$$

We introduce for $k \in \mathbf{Z}/\frac{N}{\alpha^2}\mathbf{Z}$

$$
\tilde{\chi}_k = \sum_{\xi\in\mathbf{Z}/\frac{N}{\alpha}\mathbf{Z}} \chi_{k\alpha+\xi\frac{N}{\alpha}}
$$

then, we have

$$
Z_\delta = \sum_{k\in\mathbf{Z}/\frac{N}{\alpha^2}\mathbf{Z}} \tilde{\chi}_k\overline{\tilde{\chi}_{k\omega}}
$$

with $\omega^2 \equiv 1 \pmod{\frac{2N}{\alpha^2}}$ when N is even and $\omega^2 \equiv 1 \pmod{\frac{N}{\alpha^2}}$ when N is odd. Then, there exists a δ' divisor of $\frac{N}{\alpha^2}$ such that $\alpha(\delta') = 1$ and $\omega(\delta') = \omega$. This shows that Z_δ can be rewritten as an invariant which is a sum of the form

$$
Z = \sum_i \chi_i\overline{\chi_{\sigma(i)}}
$$

where σ is a permutation of the fields. According to Verlinde and Dijkgraaf, the $\tilde{\chi}_k$'s are the characters with respect to the maximally extended chiral algebra of the model and σ is an automorphism of the fusion rules [8]. The reader will easily check that the S matrix of those $\tilde{\chi}_k$'s is given by (8) with $N' = \frac{N}{\alpha^2}$ and the same a. In fact, there is a kind of stability of the fusion rules in this case which was not present for example in the $A_1^{(1)}$ case (cf the exceptional automorphism which enables one to find the E_7 invariant out of the A_{10} one [23]). Of course, as soon as $\omega(\delta) = 1$, the partition function is a sum of moduli squared and we speak of an extended diagonal invariant. This property is easily shown to be equivalent to $n|\delta^2$. The cases where the chiral algebra cannot be extended are characterized by $\delta \wedge (n/\delta) = 1$.

3.3 Sketch of the proof.

Let us first describe how one can find the commutant.

This problem is greatly simplified by noticing that $\hat{N} \to S^\dagger \hat{N} S$ and $\hat{N} \to T^\dagger \hat{N} T$ defines a linear representation of a quotient of finite index of $SL_2(\mathbf{Z})$ on the space of complex $N \times N$ matrices $M_N(\mathbf{C})$. Hence, we only need to compute the averages of the elements of a basis of $M_N(\mathbf{C})$ under this finite quotient. This provides us with a generating family of the commutant and we still have to extract a basis. We refer the reader to [19] [17] [22] for details.

Now, we describe the second part of the proof in some detail.

We have to solve the integrality and positivity condition $\hat{N}_{\lambda,\hat{\lambda}} \in \mathbf{N}$. Thanks to the study of the commutant, the $\hat{N}$ matrix is known to be a linear combination of the

[8] In our example, σ sends u to ωu.

$\Omega^{(\delta)}$'s. Henceforth the coefficients of this linear combination verify a linear system of inequations. Our idea is to find a subset of inequations such that each one involves only a small number of coefficients. Then, we shall solve it and verify that its solutions are solutions to our integrality and positivity problem.

In order to achieve this goal, let us fix a δ and pick a row of $\hat{N}$ to which $\Omega^{(\delta)}$ contributes in $\hat{N} = \sum_\delta c_\delta \Omega^{(\delta)}$. Then, we analyse the contributions of the other $\Omega^{(\delta')}$'s. We shall choose our row such that as few $\Omega^{(\delta')}$'s as possible contribute to it. It is clear that row $\alpha(\delta)$ is the best one. We shall prove the following lemma:

Lemma 1 *When N is odd or when n/α^2 and N are even, the row $\alpha(\delta)$ contains at least one entry equal to c_δ. In the cases where N is even and $n/\alpha(\delta)^2$ is odd, there exist two divisors δ_+ and δ_- such that $\alpha(\delta_\pm) = \alpha(\delta)/2$ and $c_\delta + c_{\delta_+}, c_\delta + c_{\delta_-}$ appear as entries in the row $\lambda = \alpha(\delta)$ and all other coefficients in this row are either 0 or are greater than the two preceding ones.*

This gives us our subsystem of conditions and the reader may check that together with the unicity of the vacuum $\hat{N}_{0,0} = \sum c_\delta = 1$, this result gives the theorem.

The proof of the lemma proceeds as follows: we count the number of column indices $\bar{\lambda}$ for which $\Omega^{(\delta)}_{\alpha(\delta),\bar{\lambda}} \neq 0$ and such that $\Omega^{(\delta')}_{\alpha(\delta),\bar{\lambda}} = 0$ for any other δ'. This task is greatly simplified by the following observation: we only need to consider divisors δ' such that there exists a prime p, divisor of $\alpha(\delta)$ such that $\alpha(\delta') = \alpha(\delta)/p$ (see subsection 3.3 of [17]). We then count the number of divisors δ' with $\alpha(\delta') = \alpha(\delta)/p$ and which contribute in row $\alpha(\delta)$ at places where $\Omega^{(\delta)}$ has non-vanishing entries. This problem can be shown to be equivalent to solving certain quadratic equations over $\mathbf{Z}/p^2\mathbf{Z}$. At this point, we have to face a combinatorial subtlety: sometimes, several $\Omega^{(\delta')}$'s have non vanishing coefficients at the same place. It is however possible to handle these intersections (see subsections 3.4 and 3.5 of [17]). We can show that for N odd or N and $n/\alpha(\delta)^2$ even, the number of column indices $\bar{\lambda}$ such that only $\Omega^{(\delta)}_{\alpha(\delta),\bar{\lambda}} \neq 0$ is strictly positive. This proves the first part of the lemma. In the other case, we find zero but a more careful analysis using exactly the same kind of technology proves the second part of the lemma (see subsection 3.6 of [17]).

3.4 Physical applications

We shall mention some applications of this study: the rational Gaussian model and the $SU(N)$, E_6 and E_7 level 1 theories.

The rational Gaussian model is the theory of a free boson compactified on a circle of radius R such that $R^2/2$ is rational. Let X be this free boson, the action is given by

$$S = \frac{1}{4\pi} \int \partial X \, \bar{\partial} X \, d^2 z$$

The Hilbert space of this theory decomposes into irreducible representations of the "$U(1)_L \times U(1)_R$ current algebra" generated by ∂X, $\bar{\partial} X$.

The highest weight states relative to the $U(1)$ current algebra are created by chiral vertex operators $\exp(ipX_L(z))$. The dimension of this field is given by $h_p = p^2/2$. The quantum number associated with $U(1)$ is p.

In the case of the gaussian model, the allowed values of $(p, \bar{p})$ belong to a two-dimensional even self-dual Lorentzian lattice as a consequence of modular invariance. Therefore, this theory is not trivially rational. We need to enlarge the symmetry by imposing that there exists a $U(1)_L \times U(1)_R$ primary field which has $\bar{h} = 0$. This condition constrains the radius, we find that

$$\frac{R^2}{2} = p/q, \quad p \wedge q = 1$$

In this case, we obtain a spin pq conserved current. The extended algebra has already been identified by Moore and Seiberg [3]. They have also written the generalized characters which are theta functions. These are indexed by an integer mod $N = 2pq$ and have the following transformation law under a modular transformation:

$$S_n{}^m = \frac{1}{\sqrt{2pq}} e^{2\pi i \frac{nm}{2pq}}$$

$$T_n{}^m = \delta_n^m e^{2\pi i \frac{n^2}{4pq}}.$$

We note that the highest weight states of the irreducible representations of the extended algebra are created by vertex operators, the operator product expansion of which is known to be additive. These theories correspond to the case $a = 1$, N even and $h_n = n^2/2N$. The partition function of this model is given by

$$Z_{p,q} = \frac{1}{|\eta(\tau)|^2} \sum_{(n,m)\in \mathbf{Z}^2} q^{\frac{(pn+qm)^2}{4pq}} \bar{q}^{\frac{(pn+qm)^2}{4pq}}$$

We introduce the characters of the maximally extended chiral algebra of the theory:

$$\tilde{\chi}_k = \frac{1}{\eta(\tau)} \sum_{l\in \mathbf{Z}} q^{\frac{(k+2pql)^2}{4pq}}$$

where k is an integer modulo $2pq$. Then, we can easily rewrite $Z_{p,q}$ as

$$Z_{p,q} = \sum_{k\in \mathbf{Z}/2pq\mathbf{Z}} \tilde{\chi}_k \overline{\tilde{\chi}_{\omega k}}$$

where ω is defined as in (12) with $N = 2pq$ and $\delta = p$. Therefore $Z_{p,q}$ is one of the Z_δ's given by our theorem and is of the form given by Verlinde and Dijkgraaf's theorem.

The next example is provided by the $SU(N)$ theories at level 1. In a recent paper [10], C. Itzykson showed that the $A_N^{(1)}$ level-1 theories have $\mathbf{Z}/N\mathbf{Z}$ fusion rules. In this case, N can be even or odd, and $h_n = n(N-n)/2N$. We refer the reader to his paper for more details on the computation of the S and T matrices. Our theorem also classifies all modular invariant partition functions on the torus for rational conformal field theories with respect to the $A_N^{(1)}$ algebra at level 1. In fact, we can point out two amusing facts: the $\delta = 3$ invariant for $SU(9)$ at level 1 is the cubic root of the modular form $j(\tau)$ [10] and, using $SU(25)$ characters, the reader can also check that

$$\sum_{k=0}^{5} \chi_{5k}(\tau) = j(\tau) - 120 \tag{14}$$

Finally, E. Verlinde pointed to us that the $E_6^{(1)}$ level 1 theories have $\mathbf{Z}/3\mathbf{Z}$ fusion rules and they correspond to the $N = 3$, $a = 1$ solution of subsection 2.2 and the $E_7^{(1)}$ level 1 theories correspond to $N = 2$ and $h_n = 0, 3/4$.

Conclusion.

In this paper and in [17], we have illustrated how one can reconstruct part of a CFT starting from its fusion rules. Our particular example classifies the modular invariant partition functions of $A_N^{(1)}$ theories. This example exhibits some very interesting connections with number theory and finite group theory. For example, the problem of finding some non abelian finite group whose representation algebra could be the fusion algebra of a RCFT remains open. Another puzzling question which arises from this work is the following: given a finite dimensional unitary representation of $SL_2(\mathbf{Z})$, is it possible to find a set of modular forms f_i such that this representation gives the modular transformation laws of the $q^{h_i - \frac{c}{24}} f_i$'s? Then, we have to impose that these functions have only positive integers in their q-development. These conditions are necessary if we want a corresponding theory to exist. Finally, we have to find a chiral algebra which produces these functions as characters. These questions are the next steps of this work in the classification program of RCFT's.

The organizers of this school are thanked for their hospitality and the invitation to give this seminar. This work benefitted from many useful discussions with B. Julia, C. Itzykson and many other people. I also thank Th. Papadopoulo for his generous help with LaTex.

A Gaussian sums

Here, we recall how to compute some special sums known as Gaussian sums [25] [24]. Let us define

$$S_N(a) = \frac{1}{\sqrt{N}} \sum_{k \in \mathbf{Z}/N\mathbf{Z}} \exp\left(2\pi i \frac{ak^2}{N}\right).$$

We restrict ourselves to a invertible mod N; the other cases can trivially be reduced to this case. We refer the reader to the literature for details and proofs. We can first of all restrict the computation to the case where N is a power of a prime.

Proposition A.1 *Let a, b, c three integers with no common divisor. Then $S_{ab}(c) = S_a(bc)S_b(ac)$.*

We are reduced to computing $S_{p^\alpha}(a)$ where a and p are coprime. This can be achieved quite easily when p is odd, only $p = 2$ is more tricky. We only give the final results:

Proposition A.2 *Let p be an odd prime number and a an invertible mod p, then*

$$S_{p^\alpha}(a) = \left(\frac{a}{p}\right) S_{p^\alpha}(1)$$

Proposition A.3 *Let a be an odd integer; then*

$$S_{2^\alpha}(a) = \varepsilon(a) \left(\frac{-2^\alpha}{a}\right) S_{2^\alpha}(1)$$

where $\varepsilon(a) = 1$ *(resp, i) when* $a = 1 \pmod 4$ *(resp,* $a = -1 \pmod 4$*) and*

$$\left(\frac{x}{y}\right) = \prod_{\substack{p \text{ prime} \\ p|y}} \left(\frac{x}{p}\right)$$

is the Jacobi symbol.

Finally,

Proposition A.4 *We have*

$$S_N(1) = \frac{1+i}{2}(1 + (-i)^N).$$

With these results, one can compute all Gaussian sums quite easily. For example, we have

$$S_{4n}(\pm 1) = 1 \pm i.$$

References

[1] D. Friedan, S. Shenker: The analytic formulation of 2D CFT. Nucl. Phys. B. **281**, 509 1987.

[2] G. Moore, N. Seiberg: Polynomial equations for Rational Conformal Field Theories. Preprint IASSNS-HEP 88/18.

[3] G. Moore, N. Seiberg: Naturality in Conformal Field Theory. Preprint IASSNS-HEP 88/31.

[4] G. Moore, N. Seiberg. Classical and quantum Conformal Field Theory. Preprint IASSNS-HEP 88/35.

[5] C. Vafa: Towards classification of Conformal Field Theories. Phys. Lett. B. **206**, 421 (1988)

[6] J. Birman: *Braids, Links and Mapping class groups.* Ann. Math. Studies 82 (Princeton University Press 1974)

[7] E. Witten: Quantum Field Theory and the Jones polynomial. Preprint IQSSNS 88/33 to appear in Comm. Math. Phys.

[8] G. Segal: The definition of CFT. Oxford University Preprint.

[9] E. Verlinde: Fusion rules and modular transformations in 2D Conformal Field Theory. Nucl. Phys. B. **300** 360 (1988)

[10] C. Itzykson: Level-one Kac-Moody characters and modular invariance. In P. Binetruy, P. Sorba, R. Stora, editors, *Conformal Field Theories and related topics.* Nucl. Phys. B. (Proc. Suppl.) **5**, 150 (North Holland , Amsterdam 1988)

[11] K. Gawedzki: Théories Conformes Séminaire Bourbaki Novembre 1988.

[12] C. Vafa: Conformal field theories and punctured surfaces Phys. Lett. B., **199**, 195 1987.

[13] R. Dijkgraaf: Recent progress in rational conformal field theory. To appear in the proceedings of Les Houches XLIX-th summer school *Strings, Fields and Critical phenomena* E. Brézin, J. Zinn-Justin editors.

[14] J.M. Drouffe, C. Itzykson: *Statistical Field Theory*, to appear.

[15] P. Ginsparg: Applied Conformal Field Theory. To appear in the proceedings of Les Houches XLIX-th summer school *Strings, Fields and Critical phenomena* E. Brézin, J. Zinn-Justin editors.

[16] A.A. Belavin, A.M. Polyakov, A.B. Zamolodchikov: Infinite Conformal Symmetry in 2D-field theory. Nucl. Phys. B. **241**, 333 (1984)

[17] P. Degiovanni: $\mathbf{Z}/N\mathbf{Z}$ Conformal Field Theories. Preprint LPTENS 89/03.

[18] J.P. Serre: *Représentation linéaire des groupes finis.* (Hermann, Paris 1968)

[19] A. Cappelli, C. Itzykson, J.B. Zuber: The ADE classification of $A_1^{(1)}$ and minimal conformal field theories. Comm. Math. Phys. **113**, 1 (1987)

[20] D. Gepner, Z. Qiu: Modular invariant partition functions for parafermionic theories. Nucl. Phys. B.(FS 19) **285**, 423 (1987)

[21] A. Kato: Classification of modular invariant partition functions in 2D CFT. Modern Phys. Lett. A, Vol **2**, 585 (1987).

[22] M. Bauer, C. Itzykson: Affine characters and modular transformations. See in this volume.

[23] R. Dijkgraaf, E. Verlinde: Modular invariance and the fusion algebra. In P. Binetruy, P. Sorba, R. Stora, editors, *Conformal Field Theories and related topics.* Nucl. Phys. B. (Proc. Suppl.) **5**, 87 (North Holland , 1988)

[24] P. Cartier: Modular forms and elliptic functions. In the other volume of this school.

[25] S. Lang: *Algebraic number theory.* (Addison-Wesley, Reading 1970)

Affine Characters and Modular Transformations

M. Bauer and C. Itzykson

Service de Physique Théorique de Saclay, Laboratoire de l'Institut
de Recherche Fondamentale du Commissariat à l'Energie Atomique
F-91191 Gif-sur-Yvette Cedex, France

Abstract We describe the rational commutant of the modular action on $SU(N)$ affine characters.

1. Introduction

Partial results on the classification of rational conformal theories include the A-D-E series for the SU_2 Wess-Zumino Witten model (which applies to coset models including the minimal ones based on representations of the Virasoro algebra of central charge less than one), or $SU(N)$ at level one -reported in this volume. One might entertain the hope of an extension to all WZW theories, thought to be the fundamental bricks of all rational ones through coset constructions. We confine ourselves here to $SU(N)$ groups and study the unitary representation of (a finite factor group of) the modular group afforded by its affine characters. At a given level, call the characters $\chi_\lambda(\tau)$, where λ runs over a finite set and τ is a complex variable with positive imaginary part. A partition function on a torus specified by τ reads

$$Z(\tau,\bar{\tau}) = \sum \overline{\chi_\lambda(\tau)} Z_{\lambda,\lambda'} \chi_{\lambda'}(\tau)$$

Modular invariance requires that the matrix of non negative integers $Z_{\lambda,\lambda'}$ ($Z_{0,0} = 1$ for normalization, with 0 the vacuum sector) commutes with the unitary representation of the modular group on characters. The simplest solution is the principal invariant $Z_{\lambda,\lambda'} = \delta_{\lambda,\lambda'}$. Complementary invariants arise from automorphisms of the inhomogeneous Weyl group pertaining to the Lie algebra, conformal embeddings (which reduce to the identity when restricted to the conformal algebra) or automorphisms of the fusion rules generated by short distance expansions, formalized by Verlinde as an associative and commutative ring over integers (see below its description for $SU(N)$). It is not known if the combination of these methods yields an exhaustive list. This justifies a direct - but cumbersome - approach. We describe below the commutant of the modular representation carried by the characters in the case of the (affine) $SU(N)$ algebra. This is a technical - but significant - part of the program. Our presentation is to be expanded in a more comprehensive account in preparation.

Springer Proceedings in Physics, Vol. 47 Number Theory and Physics
Editors: J.M. Luck · P. Moussa · M. Waldschmidt

2. Affine Characters

The root lattice M of $SU(N)$ is generated by $(N-1)$ simple roots $\boldsymbol{\alpha}_i = e_i - e_{i+1}$ (the vectors $\mathbf{e}_\mu$ form an orthonormal basis in $\mathbf{R}^N$, greek indices run from 1 to N, latin indices from 1 to $N-1$). The metric is given by the Cartan matrix $g_{ij} = \boldsymbol{\alpha}_i.\boldsymbol{\alpha}_j$, equal to 2, -1 or 0 according to $\mid i-j \mid = 0, 1$ or >1. Both M and its dual, the weight lattice M^*, generated by the fundamental weights $\boldsymbol{\alpha}^i$, such that $\boldsymbol{\alpha}^i.\boldsymbol{\alpha}_j = \delta^i_j$, $\boldsymbol{\alpha}^i = g^{ij}\boldsymbol{\alpha}_j$, $g_{ij} = \mathrm{Inf}(i,j) - (ij)/N$, are invariant under the Weyl group $W \equiv S_N$, of order $N!$, generated by reflections in the hyperplanes orthogonal to the simple roots.

Specialized affine characters at level k (we also use $n = k + N$ and call it height) $\chi_{\mathbf{p}}(\tau)$, $\mathrm{Im}\tau > 0$, are indexed by a weight $\mathbf{p}$ mod nM, with a further antisymmetry in $\mathbf{p}$ with respect to W. This reduces the number of linearly independent ones to $\binom{n-1}{N-1}$ and allows us to choose $\mathbf{p}$ in a fundamental domain $B_n : p_i = (\mathbf{p}.\boldsymbol{\alpha}_i) > 0$, $\Sigma p_i < n$. Furthermore $\chi_{\mathbf{p}} = \chi_{\theta(\mathbf{p})}$, $\theta(\mathbf{p}) =^{\omega_0} (-\mathbf{p})$, $\theta^2 = 1$, and ω_0 the unique element in W such that both $\theta(\mathbf{p})$ and $\mathbf{p}$ belong to B_n (under $\omega_0\ e_{N-S} \longrightarrow e_{S+1}$, $\det\omega_0 = (-1)^{N(N-1)/2}$). For $n = N$ the theory is trivial with a unique character $\chi_{\mathbf{p}_0}(\tau) = 1$, where $(\mathbf{p}_0)_i = 1$, $\mathbf{p}_0^2 = N(N^2-1)/12$.

The modular group of 2×2 integral matrices with unit determinant, up to a sign, is generated by two transformations S and T satisfying $S^2 = (ST)^3 = I$, acting on characters as $(e(z) \equiv \exp[2i\pi z])$

$$S \qquad \chi_{\mathbf{p}}(-\tau^{-1}) = \frac{(-i)^{\frac{N(N-1)}{2}}}{(Nn^{N-1})^{1/2}} \sum_{\mathbf{p}' \in B_n} \left(\sum_{\omega \in W} \det\omega\ e\left(\frac{\mathbf{p}.^{\omega}\mathbf{p}'}{n}\right)\right) \chi_{\mathbf{p}'}(\tau)$$

$$T \qquad \chi_{\mathbf{p}}(\tau+1) = e\left(\frac{\mathbf{p}^2}{2n} - \frac{\mathbf{p}_0^2}{2N}\right) \chi_{\mathbf{p}}(\tau)$$

That these transformations generate the modular group is checked using Gauss's sums and $\chi_{\theta(\mathbf{p})} = \chi_{\mathbf{p}}$. (Due to the fact that roots have a square length multiple of 2, T is well defined). With $\mathbf{p} \in M^*$, $\mathbf{x} \in \mathbf{R}^{N-1}$ mod M and $\psi_{\mathbf{p}}(\mathbf{x}) = \sum_{\omega \in W} \det(\omega) e(\mathbf{p}.^{\omega}\mathbf{x})$, the classical Weyl formula for characters reads

$$\mathrm{ch}_{\mathbf{p}}(\mathbf{x}) = \frac{\psi_{\mathbf{p}}(\mathbf{x})}{\psi_{\mathbf{p}_0}(\mathbf{x})}$$

and

$$S_{\mathbf{p},\mathbf{p}'} = \frac{(-i)^{\frac{N(N-1)}{2}}}{(Nn^{N-1})^{1/2}} \psi_{\mathbf{p}}\left(\frac{\mathbf{p}'}{n}\right)$$

Restricting the classical characters to $\mathbf{x} \in \frac{1}{n}B_n$, the standard Clebsch-Gordan series $\mathrm{ch}_{\mathbf{p}_1}\mathrm{ch}_{\mathbf{p}_2} = \sum \mathcal{N}_{p_1,p_2}^{p_3}\mathrm{ch}_{\mathbf{p}_3}$, $\mathbf{p}_1, \mathbf{p}_2, \mathbf{p}_3 \in M^*$ with positive components $(\mathbf{p}_{1,2,3})_i$, becomes the Verlinde equation at height n

$$(V) \qquad \mathbf{p}_1, \mathbf{p}_2, \mathbf{p}_3, \mathbf{p} \in B_n \qquad \mathrm{ch}_{\mathbf{p}_1}(\frac{\mathbf{p}}{n})\mathrm{ch}_{\mathbf{p}_2}(\frac{\mathbf{p}}{n}) = \sum N_{\mathbf{p}_1\mathbf{p}_2}^{\mathbf{p}_3}\mathrm{ch}_{\mathbf{p}_3}(\frac{\mathbf{p}}{n})$$

with integral non-negative coefficients $N_{\mathbf{p}_1,\mathbf{p}_2}^{\mathbf{p}_3}$ (this requires a proof) arising from folding the previous r.h.s using the (anti-)symmetry in $\mathbf{p}_3$ with respect to the inhomogeneous Weyl group, including the reflection in the hyperplane $n - \Sigma y_i = 0$. We readily derive that

$$N_{\mathbf{p}_1\mathbf{p}_2}^{\mathbf{p}_3} = \sum_{\mathbf{p}\in B_n} \frac{S_{\mathbf{p}_1,\mathbf{p}}S_{\mathbf{p}_2,\mathbf{p}}\overline{S_{\mathbf{p}_3,\mathbf{p}}}}{S_{\mathbf{p}_0,\mathbf{p}}}$$

More symmetrically

$$N_{\mathbf{p}_1\mathbf{p}_2\mathbf{p}_3} \stackrel{\text{def}}{=} N_{\mathbf{p}_1\mathbf{p}_2}^{\theta(\mathbf{p}_3)} = \sum_{p\in B_n} \frac{S_{\mathbf{p}_1,\mathbf{p}}S_{\mathbf{p}_2,\mathbf{p}}S_{\mathbf{p}_3,\mathbf{p}}}{S_{\mathbf{p}_0,\mathbf{p}}}$$

Equation (V) implies that for fixed $\mathbf{p} \in B_n$ the vector $\mathrm{ch}_{\mathbf{p}'}(\frac{\mathbf{p}}{n})$ is an eigenvector of the commuting matrices $(N_{\mathbf{p}_1})_{\mathbf{p}''}^{\mathbf{p}'}$ with eigenvalues $\mathrm{ch}_{\mathbf{p}_1}(\frac{\mathbf{p}}{n})$. In the case of $SU(3)$ for instance, the matrix N corresponding to the weight $2\boldsymbol{\alpha}^1 + \boldsymbol{\alpha}^2$ (the fundamental 3-dimensional representation) is the incidence matrix for an oriented graph with vertices inside an equilateral triangle of side n on a triangular lattice and, as an example, at height $n = 12$ one can check that its characteristic equation of degree 55

$$\begin{aligned}\det\ (y-N) = {} & \\ & y(y-1)(y-2)(y^2-2y-2)(y^2+y+1)(y^2+2y+2)(y^2+2y+4) \\ & \times(y^4-2y^3+2y^2-4y+4)(y^4+2y^3+6y^2-4y+4)(y^6-y^3+1) \\ & \times(y^6+y^3+1)(y^{12}-34y^9+381y^6-1564y^3+2197) \\ & \times(y^{12}-14y^9+53y^6-4y^3+1)\end{aligned}$$

has indeed the roots

$$\mathrm{ch}_{2\boldsymbol{\alpha}^1+\boldsymbol{\alpha}^2}(\frac{\mathbf{p}}{n}) = e\left(\frac{2p_1+p_2}{36}\right) + e\left(\frac{-p_1+p_2}{36}\right) + e\left(\frac{-p_1-2p_2}{36}\right)$$

with p_1, p_2 positive integers, $p_1 + p_2 < 12$ and that the 54 other matrices $N_{\mathbf{p}}$ at this height are polynomials in N. We suspect that such factorizations play a role in the construction of modular invariants.

If we ignore the condition of anti-symmetry of the (affine) characters (to be restored at a later stage), drop an inessential phase and rescale all scalar products by a factor N, to deal only with integers, the study of the commutant of this (extended) representation of $SL(2,\mathbf{Z})$ (noted henceforth SL_2) requires to find matrices (with rational entries) which commute with the two unitary matrices indexed by weights in M^*/nM

$$S_{\mathbf{p},\mathbf{p}'} = \frac{1}{(Nn^{N-1})^{1/2}} e\left(\frac{\mathbf{p}.\mathbf{p}'}{Nn}\right) \qquad T_{\mathbf{p},\mathbf{p}'} = \delta_{\mathbf{p},\mathbf{p}'} e\left(\frac{\mathbf{p}^2}{2Nn} - \frac{N^2-1}{24}\right)$$

Note that as an additive group $M^*/nM = \mathbf{Z}_{nN} \times (\mathbf{Z}_n)^{N-2}$ (with $\mathbf{Z}_n$ for $\mathbf{Z}/n\mathbf{Z}$) using the fact that $\boldsymbol{\alpha}^1, \boldsymbol{\alpha}_1, ..., \boldsymbol{\alpha}_{N-2}$ generate M^*.

3. Preparation

We need some preliminary properties, some well, others less, known. Let G be a finite abelian group with two generators (and relations). Then SL_2 acts on $G \times G$ through $(g,g') \longrightarrow (g,g')\begin{pmatrix} a & b \\ c & d \end{pmatrix}$. If g and g' generate G so do their transforms. More generally we can associate to any SL_2 orbit the subgroup $H \subset G$ generated by any pair on the orbit (the converse is not true). Let $\varphi(m)$ stand for Euler's function, the order of the multiplicative group $\mathbf{Z}_m^*$ of invertible elements mod m (by convention $\mathbf{Z}_1^* = \mathbf{Z}_1$, $\varphi(1) = 1$).

Proposition 1 – G is isomorphic to $\mathbf{Z}_m \times \mathbf{Z}_{mn}$ for some positive integers m and n and there are $\varphi(m)$ orbits in $G \times G$ with associated subgroup $H = G$.

For finitely many integers $u_1, u_2, ...$, not all zero, write $[u_1, ..., u_2]$ for the *g.c.d*, i.e. positive generator of the ideal $\Sigma x_i u_i$. By convention set $[0,0,...0] = 0$. We need the following

Lemma Assume $[u,v,w] = 1$, $v \neq 0$, there exists $y \in \mathbf{Z}$ such that we have $[u + yw, v] = 1$.

For any $x \in \mathbf{Z}$ let $\delta_x = [u + xw, v] \geq 1$. Obviously $[\delta_x, w] = 1$. If $[x,u] = 1$ then $[\delta_x, u] = 1$. Indeed either $x = 0$, $u = \pm 1$ and $[\delta_x, u] = 1$, or $x \neq 0$ $[\delta_x, u] = [u + xw, v, u]$ and since x is prime to u this is also $[u,v,w] = 1$. The finite set I of positive divisors of v, of those δ_x such that $[\delta_x, u] = 1$ is non empty ($\delta_1 \in I$). Its elements are prime to u and w. The least common multiple y of all elements δ_x in I is also prime to u and w so that $\delta_y \in I$, is prime to u, divides y (by the definition of y) as well as $u + yw$ (by the definition of δ_y) hence also u. This leaves the only possibility $\delta_y \doteq 1$, concluding the proof of the lemma.

We prove the proposition in several steps.

(i) Any 2×2 integral matrix M can be written LDR, L, $R \in SL_2$, and D integral diagonal.

Write $M = \begin{pmatrix} a & b \\ c & d \end{pmatrix}$. First assume $b = 0$ and $[a, c, d] = 1$. If $d = 0$ pick v and s such that $av - cs = 1$ and the property is true in the form $M = \begin{pmatrix} a & s \\ c & v \end{pmatrix}\begin{pmatrix} 1 & 0 \\ 0 & 0 \end{pmatrix}$. If $d \neq 0$ using the lemma one can first find v then z and t such that $[av - c, d] = 1$, $(av - c)t + dz = 1$, thus we get

$$M = \begin{pmatrix} at & 1 \\ ct - dz & v \end{pmatrix}\begin{pmatrix} 1 & 0 \\ 0 & ad \end{pmatrix}\begin{pmatrix} av - c & -d \\ z & t \end{pmatrix}$$

and M is factorized as required. If $b = 0$ and $[a, c, d] > 1$ write

$$M = [a, c, d]\begin{pmatrix} a' & 0 \\ c' & d' \end{pmatrix}$$

and use the previous proof. Finally if $b \neq 0$, let x, y be such that $ax + by = [a, b] \neq 0$. Then $T = \begin{pmatrix} x & -b/[a,b] \\ y & a/[a,b] \end{pmatrix}$ belongs to SL_2, MT is lower triangular and we can apply the preceding. Note that we proved that D can be chosen of the form $\begin{pmatrix} m & 0 \\ 0 & mn \end{pmatrix}$ with $m = [a, b, c, d]$ and $ad - bc = m^2 n$.

(ii) Consider a pair (g, g') generating G and the subgroup of $\mathbf{Z}^2$ of all relations $xg + yg' = 0$. Subgroups of $\mathbf{Z}^2$ are isomorphic to $\{0\}$, $\mathbf{Z}$ or $\mathbf{Z}^2$. Since G is finite we are necessarily in the third case. The subgroup of relations is freely generated by two elements (a, c) and (b, d), such that if $xg + yg' = 0$ there exists a unique pair u, v with $\begin{pmatrix} x \\ y \end{pmatrix} = \begin{pmatrix} a & b \\ c & d \end{pmatrix}\begin{pmatrix} u \\ v \end{pmatrix}$. The matrix $M = \begin{pmatrix} a & b \\ c & d \end{pmatrix}$ is non singular and unique up to right multiplication by an invertible integral matrix (hence in SL_2 if orientation is preserved).

As a consequence of (i) on any SL_2 orbit of generators g, g' of G there exists a factorized pair h, h' and a basis of the group of relation between h and h' such that the matrix M takes the form $\begin{pmatrix} m & 0 \\ 0 & mn \end{pmatrix}$ proving the first part of the proposition that G is isomorphic to $\mathbf{Z}_m \times \mathbf{Z}_{mn}$.

(iii) If (g, g_1') and (g_2, g_2') both generate G, there exist integral matrices such that $(g_2, g_2') = (g_1, g_1')M_{12}$ and $(g_1, g_1') = (g_2, g_2')M_{21}$. Let us show that det M_{12} is well defined mod m, and that given an integer Δ, $\Delta = \det M_{12} \bmod m$, one can find a matrix M such that $\det M = \Delta$ and $(g_1, g_1')(M - M_{12}) = 0$. Since the determinant is preserved by SL_2 and any orbit of generators contains a factorized one, we can assume (g_1, g_1') factorized and $(g_1, g_1')(M - M_{12}) = 0$ iff $M - M_{12} = \begin{pmatrix} m & 0 \\ 0 & mn \end{pmatrix} P$, P an integral matrix. Thus $M \equiv M_{12} \bmod m$

and $\det M \equiv \det M_{12} \bmod m$. Let $\Delta = \det M_{12} + pm$, we search P such that $\det M = \Delta$, $M = M_{12} + \begin{pmatrix} m & 0 \\ 0 & mn \end{pmatrix} P$. Write $M_{12} = \begin{pmatrix} a & b \\ c & d \end{pmatrix}$, $M_{21} = \begin{pmatrix} a' & b' \\ c' & d' \end{pmatrix}$ $P = \begin{pmatrix} \alpha & \beta \\ \gamma & \delta \end{pmatrix}$. The equation to solve (in $\alpha, \beta, \gamma, \delta$) is

$$mn(\alpha\delta - \beta\gamma) + (d\alpha + na\delta - c\beta - nb\gamma) = p$$

Since $M_{12}M_{21}$ acts as the identity on (g_1, g_1'), $cb' + dd' \equiv 1 \bmod mn$, c and d are different from (0,0) and $[c, d, mn)] = 1$ (as we assume G not reduced to the neutral element). Let for instance $c \neq 0$ (a similar reasoning holds for $d \neq 0$). From the lemma we find δ such that $[c, d + mn\delta] = 1$. Take $\gamma = 0$. The equation reduces to

$$(d + mn\delta)\alpha - c\beta = p - na\delta$$

which admits a solution in α, β since c and $(d + mn\delta)$ are coprimes. Parenthetically the classical isomorphism $SL_2 \bmod m \sim SL_2(\mathbf{Z}_m)$ follows by applying the above reasoning to $\det M = 1 \bmod m$ and $G \sim \mathbf{Z}_m \times \mathbf{Z}_m$.

Since $M_{12}M_{21}$ acts as the identity on (g_1, g_1'), $\det M_{12} \det M_{21} \equiv 1 \bmod m$ and M_{12} is invertible mod$m$. Let (g_3, g_3') be another generating pair of G with M_{13}, M_{31} defined in an obvious way. Assume that $\det M_{13} \equiv \det M_{12} \bmod m$. Then $(g_2, g_2') = (g_3, g_3')M_{31}M_{12}$ and $\det M_{31} \det M_{12} \equiv 1 \bmod m$. We can find $M \in SL_2$ such that $(g_3, g_3')\ (M - M_{31}M_{12}) = 0$. Hence (g_2, g_2') and (g_3, g_3') are on the same orbit as soon as $\det M_{13} \equiv \det M_{12} \bmod m$ and there are at most $\varphi(m)$ generating orbits. But if Δ is invertible mod$n$ $(g_1, g_1') \begin{pmatrix} \Delta & 0 \\ 0 & 1 \end{pmatrix}$ describes the $\varphi(m)$ desired orbits as claimed in the proposition.

Any element in $G \sim \mathbf{Z}_m \times \mathbf{Z}_{mn}$ will be written as a column vector with a first entry mod$m$, second modmn. Let δ and δ' be positive divisors of m and mn respectively, Δ an element in $\mathbf{Z}_{[m/\delta, mn/\delta']}$ and $\Omega \in \mathbf{Z}^*_{[m/\delta, mn/\delta', \Delta]}$. Write the prime decomposition of m (mn) as $\prod_p p^{\alpha(p)} \left(\prod_p p^{\alpha(p)+\beta(p)}\right)$.

Proposition 2 Orbits in $G \times G$ under the (right) action of SL_2 are classified by all possible sets $(\delta, \delta', \Delta, \Omega)$. Their number is

$$m \prod_p \left((\alpha(p) + 1)(\beta(p) + 1) + \frac{\alpha(p)(1 - \beta(p))}{p} \right)$$

A pair in $G \times G$ is written as a matrix $\begin{pmatrix} \tilde{k} & \tilde{\ell} \\ \tilde{k}' & \tilde{\ell}' \end{pmatrix}$, $\tilde{k}, \tilde{\ell} \bmod m$, $\tilde{k}', \tilde{\ell}' \bmod mn$; SL_2 acts to the right. Clearly $\delta = [\tilde{k}, \tilde{\ell}, m]$ and $\delta' = [\tilde{k}', \tilde{\ell}', mn]$ are invariants of on an orbit. Set

$$\begin{pmatrix} \tilde{k} & \tilde{\ell} \\ \tilde{k}' & \tilde{\ell}' \end{pmatrix} = \begin{pmatrix} \delta & 0 \\ 0 & \delta' \end{pmatrix} \begin{pmatrix} k & \ell \\ k' & \ell' \end{pmatrix} \qquad k, \ell \bmod p = m/\delta, \quad k', \ell' \bmod q = mn/\delta'$$

and $[k,\ell,p]=1$, $[k',\ell',q]=1$. Thus $\Delta \equiv k\ell' - k'\ell \bmod [p,q]$ is SL_2 invariant. One can find integers such that $xk+y\ell \equiv 1 \bmod p$. Write $\Omega' = xk'+y\ell'$ for some representatives and $r=[\Delta,p,q]$. Then
(i) Ω' is well defined $\bmod r$
(ii) $\Omega' \in \mathbf{Z}_r^*$ i.e. one can find Ω such that $\Omega\Omega' \equiv 1 \bmod r$ (if $r=1$ take $\Omega=\Omega'=0$)
(iii) $\Omega' \bmod r$ (hence $\Omega \bmod r$) is invariant along the orbit.
Starting from a solution $x_0k+y_0\ell \equiv 1 \bmod p$, let x,y be any other solution. Thus $x = x_0+\alpha p+\beta\ell$ since x and x_0 admit k as inverse $\bmod [p,\ell]$. Consequently $(y-y_0+\alpha k)\ell \equiv 0 \bmod p$, with ℓ invertible $\bmod p/[p,\ell]$, and for some ρ and s

$$\begin{aligned} x-x_0 &= \left(\alpha + s\frac{p}{[p,\ell]}\right)\ell + \left(\beta - \frac{\ell s}{[p,\ell]}\right)p \\ y-y_0 &= -\left(\alpha + s\frac{p}{[p,\ell]}\right)k + (sk+\rho)\frac{p}{[p,\ell]} \end{aligned}$$

Since k is prime to $[p,\ell]$ choose s such that $-sk+\lambda[p,\ell]=\rho$, set

$$\alpha + s\frac{p}{[p,\ell]} = -\mu \text{ and } \gamma = \beta - \frac{\ell s}{[p,\ell]},$$

obtaining

$$x-x_0 = -\mu\ell+\gamma p \qquad y-y_0 = \mu k+\lambda p$$

and $\Omega' = xk'+y\ell' \equiv \Omega_0' + \mu\Delta \bmod [p,q] \equiv \Omega_0' \bmod r$ proving (i).
Inverting the roles of k,ℓ and k',ℓ' we find $\Omega \bmod r$, such that

$$\begin{pmatrix} k & \ell \\ k' & \ell' \end{pmatrix}\begin{pmatrix} x & x' \\ y & y' \end{pmatrix} \equiv \begin{pmatrix} 1 & \Omega \\ \Omega' & 1 \end{pmatrix} \bmod r$$

Taking determinants $\bmod r$, the ℓ.h.s. is 0, the r.h.s. $(1-\Omega\Omega')$, proving (ii).
Finally for any A in SL_2

$$\left\{\begin{pmatrix} k & \ell \\ k' & \ell' \end{pmatrix}A\right\}\left\{A^{-1}\begin{pmatrix} x & x' \\ y & y' \end{pmatrix}\right\} \equiv \begin{pmatrix} 1 & \Omega \\ \Omega' & 1 \end{pmatrix} \bmod r$$

showing that Ω (and Ω') are invariants of the orbits.

We turn to the converse and show that two pairs with the same δ,δ',Δ and Ω are on the same orbit. Reducing by δ and δ' we can assume that the pairs read $\begin{pmatrix} k & \ell \\ k' & \ell' \end{pmatrix}$ and $\begin{pmatrix} K & L \\ K' & L' \end{pmatrix}$ with $[k,\ell,p]=[K,L,p]=[k',\ell',q]=[K',L',q]=1$. From the lemma one can find representatives k',ℓ' such that $[k',\ell']=1$ and $x'k'+y'\ell'=1$ for some x',y'. By multiplying $\begin{pmatrix} k & \ell \\ k' & \ell' \end{pmatrix}$ to the right by $\begin{pmatrix} \ell' & x' \\ -k' & y' \end{pmatrix} \in SL_2$ we find an element on the same orbit with $(k',\ell') \equiv (0,1)$ mod$q$, similarly for the second matrix. Hence in the sequel we can assume $k' \equiv \ell' - 1 \equiv K' \equiv L' - 1$ mod$q$, in which case $k \equiv K \equiv \Delta$ mod$[p,q]$ and $\ell \equiv L \equiv \Omega$ mod$r$ $(= [\Delta, p, q])$. Thus with λ mod$p$, $K \equiv k + \lambda q$ mod$p$. According to the lemma it is possible to pick k and ℓ in the same classes mod$p$ such that $xk+y\ell = 1$, thus $[k,y] = 1$. Assuming a choice of λ such that $k+\lambda q \neq 0$, $[k+\lambda q, y, q] = [k,y,q] = 1$ and again from the lemma one can find u, v and μ with $u(y+\mu q) + v(k+\lambda q) = 1$. The product

$$\begin{pmatrix} x & -\ell \\ y & k \end{pmatrix} \begin{pmatrix} k+q\lambda & u \\ -(y+q\mu) & v \end{pmatrix}$$

belongs to SL_2 and transforms $\begin{pmatrix} k & \ell \\ 0 & 1 \end{pmatrix}$ into $\begin{pmatrix} k+\lambda q & \ell' \\ 0 & 1 \end{pmatrix}$ showing that we can assume $k \equiv K$ mod$p$ $(\equiv \Delta$ mod$[p,q])$. We are left with two matrices $\begin{pmatrix} k & \ell \\ 0 & 1 \end{pmatrix}, \begin{pmatrix} k & L \\ 0 & 1 \end{pmatrix}$, with $[k,\ell,p] = [k,L,p] = 1$, $k \equiv L$ mod$r (= [k,p,q])$, first lines mod$p$, second mod$q$. Understanding equivalence as right multiplication by SL_2, let us prove that for any λ mod$p$ such that $[k, \ell + q\lambda, p] = 1$, we have $\begin{pmatrix} k & \ell+\lambda q \\ 0 & 1 \end{pmatrix} \sim \begin{pmatrix} k & \ell \\ 0 & 1 \end{pmatrix}$. Pick representatives such that $[k,\ell] = 1$, $\ell + \lambda q \neq 0$, hence $[k, \ell+\lambda q, q] = [k,\ell,q] = 1$. Thus if t is the lowest common multiple of p and q, $[k, \ell+\lambda q, t] = 1$, and SL_2 acts mod$t$. Once more from the lemma we can find μ, u, v, x and y such that $(\tilde{k} = k + \mu t,\ \tilde{\ell} = \ell + q\lambda)$

$$u\tilde{k} + v\tilde{\ell} = 1 \qquad xk + y\ell = 1$$

We can choose these integers in such a way that $y \equiv v$ mod$q$. Indeed since ℓ is prime to $[k,q]$ and $(v-y)\ell \equiv 0$ mod$[k,q]$, we have $y - ek = v + fq$. Changing y into $y - ek$ and x into $x + e\ell$ we achieve our goal. Right multiplication by $\begin{pmatrix} x & -\ell \\ y & k \end{pmatrix} \begin{pmatrix} \tilde{k} & \tilde{\ell} \\ -v & u \end{pmatrix}$ in SL_2 transforms $\begin{pmatrix} k & \ell \\ 0 & 1 \end{pmatrix}$ into $\begin{pmatrix} \tilde{k} & \tilde{\ell} \\ y\tilde{k} - vk & y\tilde{\ell} + uk \end{pmatrix}$ with $\tilde{k}(= k + \mu t) \equiv k$ mod$p$, $\tilde{\ell} = k + \lambda q$, $y\tilde{k} - vk = y(k+\mu t) - vk \equiv 0$ mod$q$ and $y\tilde{\ell} + uk \equiv v\tilde{\ell} + uk \equiv v\tilde{\ell} + u\tilde{k} \equiv 1$ mod$q$. Applying $\begin{pmatrix} 1 & \mu \\ 0 & 1 \end{pmatrix}$ in SL_2 on the right, we conclude that

$$\begin{pmatrix} k & \ell \\ 0 & 1 \end{pmatrix} \sim \begin{pmatrix} k & \ell + \lambda q + \mu k \\ 0 & 1 \end{pmatrix}$$

first line mod$p$ second mod$q$, provided λ, μ are such that

$$[k,\ell,p] = [k, \ell + \lambda q + \mu k, p] = 1.$$

For arbitrary λ, μ, $\lambda q + \mu k$ mod$p$ runs over all multiples of $[k, p, q] = r$ mod$p$ thus on the same orbit as $\begin{pmatrix} k & \ell \\ 0 & 1 \end{pmatrix}$ one has all elements of the form $\begin{bmatrix} k & \ell + \sigma r \\ 0 & 1 \end{bmatrix}$ with $[k, \ell, p] = [k, \ell + \sigma r, p] = 1$, completing the proof that two pairs with the same $(\delta, \delta', \Delta, \Omega)$ are on the same orbit as

$$\begin{pmatrix} \delta\Delta & \delta\Omega \\ 0 & \delta' \end{pmatrix}$$

where we can choose $0 \le \Delta \le [m/\delta, mn/\delta'] - 1$ and $\Omega = 0$ if $[m/\delta, mn/\delta', \Delta] = 1$, otherwise we pick the smallest positive integer invertible mod$[m/\delta, mn/\delta', \Delta]$ such that $[\Delta, \Omega, m/\delta] = 1$. The counting formula follows and the proposition is proved.

4. The commutant

We return to the commutant of the group generated by S and T. To each $\mathbf{p} \in G_n = M^*/nM$ we assign an element $\mid \mathbf{p} >$ in an orthonormal basis of a finite (Hilbert) space $\mathcal{H}_n$ and unitary operators (in exponential notation)

$$Q^{\mathbf{p}} \mid \mathbf{p}' >= e\left(\frac{\mathbf{p}\cdot\mathbf{p}'}{nN}\right) \mid \mathbf{p}' > \qquad P^{\mathbf{p}} \mid \mathbf{p}' > = \mid \mathbf{p}' + \mathbf{p} >$$

Recall that scalar products have been rescaled by N. Labels in kets and operators are understood modnM, $(Q^{\mathbf{p}})^\dagger = Q^{-\mathbf{p}}$, $(P^{\mathbf{p}})^\dagger = P^{-\mathbf{p}}$ and

$$Q^{\mathbf{p}} P^{\mathbf{p}'} = e\left(\frac{\mathbf{p}\cdot\mathbf{p}'}{nN}\right) P^{\mathbf{p}'} Q^{\mathbf{p}} \qquad P^{\mathbf{p}} P^{\mathbf{p}'} = P^{\mathbf{p}+\mathbf{p}'} \qquad Q^{\mathbf{p}} Q^{\mathbf{p}'} = Q^{\mathbf{p}+\mathbf{p}'}$$

This representation of the Heisenberg group (a central extension of $G_n \times G_n$ by $\mathbf{Z}_{Nn}$) is irreducible and any operator V in $\mathcal{H}_n \otimes \mathcal{H}_n^*$ is uniquely written as

$$V = \sum_{\mathbf{p},\mathbf{p}'} P^{\mathbf{p}} Q^{\mathbf{p}'} \frac{1}{Nn^{N-1}} \mathrm{Tr}(V\, Q^{-\mathbf{p}'} P^{-\mathbf{p}})$$

From

$$S^{-1} \begin{pmatrix} Q^{\mathbf{p}} \\ P^{\mathbf{p}} \end{pmatrix} S = \begin{pmatrix} P^{\mathbf{p}} \\ Q^{-\mathbf{p}} \end{pmatrix} \qquad T^{-1} \begin{pmatrix} Q^{\mathbf{p}} \\ P^{\mathbf{p}} \end{pmatrix} T = \begin{pmatrix} Q^{\mathbf{p}} \\ e\left(-\frac{p^2}{2nN}\right) P^{\mathbf{p}} Q^{-\mathbf{p}} \end{pmatrix}$$

the group of canonical transformations generated by the adjoint action of S and T is a representation of the modular group (on $\mathcal{H}_n \otimes \mathcal{H}_n^*$). Set $\rho = e\left(\frac{1}{2Nn}\right)$, a primitive $2n$.th root of unity and use the natural projection π from G_{2n} to G_n to define the following symbols as operators on $\mathcal{H}_n$

$$(\mathbf{l},\mathbf{l}') \in G_{2n} \times G_{2n} \longrightarrow \{\mathbf{l},\mathbf{l}'\} = \rho^{\mathbf{l}.\mathbf{l}'} P^{\pi(\mathbf{l})} Q^{\pi(\mathbf{l}')}$$

These are not independent any more. Shifting $(\mathbf{l},\mathbf{l}')$ by $n(\mathbf{r},\mathbf{r}')$ with $\mathbf{r},\mathbf{r}'$ in M, they pick at most a sign

$$\{\mathbf{l}+n\mathbf{r},\mathbf{l}'+n\mathbf{r}'\} = (-1)^{((\mathbf{l}.\mathbf{r}'-\mathbf{l}'.\mathbf{r})/N)}\{\mathbf{l},\mathbf{l}'\}$$

showing that for any section $G_n \times G_n \longrightarrow G_{2n} \times G_{2n}$ we can still write

$$V = (1/(Nn^{N-1}))\Sigma\{\mathbf{l},\mathbf{l}'\}\mathrm{Tr}\; V\{\mathbf{l},\mathbf{l}'\}^{\dagger}.$$

Observe that $\rho^{\mathbf{l}^2} = \rho^{\pi(\mathbf{l})^2}$. The point is that

$$S^{-1}\{\mathbf{l},\mathbf{l}'\}S = \{\mathbf{l}',-\mathbf{l}\} \quad \text{and} \quad T^{-1}\{\mathbf{l},\mathbf{l}'\}T = \{\mathbf{l},\mathbf{l}'-\mathbf{l}\}$$

Equivalently, if L is a $(N-1)\times 2$ matrix with columns $\mathbf{l}$ and $\mathbf{l}'$ in G_{2n}

$$\{L\} \overset{S}{\longrightarrow} \left\{L\begin{pmatrix} 0 & -1 \\ 1 & 0 \end{pmatrix}\right\} \qquad \{L\} \overset{T}{\longrightarrow} \left\{L\begin{pmatrix} 1 & -1 \\ 0 & 1 \end{pmatrix}\right\}$$

As these 2×2 matrices generate SL_2 we have for any $A = \begin{pmatrix} a & b \\ c & d \end{pmatrix}$ in SL_2

$$\{L\} \overset{A}{\longrightarrow} \left\{L\begin{pmatrix} a & b \\ c & d \end{pmatrix}\right\}$$

Orbits ω of SL_2 on $G_{2n} \times G_{2n}$ are finite, the decomposition $G_{2n} \sim \mathbf{Z}_{2nN} \times (\mathbf{Z}_{2n})^{N-2}$ showing that SL_2 is only effective mod$2nN$. From section 3 the adjoint action provides therefore a (reducible) representation of $SL(2;\mathbf{Z}_{2nN})$. The operators

$$I_\omega = \sum_{(\mathbf{l},\mathbf{l}')\in\omega} \{\mathbf{l},\mathbf{l}'\}$$

obviously generate the commutant. We want to investigate which among them are linearly independent. Let $\dim_n$ be the dimension over $\mathbf{C}$ of the commutant and $n = \prod_p p^{\alpha(p)}$ the prime decomposition of n.

Proposition 3

(i) If N is odd the commutant admits a basis of matrices in one to one correspondence with SL_2 orbits in $G_n \times G_n$, while if N is even, we need only even orbits (generated by pairs $(\mathbf{p},\mathbf{p}')$ with $\mathbf{p}^2 \equiv \mathbf{p}'^2 \equiv$ mod4). As a result

$$\dim_2(n) \;= \sigma(n) = \prod_p (\alpha(p)+1)$$

$$\dim_3(n) \quad = 2n(\alpha(3)+1)\prod_{p\neq 3}(\alpha(p)+1+p^{-1}\alpha(p))$$

(ii) By a linear transformation the commutant admits a basis (over $\mathbf{C}$) consisting of matrices with integral entries. This is also a basis over $\mathbf{Q}$.

(iii) Fix $(\mathbf{p},\mathbf{p}') \in G_n \times G_n$ and pick any preimages $(\mathbf{l},\mathbf{l}')$ in $G_{2n}\times G_{2n}$. From $\rho^{\mathbf{l}^2} = \rho^{\pi(\mathbf{l})^2}$ it follows that for $A = \begin{pmatrix} a & b \\ c & d \end{pmatrix} \in SL_2$

$$\rho^{(a\mathbf{l}+c\mathbf{l}').(b\mathbf{l}+d\mathbf{l}')-\mathbf{l}.\mathbf{l}'} = \rho^{ab\mathbf{p}^2+cd\mathbf{p}'^2+2bc\mathbf{p}.\mathbf{p}'}$$

Hence if ω is an orbit in $G_{2n}\times G_{2n}$ containing $(\mathbf{l},\mathbf{l}')$ the operator $I_\omega\rho^{-\mathbf{l}.\mathbf{l}'}$ only depends on the orbit of the projection $(\pi(\mathbf{l}),\pi(\mathbf{l}'))$. If two pairs on the same orbit ω with $\pi(\mathbf{l}_1,\mathbf{l}_1') = \pi(\mathbf{l}_2,\mathbf{l}_2')$ are such that $\mathbf{l}_1\cdot\mathbf{l}_1' \neq \mathbf{l}_2\cdot\mathbf{l}_2' \bmod 2nN$ (but $\mathbf{l}_1\cdot\mathbf{l}_1' = \mathbf{l}_2{\cdot}\mathbf{l}_2' \bmod nN$) then $I_\omega = 0$. On the other hand if all pairs with equal projection have equal scalar product $\bmod 2nN$, the same will be true all along their orbit. Moreover the operators $P^{\mathbf{p}}Q^{\mathbf{p}'}$ span $\mathcal{H}_n\otimes\mathcal{H}_n^*$ and disjoint orbits in $G_n\times G_n$ involve linearly independent sets of operators. It is clear that a basis of the commutant corresponds to admissible orbits $\mathcal{O}$ of SL_2 on $G_n\times G_n$ for which when lifting $(\mathbf{p},\mathbf{p}')\in\mathcal{O}$ to $(\mathbf{l},\mathbf{l}')\in G_{2n}\times G_{2n}$ all pairs of their orbit $\omega(\pi(\omega)=\mathcal{O})$ that project on $(\mathbf{p},\mathbf{p}')$ have equal scalar product $\bmod 2nN$.

Recall that M^* is generated by $\boldsymbol{\alpha}^1,\boldsymbol{\alpha}_1,\ldots,\boldsymbol{\alpha}_{N-2}$. Consider a pair $(\mathbf{l},\mathbf{l}')$ and its transform by $A = \begin{pmatrix} a & b \\ c & d \end{pmatrix} \in SL_2$. Write $\mathbf{l}$ (resp.$\mathbf{l}'$) as $x_0\boldsymbol{\alpha}^1 + \sum_1^{N-2} x_i\boldsymbol{\alpha}_i$ $(y_0\boldsymbol{\alpha}^1 + \sum_1^{N-2} y_i\boldsymbol{\alpha}_i)$. The condition of equal projection translates into

$$\begin{aligned}(x_0,y_0)\left[\begin{pmatrix} a & b \\ c & d \end{pmatrix} - I\right] &\equiv 0 \bmod nN \\ (x_i,y_i)\left[\begin{pmatrix} a & b \\ c & d \end{pmatrix} - I\right] &\equiv 0 \bmod n \qquad 1\le i\le N-2\end{aligned}$$

Thus $(a\mathbf{l}+c\mathbf{l}')(b\mathbf{l}+d\mathbf{l}')-\mathbf{l}\cdot\mathbf{l}' \equiv (N-1)[(ax_0+cy_0)\,(bx_0+dy_0)\,-x_0y_0] \bmod 2Nn$. The r.h.s. is a multiple of $(N-1)Nn$ therefore zero mod $2Nn$ if N is odd. If N is even we can choose a matrix A is SL_2 such that

$$\begin{pmatrix} a & b \\ c & d \end{pmatrix} \equiv \begin{pmatrix} 1 & \beta nN \\ \gamma nN & 1 \end{pmatrix} \quad \bmod 2Nn$$

satisfying the condition of equal projection, for which the r.h.s. of the previous equation is equal to $nN(\beta x_0^2+\gamma y_0^2) \bmod 2Nn$. Since β and γ are arbitrary, we conclude that unless x_0 and y_0 are both even the corresponding invariant will vanish. Conversely if N, x_0, y_0 are all even $A\in SL_2$ satisfying the projection condition, it is readily found that $(ax_0 + cy_0)(bx_0 + dy_0) \equiv x_0 y_0 \bmod 2Nn$. The conditions $x_0 \equiv y_0 \equiv 0 \bmod 2$ are equivalent to $x_0^2 \equiv y_0^2 \equiv 0 \bmod 4$, hence ($N$ is even) to $\mathbf{l}^2 \equiv \mathbf{l}'^2 \equiv 0 \bmod 4$, and since $\mathbf{l}^2 \equiv \pi(\mathbf{l})^2 \bmod 2nN$, this is also $\pi(\mathbf{l})^2 \equiv \pi(\mathbf{l}')^2 \bmod 4$ (note that if $\mathbf{p} \in G_n$, $\mathbf{p}^2$ is well defined mod $2nN$, a multiple of 4 if N is even, and the above reasoning implies that if $(\mathbf{p}, \mathbf{p}') \in G_n \times G_n$, and $\mathbf{p}^2 \equiv \mathbf{p}'^2 \equiv 0 \bmod 4$ this is also true for any other pair along their SL_2−orbit).

To summarize, for any orbit $\mathcal{O}$ in $G_n \times G_n$ if N is odd or any even orbit if N is even, we lift a pair $(\mathbf{p}, \mathbf{p}') \in \mathcal{O}$ to a pair $(\mathbf{l}, \mathbf{l}') \in G_{2n} \times G_{2n}$ generating ω and use I_ω. Distinct choices $\mathcal{O} \neq \mathcal{O}'$ involve disjoint sets of independent monomials $P^{\mathbf{p}} Q^{\mathbf{p}'}$, which insures linear independence.

In general it is quite hard to even enumerate the orbits, however for $N = 2$ or 3, one derives explicit answers from proposition 2 giving the values for $\dim_2(n)$ (previously known) and $\dim_3(n)$ tabulated below (if n is prime it reduces to $2(2n+1)$)

n	1	2	3	4	5	6	7	8	9	10	11	12	13	14
dim_3	2	10	12	32	22	60	30	88	54	110	46	192	54	150

n	15	16	17	18	19	20	21	22	23	24	25	26	27
dim_3	132	224	70	270	78	352	180	230	94	528	170	270	216

(iv) We are interested in integral valued elements of the commutant. While the previous construction is canonical (up to overall phases which can be chosen as $2Nn$-th roots of unity) we have not been as successful for integral matrices. The proposition does not claim that any integral matrix in the commutant is expandable on such a basis with integral coefficients but (using the theorem on incomplete bases) that its expansion involves at most rational coefficients (with added work one could find a largest common multiple for all denominators).

Consider an orbit $\mathcal{O}$ in $G_n \times G_n$ (an even one if N is even). Let ω be a covering orbit in $G_{2n} \times G_{2n}$ with an associated subgroup $G \subset G_{2n}$ generated by any pair in ω, G isomorphic to $\mathbf{Z}_u \times \mathbf{Z}_{uv}$. For $(\mathbf{l}, \mathbf{l}') \in \omega$ and $\mathbf{x}$ and $\mathbf{x}'$ in G, we can find an integral matrix such that $(\mathbf{x}, \mathbf{x}') = (\mathbf{l}, \mathbf{l}')M$. We set $B_\omega(\mathbf{x}, \mathbf{x}') = \det M$ defined mod$u$ (even though $\mathbf{x}, \mathbf{x}'$ need not generate G), invariant if we pick another pair on ω or change $\mathbf{x}, \mathbf{x}'$ into another pair of the orbit ω' it generates in $G \times G \subset G_{2n} \times G_{2n}$, and bilinear in both rows of M, hence in $\mathbf{x}$ and $\mathbf{x}'$. For fixed G and varying ω we get $\varphi(u)$ different B_ω's (we could write $B_\omega(\omega')$).

In $G_{2n} \sim \mathbf{Z}_{2nN} \times (\mathbf{Z}_{2n})^{N-1}$ the order of any element divides $2nN$, consequently for the subgroup $G \sim \mathbf{Z}_u \times \mathbf{Z}_{uv}$, the integer u divides $2nN$. For integral t(mod$u$) let us compute the matrix elements of the well-defined operator

$$\mathcal{I}_{\omega,t} = \frac{1}{|G|} \sum_{\mathbf{x},\mathbf{x}' \in G} \rho^{(2nNt/u) B_\omega(\mathbf{x},\mathbf{x}')} \{\mathbf{x}, \mathbf{x}'\}$$

With $\mathbf{p}, \mathbf{p}' \in G_n$ we have $\langle \mathbf{p}|\{\mathbf{x}, \mathbf{x}'\}|\mathbf{p}'\rangle = \delta_{\mathbf{p},\mathbf{p}'+\pi(\mathbf{x})}\rho^{\mathbf{x}.\mathbf{x}'+2\mathbf{x}'.\mathbf{p}'}$. The sum

$$\sum_{\mathbf{x}'\in G} \rho^{\mathbf{x}.\mathbf{x}'+2\mathbf{x}'.\mathbf{p}'+(2nNt/u)B_\omega(\mathbf{x},\mathbf{x}')}$$

vanishes if for some $\mathbf{y}' \in G$ the exponent is non vanishing mod$2nN$. This is seen by writing

$$\sum_{\mathbf{x}'\in G} \equiv \frac{1}{2nN} \sum_{m \bmod 2nN} \sum_{\mathbf{x}'+m\mathbf{y}'\in G}$$

and inverting the orders of the sums (using the bilinear property of B_ω). The average over m yields zero. If the exponent vanishes mod$2nN$ for any $\mathbf{x}' \in G$ the sum yields $\mid G \mid$. As a consequence for any $\omega, t, \mathcal{I}_{\omega,t}$ has integral matrix elements.

If ω and ω' generate the same subgroup G and ω'' is a SL_2–orbit in $G \times G$ we have $B_\omega(\omega'') \equiv B_\omega(\omega')B_{\omega'}(\omega')$ mod$u$ by definition. Hence $\mathcal{I}_{\omega,t} = \mathcal{I}_{\omega',tB_\omega(\omega')}$. Grouping orbits we can write

$$\mathcal{I}_{\omega,t} = \frac{1}{\mid G \mid} \sum_{\omega'\subset G\times G} \rho^{(2nNt/u)B_u(\omega')} I_{\omega'}$$

This relation can be inverted. To an abelian group H we can assign an invariant m equal to the maximal length of a chain of distinct subgroups (Jordan-Hölder),

$$H_0 \equiv \{0\} \subsetneq H_1 \subsetneq \ldots \subsetneq H_m \equiv H\ .$$

To each orbit ω we associate $m(\omega)$, the invariant of the subgroup $G \subset G_{2n}$ it generates. For $\omega_0 = \{(0,0)\}$, $\mathcal{I}_{\omega_0,0} = \mathcal{I}_{\omega_0}$. Assume recursively that we can express all I_ω's with $m(\omega) \leq m$ as linear combinations of $\mathcal{I}_{\omega',t}$'s $(m(\omega') \leq m)$. If $m(\omega) = m + 1$, the associated G contains $\varphi(u)$ generating orbits $(u \equiv u(G))$ of length $m + 1$, all others (generating subgroups of G) have length smaller or equal to m, the corresponding I's being therefore expressible in terms of $\mathcal{I}$'s. In the above relation we bring those I's to the ℓ.h.s., then Fourier transform both sides in $t \in \mathbf{Z}_u$, getting expressions for the I_ω's such that ω generates G $(m(\omega) = m + 1)$ in terms of $\mathcal{I}$'s, or zero, hence relations (if any) or identities. This completes the proof by induction taking a maximal free subset of the $\mathcal{I}$'s.

C.I. would like to thank his hosts in Japan for making him aware of a work by S. Tanaka, "Irreducible representations of the binary congruence group modp^λ", J. Math. Kyoto Univ. **7**, 123-132 (1967) akin to part of the present work.

Conformal Field Theory on a Riemann Surface

H. Ooguri

Enrico Fermi Institute and Department of Physics,
University of Chicago, 5640 S. Ellis Ave., Chicago, IL60637, USA

Abstract. One of the motivations for studying conformal field theory on a Riemann surface is to have a better understanding of the structure of quantum string theory. Recent research has revealed that there are deep connections between conformal field theory and number theory, which may have profound implications to string theory. The equivalence of bosons and fermions in two dimensions is such an example; there an addition theorem of theta-function plays an important role.

1. Bosonization

In this section, I would like to explain the equivalence of bosons and fermions on a two-dimensional plane. In this simple topology, the operatorial description of a free fermion system is available, and one can establish the equivalence between bosons and fermions as an operatorial identity. In later sections, we extend this idea to a general Riemann surface.

Consider a set of fermion operators ψ_n, ψ_m^+ $(n, m \in \mathbf{Z} + \frac{1}{2})$ satisfying the following anti-commutation relations:

$$\begin{aligned} \{\psi_n, \psi_m^+\} &= \delta_{n+m,0} \, , \\ \{\psi_n, \psi_m\} &= 0 \, , \\ \{\psi_n^+, \psi_m^+\} &= 0 \, . \end{aligned} \tag{1}$$

Starting with a vacuum state $|0\rangle$

$$\psi_n|0\rangle = 0 \, , \; \psi_n^+|0\rangle = 0 \quad (n \geq \frac{1}{2}) \, , \tag{2}$$

the Fock space of free fermions is spanned by states of the form

$$|n_1, \ldots, n_N; m_1, \ldots, m_M\rangle = \psi_{-n_1} \ldots \psi_{-n_N} \psi_{-m_1}^+ \ldots \psi_{-m_M}^+ |0\rangle \, . \tag{3}$$

This Fock space is endowed with an inner product such that hermitian conjugates of ψ_n and ψ_m^+ are ψ_{-n} and ψ_{-m}^+ respectively. It is convenient to consider the fermion field operators $\psi(z)$ and $\psi^+(z)$ defined as

$$
\begin{aligned}
\psi(z) &= \sum_{n\in\mathbf{Z}+\frac{1}{2}} \psi_n z^{-n-1/2} \\
\psi^+(z) &= \sum_{n\in\mathbf{Z}+\frac{1}{2}} \psi_n^+ z^{-n-1/2} ,
\end{aligned} \tag{4}
$$

where z is a complex number. One can then easily compute the matrix element $\langle 0|\psi(z_1)\ldots\psi(z_N)\psi^+(w_N)\ldots\psi^+(w_1)|0\rangle$ to show that it is a rational function of z_i and w_i $(i = 1,\ldots,N)$ having poles at $z_i = w_j$ and totally antisymmetric in z_i's and w_i's. These are the properties we expect for correlation functions of free fermions on the complex plane.

One of the basic strategies of conformal field theory is to examine its infinite dimensional symmetry and to extract non-trivial operatorial identities from it. On the Fock space of free fermions, we can define a U(1) current operator $J(z)$ and an energy-momentum tensor operator $T(z)$ as

$$
\begin{aligned}
J(z) &= \sum_{n\in\mathbf{Z}} J_n z^{-n-1} , \quad J_n = \sum_{m\in\mathbf{Z}+\frac{1}{2}} : \psi_m \psi_{n-m}^+ : \\
T(z) &= \sum_{n\in\mathbf{Z}} L_n z^{-n-1} , \quad L_n = \sum_{m\in\mathbf{Z}+\frac{1}{2}} \frac{1}{2}n : \psi_m \psi_{n-m}^+ : ,
\end{aligned} \tag{5}
$$

where the normal ordering symbol $: \psi_n \psi_m^+ :$ in the above is defined as

$$
: \psi_n \psi_m^+ : = \begin{cases} \psi_n \psi_m^+ & (\text{if } m \geq 1/2) \\ -\psi_m^+ \psi_n & (\text{if } m < 1/2) . \end{cases} \tag{6}
$$

These operators make a closed algebra

$$
\begin{aligned}
[L_n, L_m] &= (n-m)L_{n+m} + \frac{1}{12}(n^3 - n)\delta_{n+m,0} \\
[L_n, J_m] &= -mJ_{n+m,0} , \quad [J_n, J_m] = n\delta_{n+m,0} .
\end{aligned} \tag{7}
$$

To construct a representation of this algebra, we start with a highest weight state $|h,q\rangle$ such that

$$
\begin{aligned}
L_0|h,q\rangle &= h|h,q\rangle , \quad J_0|h,q\rangle = q|h,q\rangle \\
L_n|h,q\rangle &= 0 \quad , \quad J_n|h,q\rangle = 0 \ (n \geq 0) .
\end{aligned} \tag{8}
$$

A highest weight representation is generated by acting L_{-n} and J_{-n} on $|h,q\rangle$. The Fock space of free fermions is decomposed into irreducible representation spaces of the algebra (7). In fact there are infinitely many states in the Fock space satisfying the highest weight condition (8). The following are examples of the highest weight states and their highest weight:

$$
\begin{aligned}
|0\rangle \quad (h &= 0,\ q = 0) \\
\psi_{-1/2}|0\rangle \quad (h &= \frac{1}{2},\ q = 1) \\
\psi^{+}_{-1/2}|0\rangle \quad (h &= \frac{1}{2},\ q = -1)\ .
\end{aligned}
\tag{9}
$$

In the algebra (7) it sometimes happens that a single representation space contains more than one highest weight state. For example, when a highest weight (h,q) satisfies $h = q^2/2$, it is easy to show that $|\eta\rangle = (L_{-1} - qJ_{-1})|h,q\rangle$ is also a highest weight state. However, such states are not allowed to exist in the Fock space where the algebra (7) is realized. Since hermitian conjugates of L_n and J_m are L_{-n} and J_{-m} respectively, the square norm of the state $|\eta\rangle$ is $\langle h,q|(L_1 - qJ_1)(L_{-1} - qJ_{-1})|h,q\rangle$ and it vanishes because $|\eta\rangle$ is a highest weight state. The state with zero norm is impossible in the Fock space where the inner product is non-degenerate, and the state $|\eta\rangle$ itself must vanish. Thus we have obtained the identity

$$
(L_{-1} - qJ_{-1})\ |h = q^2/2\ ,q\rangle \ = \ 0\ . \tag{10}
$$

The highest weight states listed in eq.(9) satisfy the condition $h = q^2/2$. For example $L_{-1}|0\rangle$ should vanish identically. The other two examples have more important implications. Due to the property (2) of the vacuum state, $\psi_{-1/2}|0\rangle$ ($\psi^{+}_{-1/2}|0\rangle$) can also be written as $\psi(z=0)|0\rangle$ ($\psi^{+}(z=0)|0\rangle$). It is also easy to derive the following commutation relation:

$$
[L_n, \psi(z)] \ = \ (z^{n+1}\frac{d}{dz} + \frac{n+1}{2}z^n)\psi(z) \tag{11}
$$

and a similar equation for $\psi^{+}(z)$. We then obtain the following identity:

$$
\partial_z \psi(z=0)|0\rangle \ = \ [L_{-1}, \psi(0)]|0\rangle \ = \ L_{-1}\psi(0)|0\rangle \ = \ J_{-1}\psi(0)|0\rangle\ . \tag{12}
$$

It is a general phenomenon in conformal field theory that, if there is a relation among state vectors, there is a corresponding relation among operators. In our case the corresponding operatorial relation is

$$
\partial_z \psi(z) \ =: \ J(z)\psi(z)\ :\ . \tag{13}
$$

The normal ordering symbol : : on the right hand side is the same as the one we have defined in the above. By multiplying both sides by the vacuum state $|0\rangle$ and by setting $z = 0$, the above equation reduces to eq.(12). This equation is a first order differential equation in z and formally integrable as

$$\psi(z) = : \exp(\int_{z_0}^{z} dw J(w))\psi(z_0) : . \tag{14}$$

Thus the fermion field operator is expressed in terms of the bosonic operator $J(z)$. (We call $J(z)$ bosonic because it satisfies the commutation relation (7).) This relation is called the boson-fermion correspondence.

2. Bosonization on a Riemann Surface

We have derived the bosonization rule (14) on a plane using the representation theory of the algebra (7). Since we can also consider free fermions on a general Riemann surface Σ, it is natural to ask whether this relation holds on Σ [1]. For a general Riemann surface, we understand eq.(14) as relations among correlation functions. We compute correlation functions of free fermions and currents on Σ, and show that eq.(14) holds [2,3].

Let us first consider a correlation function of free fermions $\langle\psi(z_1)\ldots\psi^+(w_1)\rangle$. We assume that it is given by a functional integral

$$\begin{aligned}&\langle\psi(z_1)\ldots\psi(z_N)\psi^+(w_N)\ldots\psi^+(w_1)\rangle = \\ &\int[d\psi, d\psi^+]\psi(z_1)\ldots\psi(z_N)\psi^+(w_N)\ldots\psi^+(w_1)\exp(\int_\Sigma d^2z\psi^+\partial_{\bar{z}}\psi) ,\end{aligned} \tag{15}$$

where the integral is taken over Grassmannian valued fields $\psi(z)$ and $\psi^+(z)$ transforming as half-forms on Σ. We also assume that they are quasiperiodic, i.e. they are multiplied by phase factors when they go around homologically non-trivial cycles on Σ. Due to the Wick theorem for free fermions, the 2N-point correlation function in the above is expressed in terms of two-point functions as

$$\langle\psi(z_1)\ldots\psi(z_N)\psi^+(w_N)\ldots\psi^+(w_1)\rangle = \mathcal{Z}\cdot\det_{i,j}\left(\frac{\langle\psi(z_i)\psi^+(w_j)\rangle}{\mathcal{Z}}\right) , \tag{16}$$

where $\mathcal{Z}$ is the zero point function.

The two-point function $\langle\psi(z)\psi^+(w)\rangle$ defined as in the above has a unique pole at $z = w$. Its pole residue is equal to the zero-point function $\mathcal{Z}$. This fact combined with its periodicity on Σ uniquely characterize the two-point function. It can be expressed in terms of the Riemann theta-function as

$$\langle\psi(z)\psi^+(w)\rangle = \mathcal{Z}\cdot\frac{\vartheta\begin{bmatrix}\theta\\\phi\end{bmatrix}(\int_w^z\vec{\omega}(u)du|\Omega)}{\vartheta\begin{bmatrix}\theta\\\phi\end{bmatrix}(\vec{0}|\Omega)\cdot E(z,w)} . \tag{17}$$

I explain the notations in the above. On the Riemann surface Σ, we choose a standard basis of homology cycles $\{\alpha_i, \beta_j\}_{i=1,\ldots,g}$ (G is the genus of Σ); α_i and β_i intersect once and other pairs do not. With respect to this homology basis, we define a basis of holomorphic one-forms on Σ, $\{\omega^i(z)\}_{i=1,\ldots,g}$, normalized as

$$\oint_{\alpha_i} \omega^j(z)dz = \delta_{ij} . \tag{18}$$

Ω is the $g \times g$ period matrix, whose matrix element Ω_{ij} is given by a contour integral of $\omega^j(z)$ around the cycle β_i.The Riemann theta function $\vartheta \begin{bmatrix} \theta \\ \phi \end{bmatrix} (\vec{z}|\Omega)$, $(\vec{\theta}, \vec{\phi} \in \mathbf{R}^g, \vec{z} \in \mathbf{C}^g)$ is defined as

$$\vartheta \begin{bmatrix} \theta \\ \phi \end{bmatrix} (\vec{z}|\Omega) = \sum_{\vec{n} \in \mathbf{Z}^g} \exp \left[2\pi i \left(\frac{1}{2} \left(\vec{n} + \vec{\theta}\right)^t \Omega \left(\vec{n} + \vec{\theta}\right) + \left(\vec{n} + \vec{\theta}\right)^t \left(\vec{z} + \vec{\phi}\right) \right) \right] . \tag{19}$$

By choosing a pair of g-vectors, $\vec{\delta}_1, \vec{\delta}_2 \in \frac{1}{2}\mathbf{Z}^g/\mathbf{Z}^g$, such that $4\vec{\delta}_1 \cdot \vec{\delta}_2$ is an odd integer, we define the prime form $E(z,w)$ as

$$E(z,w) = \frac{\vartheta \begin{bmatrix} \delta_1 \\ \delta_2 \end{bmatrix} (\int_w^z \vec{\omega}(u)du|\Omega)}{\sqrt{\zeta(z)}\sqrt{\zeta(w)}} \tag{20}$$

$$\sqrt{\zeta(z)} = \sum_{i=1}^{g} \omega^i(z) \frac{\partial}{\partial \delta_2^i} \vartheta \begin{bmatrix} \delta_1 \\ \delta_2 \end{bmatrix} (0|\Omega) . \tag{21}$$

The prime form has a unique zero at $z = w$ and does not depend on the choice of $\vec{\delta}_1$ and $\vec{\delta}_2$. The two-point function (17) given by assembling these elements has a unique pole at $z = w$ with a pole residue $\mathcal{Z}$. When the fermion $\psi(z)$ goes around the homology cycle α_i (β_i), the two-point function $\langle\psi(z)\psi^+(w)\rangle$ is multiplied by a phase factor $e^{2\pi i\theta_i}$ $\cdot(e^{-2\pi i\phi_i})$.

Let us now discuss correlation functions involving currents. The current $J(z)$ is defined as a field bilinear in fermions:

$$\langle J(z)\ldots\rangle = \lim_{w \to z} \left(\langle\psi(z)\psi^+(w)\ldots\rangle - \frac{1}{z-w}\langle\ldots\rangle \right) . \tag{22}$$

Thus it has the following operator product expansions with other fields:

$$J(z)\psi(w) \sim \frac{1}{z-w}\psi(w) , \quad J(z)\psi^+(w) \sim \frac{1}{z-w}\psi^+(w) ,$$
$$J(z)J(w) \sim \frac{1}{(z-w)^2} \quad (z \sim w) .$$

We assume that the current $J(z)$ is holomorphic elsewhere and behaves as a one-form on Σ. Now I am going to show that these properties uniquely determine correlation functions of the currents. A one-form on Σ can be expressed as a sum of two terms, a derivative of some scalar function ϕ on Σ, and a linear combination of the holomorphic one-forms:

$$\begin{aligned} &\langle J(z)J(w_1)\dots J(w_N)\psi(u_1)\dots\psi(u_M)\psi^+(v_M)\dots\psi^+(v_1)\rangle \\ &= \partial\phi(z;w_1,\dots,v_1)+\sum_{i=1}^{g}\omega^i(z)p_i(w_1,\dots,u_1)\ . \end{aligned} \tag{24}$$

As a function of z, ϕ is uniquely determined by its poles and single-valuedness on Σ, and expressed using the prime form $E(z,w)$ as

$$\begin{aligned} \phi(z) &= \\ &\sum_{i=1}^{N}\partial_{w_i}\log(E(z,w_i))\langle J(w_1)\dots J(w_{i-1})J(w_{i+1})\dots J(w_N)\psi(u_1)\dots\psi^+(v_1)\rangle \\ &+\sum_{i=1}^{M}\log\left(\frac{E(z,u_i)}{E(z,v_i)}\right)\langle J(w_1)\dots J(w_N)\psi(u_1)\dots\psi^+(v_1)\rangle\ . \end{aligned} \tag{25}$$

To compute p^i, we consider a contour integral of the current $J(z)$ around the α_i-cycle. Due to the single-valuedness of $\phi(z)$ and normalization of the holomorphic one-forms, we obtain

$$p^i = \langle\oint_{\alpha_i} dz J(z)J(w_1)\dots\psi^+(v_1)\rangle\ . \tag{26}$$

The fermion $\psi(z)$ is quasi-periodic on Σ and multiplied by a phase factor $e^{2\pi i\phi_i}$ ($e^{-2\pi i\theta_i}$) when it goes around the β_i (α_i)-cycle (ϕ_i and θ_i appear in the correlation function of the fermions as in eq.(17).) Because of the operator product expansion

$$J(z)\psi(w)\sim\frac{1}{z-w}\psi(w)\ ,\ J(z)\psi^+(w)\sim\frac{-1}{z-w}\psi^+(w)\ , \tag{27}$$

the insertion of the contour integral of the current along the α_i-cycle changes the periodicity of the fermion around its dual cycle β_i. Thus we obtain

$$p^i = \frac{\partial}{\partial\phi_i}\langle J(w_1)\dots J(w_N)\psi(u_1)\dots\psi^+(v_1)\rangle\ . \tag{28}$$

By substituting eqs.(25) and (28) into eq.(24), we obtain the recurrence relation which relates a correlation function with $(N+1)$ currents to ones with N and $(N-1)$ currents. Thus arbitrary correlation functions containing currents can be derived inductively from correlation functions without current insertion.

So far we have not discussed the zero-point function $\mathcal{Z}(\theta,\phi)$. Due to the recurrence relation in the above, its ϕ-derivatives are related to the one-point function of the current as

$$\langle J(z)\rangle = \sum_i \omega^i(z)\frac{\partial}{\partial\phi_i}\mathcal{Z}(\theta,\phi) . \tag{29}$$

On the other hand the current is bilinear in fermion fields (22), so the one-point function of the current is related to the known two-point function (17) of fermions

$$\begin{aligned}\langle J(z)\rangle &= \lim_{w\to z}\left(\langle\psi(z)\psi^+(w)\rangle - \frac{1}{z-w}\right)\\ &= \mathcal{Z}\cdot\frac{\partial}{\partial\phi_i}\log\left(\vartheta\begin{bmatrix}\theta\\ \phi\end{bmatrix}(\vec{0}|\Omega)\right) .\end{aligned} \tag{30}$$

Comparing the above two equations, we find

$$\frac{\partial}{\partial\phi_i}\left(\frac{\mathcal{Z}(\theta,\phi)}{\vartheta[^\theta_\phi](\vec{0}|\Omega)}\right) = 0 , \tag{31}$$

i.e. the ϕ-dependence of the zero-point function $\mathcal{Z}$ is given by the theta-constant.

By assembling the above results, we can compute correlation functions of the currents. For example, their two-point function is given by

$$\frac{1}{\mathcal{Z}}\langle J(z)J(w)\rangle = \partial_z\partial_w\log(E(z,w)) + \sum_{i,j}\omega^i(z)\omega^j(z)\frac{1}{\vartheta[^\theta_\phi]}\frac{\partial^2}{\partial\phi_i\partial_j}\vartheta\begin{bmatrix}\theta\\ \phi\end{bmatrix} . \tag{32}$$

Now we are ready to verify the bosonization rule (14) on Σ. Previously the 2N-point function of fermions is computed using the Wick theorem as in eq.(16). On the other hand, the bosonization rule implies

$$\langle\psi(z_1)\ldots\psi(z_N)\psi^+(w_N)\ldots\psi^+(w_1)\rangle = \langle\prod_{i=1}^N\exp(\int_{w_i}^{z_i}J(z)dz)\rangle . \tag{33}$$

The right hand side can be computed by expanding the exponentials in powers of currents and using the correlation functions of the currents given in the above. The result is

$$\langle\prod_{i=1}^N\exp(\int_{w_i}^{z_i}J(z)dz)\rangle = \mathcal{Z}\cdot\frac{\prod_{i<j}E(z_i,z_j)E(w_j,w_i)}{\prod_{i,j}E(z_i,w_j)}\frac{\vartheta[^\theta_\phi]\left(\sum_i\int_{w_i}^{z_i}\vec{\omega}(z)dz|\Omega\right)}{\vartheta[^\theta_\phi](\vec{0}|\Omega)} . \tag{34}$$

Due to the addition theorem of the theta-function, this expression is cast into the form

$$\langle\psi(z_1)\ldots\psi(z_N)\psi^+(w_N)\ldots\psi^+(w_1)\rangle = \mathcal{Z}\cdot\det_{ij}\left(\frac{\vartheta[\frac{\vec\theta}{\vec\phi}]\left(\int_{w_j}^{z_i}\vec\omega(u)du|\Omega\right)}{\vartheta[\frac{\vec\theta}{\vec\phi}](0|\Omega)E(z_i,w_j)}\right), \tag{35}$$

which is equal to the left hand side of eq.(33). Thus the bosonization rule is valid on Σ. The addition theorem of this kind plays an important role in number theory (I refer the reader to the lectures by Bost).

3. Infinite Dimensional Grassmannian Manifold

To appreciate the geometrical significance of the addition theorem, let us consider the following generating functional for correlation functions of currents [4]. Choose an arbitrary point p on Σ and a coordinate z around p such that $z(p) = \infty$. Considering a formal power series of z, $\xi(z;x) = \sum_{n=1}^{\infty} x_n z^{-n}$, a generating function is defined as

$$\tau(x_1, x_2, \ldots) = \langle\exp(\oint_p \frac{dz}{2\pi i}\xi(z;x)J(z))\rangle\ . \tag{36}$$

By expanding the right hand side in powers of $J(z)$ and using the correlation functions of the currents, one can evaluate $\tau(x)$ as

$$\tau(x) = \mathcal{Z}\cdot\exp\left(\sum_{n,m=1}^{\infty} Q_{nm}x_n x_m\right)\frac{\vartheta[\frac{\theta}{\phi}]\left(-\sum_{n=1}^{\infty}\vec A_n x_n|\Omega\right)}{\vartheta[\frac{\theta}{\phi}](0|\Omega)}, \tag{37}$$

where Q_{nm} and $\vec A_n$ are the Taylor expansion coefficients of the prime form $E(z,w)$ and the holomorphic one-forms $\omega^i(z)$ near ∞:

$$\begin{aligned}\partial_z\partial_w\log(E(z,w)) &= \frac{1}{(z-w)^2} + 2\sum_{n,m\geq 1} Q_{nm}z^{-n-1}w^{-m-1}\\ \vec\omega(z) &= -\sum_{n\geq 1}\vec A_n z^{-n-1}\ .\end{aligned} \tag{38}$$

It is easy to check that correlation functions of currents are derived from $\tau(x)$ as

$$\langle J(z_1)\ldots J(z_N)\rangle = \mathcal{J}(z_1;x)\ldots\mathcal{J}(z_N;x)\tau(x)\Big|_{x_n=0,\,n=1,2,\ldots}, \tag{39}$$

where

$$\mathcal{J}(z;x) = \partial_z\xi(z;x) + \sum_{n\geq 1}\frac{\partial}{\partial x_n}z^{-n-1}\ . \tag{40}$$

We are going to see that the function $\tau(x)$ satisfies an intriguing identity. Because of the boson-fermion correspondence (33), the function $\tau(x)$ also generates correlation functions of fermions.

Defining a vertex operator $V(z; x_0, x_1, x_2, \ldots)$ as

$$V(z;x) = \exp(\sum_{n\geq 0} x_n z^n) \exp\left(\log(z)\frac{\partial}{\partial x_0} - \sum_{n\geq 1} \frac{1}{n} z^{-n} \frac{\partial}{\partial x_n}\right) \tag{41}$$

the 2N-point correlation function of free fermions in the presence of the current is given by

$$\begin{aligned} &\langle \exp(\oint_p \frac{dz}{2\pi i}\xi(z;x)J(z))\psi(z_1)\ldots\psi(z_N)\psi^+(w_N)\ldots\psi^+(w_1)\rangle \\ &= V(z_1;x)\ldots V(z_N;x)V(w_N;-x)\ldots V(w_1;-x)\tau(x) \\ &= \frac{\prod_{i<j}(z_i - z_j)(w_j - w_i)}{\prod_{i,j}(z_i - w_j)} \cdot \tau\left(x - \sum_{i=1}^{N}([z_i] - [w_i])\right) \\ &\text{with} \quad [z]_n = \frac{1}{n} z^{-n} . \end{aligned} \tag{42}$$

The two-point function is

$$\langle \exp(\oint_p \frac{dz}{2\pi i}\xi(z;x)J(z))\psi(z)\psi^+(w)\rangle = \frac{1}{z-w} \cdot \tau(x - [z] + [w]) . \tag{43}$$

According to the Wick theorem (as in eq.(16)), the 2N-point function is a product of two-point functions. This implies the following identity for $\tau(x)$:

$$\begin{aligned} &\frac{\prod_{i<j}(z_i - z_j)(w_j - w_i)}{\prod_{i,j}(z_i - w_j)} \cdot \frac{\tau(x - \sum_{i=1}^{N}([z_i] - [w_i]))}{\tau(x)} \\ &= \det_{i,j}\left(\frac{1}{z_i - w_j} \frac{\tau(x - [z_i] + [w_j])}{\tau(x)}\right) . \end{aligned} \tag{44}$$

To understand the geometric meaning of this identity, let us introduce the following function:

$$\zeta(w_1, \ldots, w_N) = \prod_{i<j}(w_i - w_j) \cdot \tau(\sum_{i=1}^{N}[w_i]) . \tag{45}$$

Substituting $x = \sum_{i=1}^{N}[z]$ in eq.(44) and rearranging the equation, we obtain

$$\sum_{k=1}^{N+1}(-1)^k \zeta(z_1, \ldots, z_{N-1}, w_k)\zeta(w_1, \ldots, w_{k-1}, w_{k+1}, \ldots, w_{N+1}) = 0 . \tag{46}$$

This equation is reminiscent of the Plücker relation for a Grassmannian manifold.

A Grassmannian manifold $Gr(m,n)$ is a space of m-dimensional hyperplanes in the n-dimensional Euclidean space $(m < n)$. The hyperplane is defined by a basis of its tangent vectors, $\zeta_1^i, \ldots, \zeta_m^i$ (i is a vector index $i = 1, \ldots, n$). Since the choice of a basis is rather arbitrary, ζ_j^i itself does not make a good coordinate of $Gr(m,n)$. To define coordinates on $Gr(m,n)$ we consider

$$\zeta^{i_1 \ldots i_m} = \det \begin{pmatrix} \zeta_1^{i_1} & \ldots & \zeta_m^{i_1} \\ \vdots & & \vdots \\ \zeta_1^{i_m} & \ldots & \zeta_m^{i_m} \end{pmatrix} . \tag{47}$$

This defines a well-defined map of $Gr(m,n)$ into a projective space $P(\wedge^m \mathbf{C}^n)$ in a sense that the image of the hyperplane is independent of a choice of its basis ζ_j^i. Although this map is not surjective, it is possible to specify the image of $Gr(m,n)$ in the projective space. The image is defined by the following set of equations known as the Plücker relation:

$$\sum_{k=1}^{m+1} (-1)^k \zeta^{i_1 \ldots i_{m-1} j_k} \zeta^{j_1 \ldots j_{k-1} j_{k+1} \ldots j_{m+1}} = 0 . \tag{48}$$

In view of the similarity between eqs.(46) and (48), one might suspect that the generating function $\tau(x)$ gives coordinates for some Grassmannian manifold. In fact it was shown by Sato [5] that eq.(46) defines an image of an infinite dimensional Grassmannian manifold into a projective space. Subsequently, Date, Jimbo, Kashiwara and Miwa [6] proved that, when $\tau(x)$ satisfies eq.(46), there exists a state vector $|\tau\rangle$ in the Fock space of free fermions such that $\tau(x)$ is expressed as

$$\tau(x) = \langle 0| \exp\left(\sum_{n=1}^{\infty} x_n J_n \right) |\tau\rangle . \tag{49}$$

Therefore, for each Riemann surface Σ (with some additional structure like periodicity of fermions, choice of a point p, a coordinate z around p), we can associate a state vector $|\tau\rangle$ which contains all the information on correlation functions of fermions on Σ. Using this result, one can develop an operator formalism for free fermions on a Riemann surface [7,8], which has proved useful in studying quantum string theory.

Acknowledgement

This work is supported in part by the United States Department of Energy under grant number DE-AC-02-76ER02220.

References

For more discussions on the boson-fermion correspondence on a Riemann surface see for example

[1] L. Alvarez-Gaumé, J. B. Bost, G. Moore, P. Nelson and C. Vafa, Phys. Lett. **B 178** (1986) 41; J. B. Bost and P. Nelson, Phys. Rev. Lett. **57** (1986) 795.

[2] E. Verlinde and H. Verlinde, Nucl. Phys. **B 288** (1987) 357.

[3] T. Eguchi and H. Ooguri, Phys. Lett. **B 187** (1987) 127.

The function $\tau(x)$ first appeared in the context of solvable non-linear partial differential equations, see for example

[4] R. Hirota, in Lect. Notes in Math. **515** (Springer 1976).

It was noticed by Sato that the equation for $\tau(x)$ is the Plücker relation, as explained in

[5] M. Sato and Y. Sato, in Non-linear Partial Differential Equations in Applied Science (North-Holland 1983).

[6] E. Date, M. Jimbo, M. Kashiwara and T. Miwa, in Proc. Int. Symp. on Non-linear Integrable Systems (World Scientific 1983).

Operator formalism of free fermions on a Riemann surface was discussed in

[7] N. Ishibashi, Y. Matsuo and H. Ooguri, Mod. Phys. Lett. **A2** (1987) 119.

[8] L. Alvarez-Gaumé, C. Gomez and C. Reina, Phys. Lett. **B 190** (1987) 55.

Modular Invariance of Field Theories and String Compactifications

A. Shapere

The Institute for Advanced Study, Olden Lane, Princeton, NJ 08540, USA

Abstract. There is a large class of Gaussian lattice models which are self-dual with respect to inversion of coupling constants. When a theta term is added, this duality may extend to an invariance under an action of an infinite discrete modular group on the coupling parameter space. In particular, 4-dimensional Abelian lattice gauge theories possess a $Sp(2k, \mathbf{Z})$ modular symmetry, and string compactifications on d-dimensional tori are invariant under $SO(16+d, d; \mathbf{Z})$. We review these results and their applications, and discuss the classification of self-dual string compactifications.

1. Introduction

Duality, in the sense of Kramers and Wannier [1], is an equivalence between statistical models at high and low temperatures. A model is said to be self-dual if its partition function is left invariant by a duality transformation. Typically, the effect of a duality transformation is to interchange two physical variables; in the Ising model, duality exhanges order and disorder parameters, while in gauge theories it exhanges electric and magnetic charge. An interesting extension of Kramers–Wannier duality was first pointed out by Cardy [2]. He found, in a particular class of models [3], a form of self-duality that leads to an astonishingly rich structure over coupling parameter space. In these models, the partition function is invariant under two types of transformations, which close on an infinite discrete group.

Many self-dual models may be extended by adding a topological "theta term" to the action. In general, these terms are total derivatives, like $\theta F_{\mu\nu}\widetilde{F}_{\mu\nu}$. As such, they have no effect on the classical equations of motion, and their contribution to the action is only nonzero in the presence of fields with non-trivial global topology. This contribution is in general quantized, so that the dependence of the physics on the coefficient θ of such a term is periodic.

Theta terms are most familiar in the context of non-Abelian gauge theories, where they count winding numbers of mappings of the three-sphere at infinity into the gauge group. They may also appear in abelian gauge theories with monopoles. Cardy studied these theories on a lattice, and found an extension of the notion of self-duality to the case $\theta \neq 0$. Namely, for gauge group $\mathbf{Z}_N$, the

Springer Proceedings in Physics, Vol. 47 **Number Theory and Physics**
Editors: J.M. Luck · P. Moussa · M. Waldschmidt

partition function $Z[g,\theta]$ is invariant under the transformations

$$S: \tau \to -1/\tau \tag{1.1}$$

$$T: \tau \to \tau + 1 \tag{1.2}$$

where τ is a complex coupling parameter

$$\tau = \frac{\theta}{2\pi} + i\frac{2\pi}{Ng^2} \tag{1.3}$$

and g is the gauge coupling. When $\theta = 0$ the second transformation reduces to the duality transformation $g \to 2\pi/Ng$. Together, these two transformations generate the modular group $SL(2,\mathbf{Z})$.

Recently, these results were extended in several ways [5]. It was shown that more complicated four–dimensional abelian gauge theories, with gauge group $\mathbf{Z}_N \times \cdots \times \mathbf{Z}_N$, admit an action of $Sp(2k,\mathbf{Z})$ (where k is the number of factors of $\mathbf{Z}_N$). Self-dual models in d dimensions may also be constructed, with concommitantly generalized theta terms.

For the case of $d = 2$, this generalized duality symmetry has some interesting applications to toroidal compactifications in string theory. Bosonic and heterotic string compactifications on a circle of radius R are self-dual with respect to the transformation $R \to 1/2R$, if R is measured in Planck units. It turns out that the generalization to d-dimensional toroidal compactifications in the presence of arbitrary constant background fields G and B is

$$B + G \to \tfrac{1}{4}(B+G)^{-1} \tag{1.4}$$

where the matrix inverse is indicated on the right hand side. This transformation, together with periodicities in B and $GL(d,\mathbf{Z})$ changes of basis, leads to an action of a discrete modular group $SO(d,d;\mathbf{Z})$ (actually $SO(16+d,d;\mathbf{Z})$ for heterotic strings) on the moduli space of toroidal compactifications. Here, modular invariance gives us information about the global geometry of the space of compactifications and is useful for finding special symmetric backgrounds.

2. Self–Dual Models with Theta Terms

The self-dual gauge model studied by Cardy and Rabinovici [3] is an Abelian lattice model with gauge group $\mathbf{Z}_N$, with partition function

$$Z = \mathrm{Tr}_{\{\phi,n,s\}} \exp\Big[-\frac{1}{2g^2}\sum_P(\Delta_\mu\phi_\nu - \Delta_\nu\phi_\mu - 2\pi s_{\mu\nu})^2 + iN\sum_L n_\mu\phi_\mu$$
$$+ \frac{iN\theta}{32\pi^2}\sum_{r,r'} f(r-r')\,\epsilon_{\mu\nu\lambda\sigma}(\Delta_\mu\phi_\nu - \Delta_\nu\phi_\mu - 2\pi s_{\mu\nu})(r)$$

$$\cdot (\Delta_\lambda \phi_\sigma - \Delta_\sigma \phi_\lambda - 2\pi s_{\lambda\sigma})(r') \Big] . \tag{2.1}$$

ϕ_μ is a gauge field defined on the links of a four-dimensional lattice, with values from 0 to 2π; n_μ and $s_{\mu\nu}$ are integer-valued fields defined respectively on links and plaquettes. The sum over n_μ has the effect of constraining ϕ_μ to take its values in $\mathbf{Z}_N$, $\phi_\mu = \frac{2\pi}{N} k_\mu$. This sum is itself constrained by $\Delta_\mu n_\mu = 0$, since ϕ_μ can only couple to a conserved current. We interpret n_μ as the electric current of a charged field. Finally, f is a smearing function that interpolates between a plaquette and its nearest perpendicular neighbors. In this section, we make an approximation by ignoring f. (Later on, we will eliminate the need for a smearing function by placing dual fields on dual lattices.)

The first term in the action (2.1) is a Villain–type [4] kinetic term for the gauge field; it may be cast in a more recognizable form by writing

$$agF_{\mu\nu} \equiv \Delta_\mu \phi_\nu - \Delta_\nu \phi_\mu - 2\pi s_{\mu\nu} \tag{2.2}$$

where a is the lattice spacing; the kinetic term then becomes $-\frac{1}{2}F^2$ in the continuum limit, when the sum is replaced by an integral. Treating ϕ_μ as an angle and separating out $s_{\mu\nu}$ amounts to allowing multivalued configurations for the gauge field; the sum on $s_{\mu\nu}$ corresponds to a sum over topologically nontrivial configurations with magnetic monopoles. In fact, the monopole current is $m_\mu = \frac{1}{2}\epsilon_{\mu\nu\lambda\sigma}\Delta_\nu s_{\lambda\sigma}$, so that the final term couples m_μ directly to the gauge field. When summed over the lattice, the $\Delta\phi\Delta\phi$ pieces in this term vanish, and what is left of the sum is always a multiple of $16\pi/N$. Hence Z is periodic in $\theta \to \theta + 2\pi$. This last term, the theta term, approximates $F_{\mu\nu}\widetilde{F}_{\mu\nu}$.

When $\theta = 0$, the partition function (2.1) may be shown to be self-dual under $g \to 2\pi/Ng$. One way to show this is to integrate out the gauge field; the resulting action describes an electromagnetic Coulomb gas for electric and magnetic charges quantized in units of Ng and $2\pi g^{-1}$, respectively. Physically, the effect of the duality transformation is simply to interchange the electric and magnetic currents n_μ and m_μ [3][6]. Thus, correlation functions at gauge coupling g are equal to the corresponding correlation functions with electric and magnetic operators interchanged, at gauge coupling $2\pi/Ng$.

The appropriate generalization of duality to nonzero θ is not at first glance obvious, but after a bit of straightforward algebra, Eq.(1.1) is shown to be the correct transformation. The crucial point is that this duality does not commute with $\theta \to \theta + 2\pi$, and the failure of these symmetries to commute generates an infinite discrete symmetry group of the partition function. Indeed, in the notation of Eqs.(1.1)–(1.2), a general group element may be represented as

$$T^{a_n} S T^{a_{n-1}} S \cdots S T^{a_1} \tag{2.3}$$

where the a_i are all integers. The effect of this transformation on τ is then most readily expressed in the form of a continued fraction:

$$\tau \rightarrow a_n - \cfrac{1}{a_{n-1} - \cfrac{1}{\cdots - \cfrac{1}{a_1 - \tau}}} \tag{2.4}$$

A transformation of this type may always be collapsed to the form

$$\tau \rightarrow \frac{a\tau + b}{c\tau + d} \qquad ad - bc = 1; \qquad a, b, c, d \in \mathbf{Z} \tag{2.5}$$

Conversely, any such transformation may be expanded uniquely as a continued fraction (2.4). The transformations (2.5) form a representation of the modular group $SL(2, \mathbf{Z})$ on the coupling parameter space. (In fact, this representation of $SL(2, \mathbf{Z})$ is two–to–one, since changing all the signs of a, b, c, d gives a different matrix but the same mapping.)

The existence of this large group of symmetries has immediate implications for the phase structure of this model. If we suppose that there is a unique phase transition in g when $\theta = 0$, then by Eq. (2.4) the critical point must be at $g = (2\pi/N)^{1/2}$, or $\tau = i$. Away from the imaginary τ axis (when $\theta \neq 0$), the duality transformation $\tau \rightarrow -1/\tau$ interchanges the interior and exterior of the unit circle. So until the phase boundary bifurcates or terminates, it must coincide with an arc of the unit circle. Then by periodicity in θ, we obtain a whole sequence of arcs. The simplest possibility is that these arcs are joined at $\theta = (2n+1)\pi$. $SL(2, \mathbf{Z})$ now does the rest, leading to the complicated phase diagram of Figure 1. As discussed in detail in [2], this phase diagram represents a beautiful realization of what 't Hooft calls oblique confinement [7].

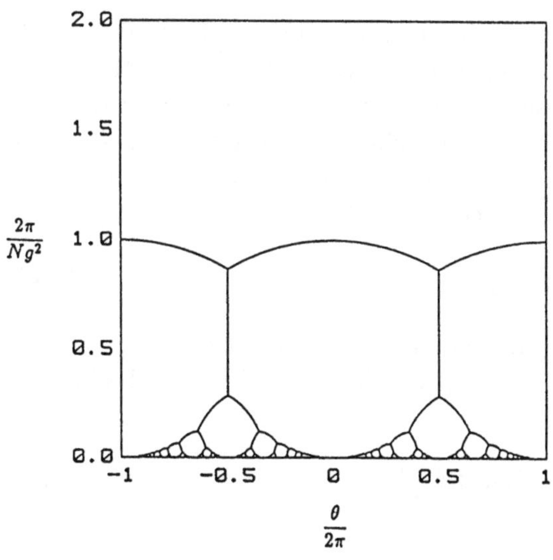

Figure 1. Phase diagram of an Abelian lattice gauge theory with magnetic charges. In the strong coupling limit, there is a phase for every rational value of $\theta/2\pi$.

3. Three Generalizations

In this section we shall make three generalizations of the model (2.1) of Cardy and Rabinovici. We shall successively construct self-dual models of arbitrary dimensionality and tensoriality, then generalized theta terms, and finally models with arbitrary Abelian gauge group. In all cases, we shall see that duality is an element of order 2 of an infinite modular group, whose exact form depends on both the dimensionality of the underlying spacetime and the number of simple factors k in the gauge group. Specifically, the modular group is either $Sp(2k, \mathbf{Z})$ or $SO(k, k; \mathbf{Z})$, depending on whether the $k \times k$ matrix of theta parameters is symmetric or antisymmetric.

The gauge models we have discussed so far are self-dual only in four dimensions. But it is easy to construct other types of self-dual models of arbitrary dimensionality. Roughly speaking, these models contain p-form and $(d-p)$-form fields, which get mapped into each other under duality. In the special case where d is even and $p = d/2$, the model may be taken to contain just one self-dual p form field. For example, when $d = 4$, we recover the self-dual gauge model of Sec. 2, with 2-form $F_{\mu\nu}$. We shall refer to p as the *rank* of a model.

To be specific, let us consider the following generalization of the Cardy–Rabinovici partition function to d dimensions [6]:

$$Z_p = \mathrm{Tr}\ \exp\left[-\frac{1}{2g^2}\sum_{S_p}(\Delta_{\mu_1}\phi_{\mu_2\ldots\mu_p} - 2\pi s_{\mu_1\ldots\mu_p})^2 + iN\sum_{S_{p-1}} n_{\mu_1\ldots\mu_{p-1}}\phi_{\mu_1\ldots\mu_{p-1}}\right] \tag{3.1}$$

Here, ϕ, s, and n are totally antisymmetric fields of rank $p-1$, p, and $p-1$ respectively, and Δ is shorthand for the antisymmetrized lattice derivative. The trace is over continuous values of ϕ and integer values of q and n, and the two sums in the exponent are over p and $p-1$ simplices of the lattice. As in the model of Eq.(2.1), the sum over n restricts ϕ to the values $\frac{2\pi}{N}$. Finally, in order that ϕ couple to a conserved current, we require that

$$\Delta_{\mu_1} n_{\mu_1\ldots\mu_{p-1}} = 0 \tag{3.2}$$

In [5], it is shown that the partition function Z_p is dual to $\widetilde{Z}_{d-p}$, at the coupling $\tilde{g} = 2\pi/Ng$. Hence, we can obtain a self-dual model by combining Z_p and Z_{d-p}. That is, the model

$$Z_{SD}[g] = Z_p[g]Z_{d-p}[g] \tag{3.3}$$

is self-dual under $g \longrightarrow 2\pi/Ng$.

Eq.(3.3) gives us a trivial class of self-dual models in d dimensions. The models are trivial because they contain no coupling between the dual fields.

Theta terms, as we shall define them, provide a natural coupling of p-form fields to dual $(d-p)$-form fields.

The familiar gauge theory theta term is an example; it couples a two-form to a $(4-2)$-form. It is a "topological" term because it depends only on the homology class of the 4-form $F_{\mu\nu}\tilde{F}_{\mu\nu}$, so that its integral over spacetime is quantized. A direct consequence of this is that the physics only depends on the value of θ *modulo* 2π.

Let us now add to the action of Eq. (3.1) the term

$$\frac{iN\theta}{4\pi^2}\sum_{S_p}\sum_{S_{d-p}}\frac{1}{p!(d-p)!}\epsilon_{\mu_1\ldots\mu_d}\left(\Delta_{\mu_1}\phi_{\mu_2\ldots\mu_p}(r)-2\pi s_{\mu_1\ldots\mu_p}(r)\right)$$
$$\cdot\left(\Delta_{\mu_{d-p+1}}\tilde{\phi}_{\mu_{d-p+2}\cdots\mu_d}(r)-2\pi\tilde{s}_{\mu_{d-p+1}\cdots\mu_d}(r)\right) \tag{3.4}$$

(Note that if ϕ and $\tilde{\phi}$ are defined on dual lattices, then no interpolating function f is necessary.) We shall henceforth abbreviate this as

$$\frac{iN\theta}{4\pi^2}\sum(\Delta\phi-2\pi s)\wedge(\Delta\tilde{\phi}-2\pi\tilde{s}) \tag{3.5}$$

As is the case for gauge models, the summation of the summand over the lattice is quantized in units of $4\pi^2/N$, so that $\theta\rightarrow\theta+2\pi$ is a symmetry of the partition function. In addition, the partition function is invariant under precisely the same duality transformation (1.1) as Cardy and Rabinovici's gauge model. So all of the analysis of [2] applies unaltered to this class of models.

For our third generalization, we now consider the case of an arbitrary finite Abelian gauge group. This is a very straightforward extension of the above analysis, and amounts to considering fields ϕ, s, and n which take their values in a lattice. The lattice will be defined by basis vectors e_i and periodicities N_i, and the model will have $\mathbf{Z}_{N_1}\times\cdots\times\mathbf{Z}_{N_k}$ symmetry. The partition function (3.1) generalizes to

$$Z_p=\operatorname{Tr}\exp\left[-\tfrac{1}{2}\sum_{S_p}G_{ij}\left(\Delta\phi^i-2\pi s^i\right)\left(\Delta\phi^j-2\pi s^j\right)+i\sum_{S_{p-1}}N^in^i\phi^i\right] \tag{3.6}$$

As before, the integer-valued fields s^i are responsible for the periodicity of the ϕ^i, and the sum over n^i ensures that each ϕ^i is $2\pi/N_i$ times an integer. The new feature is the replacement of the coupling constant g^{-2} by a coupling matrix G_{ij}, which is interpreted as defining a lattice with basis $\{e_i\}$ such that $G_{ij}=e_i\cdot e_j$. Let us now take $N_1=\cdots=N_k=N$. By Eq. (2.4), when G is diagonal, duality maps G to $\frac{N^2}{4\pi^2}G^{-1}$. (It is easy to extend this to the case of unequal N_i.)

To self-dual theories containing both p and $d-p$ forms, which take values in dual lattices, we may add a generalized theta term of the form

$$i\sum_{S_p,\ S_{d-p}}B_{ij}(\Delta\phi-s)^i\wedge(\Delta\tilde{\phi}-\tilde{s})^j \tag{3.7}$$

B_{ij} will be symmetric or antisymmetric according to whether p-forms and $d-p$ forms commute or anticommute, *i.e.*, whether $p \cdot (d-p)$ is even or odd. Because of the topological nature of this term, the full partition function will be invariant under periodic shifts of any component of B.

Then the full generalization of the duality relation (1.1) is

$$\frac{2\pi}{N}\begin{pmatrix} G & iB \\ iB^\top & G \end{pmatrix} \rightarrow \frac{N}{2\pi}\begin{pmatrix} G & iB \\ iB^\top & G \end{pmatrix}^{-1} \tag{3.8}$$

(we take $N_1 = \cdots = N_k$ here and hereafter). There are two further discrete symmetries: first, each component B_{ij} (together with $B_{ji} = \pm B_{ij}$) is an angular variable, with periodicity under

$$B_{ij} \rightarrow B_{ij} + \frac{N}{2\pi} \tag{3.9}$$

In addition, the torus defined by G_{ij} is unchanged if we choose a different set of basis vectors $\{e_i{}'\}$, related to the original basis by a $GL(d, \mathbf{Z})$ transformation. The action is therefore invariant if

$$G \rightarrow a^\top G a \qquad\qquad B \rightarrow a^\top B a \tag{3.10}$$

where $a \in GL(d, \mathbf{Z})$.

These three invariances, duality, periodic shifts of B_{ij}, and lattice symmetry, fail to commute just as before, generating a large discrete symmetry group. In fact, when B is symmetric (*e.g.*, for four-dimensional gauge theories), this group is none other than the symplectic modular group $Sp(2k, \mathbf{Z})$ [8]. (For $k = 1$ this is equal to $SL(2, \mathbf{Z})$.) The parameter space of this class of models, on which $Sp(2k, \mathbf{Z})$ acts, consists of symmetric matrices with positive definite imaginary part (corresponding to a positive definite metric G_{ij}). This space is known to mathematicians as the Siegel upper half plane. The group that results when B is antisymmetric (*e.g.*, for two-dimensional sigma models) is $SO(k, k; \mathbf{Z})$.

As a final generalization, we should mention that these discrete symmetries hold regardless of the underlying spacetime metric. This is because interchanging p-forms and $(d-p)$-forms, is a purely geometric operation, implemented by the Hodge star operator.

4. Toroidal String Compactifications

Thus far, our discussion has been restricted to lattice models. It is natural to ask whether continuum field theories can be self-dual as well. There are two continuum limits that need to be taken — the lattice spacing $a \rightarrow 0$ and the group $\mathbf{Z}_N \rightarrow U(1)$ — and in general a renormalization group analysis must be applied. However, if the continuum limit is a conformally invariant field theory, then there is nothing to worry about, and we expect this field theory to be self–dual. This is what happens for our 2-dimensional models, if the continuum limit happens to be a bosonic or heterotic string theory: these string theories, toroidally compactified, are self–dual in precisely the sense of Section 3. In this section, we shall explicitly show that the moduli space of background field configurations of the heterotic string possesses an invariance under $SO(16+d,d;\mathbf{Z})$, which includes the well–known $R \to 1/2R$ duality symmetry of circle compactifications with radius R [9][10]. This invariance will also turn out to be quite useful, in forming a picture of the global geometry of moduli space and in locating special symmetric compactifications.

Toroidal compactifications of heterotic strings may be viewed in two ways: as compactifications of the independent left- and right-movers on tori of Lorentzian self-dual lattices in $R^{16+d,d}$ [11], or as compactifications from 10 dimensional spacetime down to $D = 10-d$ dimensions in the presence of background B_{ij} and A_{iI} fields [12]. (Lower-case roman indices here run from $D+1$ to 10, upper-case from 11 to 26.) In the first picture, duality is a mapping between Lorentzian lattices which sends lattice vectors

$$(L,\tilde{L}) \to (L,-\tilde{L}) \tag{4.1}$$

This preserves lattice self-duality since the norm $L^2 - \tilde{L}^2$ is unchanged. It is easy to see the effect of this mapping on string theory amplitudes. The cosmological constant, for example, is invariant: to all orders, the contribution of winding states to the partition function depends only on inner products of L's with L's and of $\tilde{L}$'s with $\tilde{L}$'s. There is a direct correspondence between Lorentzian self-dual lattices and configurations (G,B,A) [12]. So to establish a duality relation expressed in terms of background fields, it suffices to show that the corresponding Lorentzian lattices are related as in (4.1).

The background fields make their appearances in the worldsheet action as follows [12]:

$$S_{WS} = \int d\sigma dt \left(\sqrt{h}h^{ab}G_{ij}\partial_a X^i \partial_b X^j + i\epsilon^{ab}B_{ij}\partial_a X^i \partial_b X^j + i\epsilon^{ab}A_{iI}\partial_a X^i \partial_b X^I\right) \tag{4.2}$$

where h is the worldsheet metric. At least when $A=0$, this bears a marked resemblance to the action (3.6) when $d=2$.

We first examine the case $B = A = 0$. Consider a compactification from 10 to D dimensions on a $10-D$ dimensional torus, which is specified by a background metric G_{ij}. In terms of the basis vectors e_i for the torus and their duals e_i^*, $G_{ij} = e_i \cdot e_j$ and the momenta and winding vectors are

$$p = m_i e_i^* \tag{4.3}$$

$$w = n_i e_i \tag{4.4}$$

According to [12], winding vectors in the Lorentzian lattice picture are related to momenta p and winding vectors w in the background field picture *via*

$$L^i = n^i + \tfrac{1}{2}G^{ik}m_k \tag{4.5}$$
$$\tilde{L}^i = -n^i + \tfrac{1}{2}G^{ik}m_k \tag{4.6}$$

where L and $\tilde{L}$ live in $\mathbb{R}^d$ with the metric G_{ij}. The transformation $2G \to \frac{1}{2}G^{-1}$ is equivalent to $e_i \leftrightarrow \frac{1}{2}e_i^*$, which exchanges the momentum and winding vector lattices $\frac{1}{2}p$ and w. Under this transformation, $(L, \tilde{L}) \to (L, -\tilde{L})$.

For $B \neq 0$, the correspondence (4.5)-(4.6) generalizes to

$$L^i = n^i + \tfrac{1}{2}G^{ik}m_k - G^{ij}B_{jk}n^k \tag{4.7}$$
$$\tilde{L}^i = -n^i + \tfrac{1}{2}G^{ik}m_k - G^{ij}B_{jk}n^k \tag{4.8}$$

and the duality relation becomes

$$2(B+G) \to \tfrac{1}{2}(B+G)^{-1} \tag{4.9}$$

This symmetry may be verified explicitly by checking its effect on the the inner products of the corresponding Lorentzian lattice vectors:

$$\{L_1 \cdot L_2\} \to \{L_1 \cdot L_2\} \tag{4.10}$$
$$\{L_1 \cdot \tilde{L}_2\} \to \{-L_1 \cdot \tilde{L}_2\} \tag{4.11}$$
$$\{\tilde{L}_1 \cdot \tilde{L}_2\} \to \{\tilde{L}_1 \cdot \tilde{L}_2\} \tag{4.12}$$

In fact, duality does not preserve these inner products individually, but only the set of inner products. It has the effect of interchanging the summations over m and n, as defined in eqs.(4.7-8).

Finally, we turn to the case of nonzero A. Upon inspecting the action (4.2), one might naively expect that the correct duality transformation is obtained by inverting the matrix

$$M = 2\begin{pmatrix} B+G & A \\ 0 & \Gamma \end{pmatrix} \tag{4.13}$$

where Γ is the metric for the internal 16-dimensional torus (*e.g.*, the $E_8 \times E_8$ matrix). However, the situation is now complicated by the constraint $\partial\widetilde{X}^I = 0$, and it turns out that the correct matrix to invert is instead

$$M' = 2\begin{pmatrix} B+G+\frac{1}{4}A^I A_I & A \\ 0 & \Gamma \end{pmatrix} \tag{4.14}$$

One can verify directly, by a straightforward computation, that

$$M' \to M'^{-1} \tag{4.15}$$

is a duality transformation by studying the inner products of Lorentzian lattice vectors expressed in terms of background fields [5].

Because of the correspondence between Lorentzian lattices and background fields, it is easy to see that the group of generalized duality transformations on B and G is $SO(d, d; \mathbf{Z})$, when $A = 0$. For nonzero A, the group extends to $SO(16 + d, d; \mathbf{Z})$. This is just the group of invariant changes of basis of a Lorentzian lattice, plus the duality $(L, \tilde{L}) \to (L, -\tilde{L})$. The latter is not in general a symmetry of the lattice; it would be worth knowing whether there are any other independent, nongeometric symmetries.

One consequence of spacetime modular invariance is that toroidal compactifications never get too small. That is, any compactification is equivalent, under some modular transformation, to a compactification in a fundamental domain of $SO(16+d, d; \mathbf{Z})$ located a finite distance (of order unity) from the the $G_{ij} = 0$ hyperplane (provided such a domain exists, which seems reasonable to expect).

5. Two-Dimensional Compactifications

The moduli space of two-dimensional compactifications (with $A = 0$) is simple enough to study in detail, yet complex enough to contain a surprising amount of structure, much of which is due to its global symmetries. We shall show that this moduli space is isomorphic to the direct product of two one-complex-dimensional Siegel upper half planes, with modular group $SL(2, \mathbf{Z}) \times SL(2, \mathbf{Z})$. Then, we shall use this information to find four compactifications of enhanced symmetry.

The four real parameters spanning the space of two-dimensional compactifications may be expressed in terms of two complex numbers as follows [13]:

$$\rho = 2(B_{12} + i\sqrt{G}) \tag{5.1}$$

$$\eta = \frac{G_{12}}{G_{22}} + i\frac{\sqrt{G}}{G_{22}} \tag{5.2}$$

Under duality (4.9), $\rho \to -1/\rho$, with η unchanged, while switching the 1 and 2 components sends η to $-1/\eta$ and leaves ρ unchanged. In addition to two dualities, there are periodicities in both η and ρ. The periodicity $\eta \to \eta + 1$ amounts to a shift of the basis vectors e_i defining the matrix $G_{ij} = e_i e_j$. These transformations define an explicit action of

$$SL(2, \mathbf{Z}) \times SL(2, \mathbf{Z}) \simeq SO(2, 2; \mathbf{Z}) \tag{5.3}$$

on our moduli space, which is the product of two upper–half complex planes. Modding out by this action leaves a direct product of two fundamental domains of $SL(2, \mathbf{Z})$.

The modular group does not act freely on moduli space; in general there will be isolated points that are fixed by subgroups of the full modular group. These special compactifications possess extra discrete symmetry, and may also have enhanced gauge symmetries. Furthermore, if the cosmological constant Λ is nonzero as in the $O(16) \times O(16)$ heterotic string, then isolated fixed points are guaranteed to be extrema of Λ.

For two-dimensional compactifications with $A = 0$, we can locate these points immediately. There are just two fixed points per fundamental domain (excluding infinity), one of order two and one of order three, at i and $\omega \equiv \frac{1}{2} + i\frac{\sqrt{3}}{2}$. Thus we can find a total of four fixed points (ρ, η) in the direct product of two fundamental domains: (i,i), (i,ω), (ω,i), and (ω,ω). The first and last compactifications lead to enhancement of the gauge symmetry groups, by $SU(2) \times SU(2)$ and $SU(3)$, respectively. The second and third also have enhanced $SU(3)$ symmetry, although it takes more work to see this [15] . The fact that the latter two compactifications have irrational mass spectra suggests that rational conformal theories may not necessarily play a privileged dynamical role. At all four of these points, the cosmological constant Λ of the $O(16)\times O(16)$ heterotic string is extremized. We have looked for other extrema of Λ_{1-loop} numerically; these four seem to be the only ones in a fundamental domain.

6. Further Directions

Modular invariance of string compactifications is a powerful principle. We have just seen how it may be used to locate special two–dimensional compactifications. More generally, we would like to be able to enumerate all of the points fixed by discrete subgroups of $SO(16 + d, d; \mathbf{Z})$, for several reasons. Isolated fixed points are always extrema of the vacuum energy, and as such are candidate vacua for ground states with broken supersymmetry (modulo the cosmological constant problem). Also, based on our experience, these special points often seem to possess enhanced gauge symmetry. Finally, at a fixed point one may always mod out the compactification by the discrete subgroup of $SO(16+d, d; \mathbf{Z})$ that fixes it, to obtain an orbifold. The classification of fixed point subspaces is closely related to the classification of orbifold compactifications.

Modular invariance also has important consequences in the low–energy limit. It places strong restrictions on the form of the effective action describing the interactions of massless fields [16]. In the supersymmetric case, the low–energy action is additionally constrained by analyticity requirements, leading to a beautiful connection between supergravity and the theory of modular forms.

I would like to thank the organizers for a uniquely memorable conference. This work was supported in part by the Department of Energy under Grant. No. DE-AC02-76ERO-2220.

References

[1] H.A. Kramers and G.H. Wannier, *Phys.Rev.* **60**, (1941) 252.

[2] J. Cardy, *Nucl. Phys.* **B205** (1982) 17.

[3] J. Cardy and E. Rabinovici, *Nucl. Phys.* **B205** (1982) 1.

[4] J.V. José, L.P. Kadanoff, S. Kirkpatrick, D.R. Nelson, *Phys. Rev.* **16** (1977) 1217.

[5] A. Shapere and F. Wilczek, "Self-Dual Models with Theta Terms," *Nucl. Phys.* **B320** (1989) 669.

[6] R. Savit, *Rev. Mod. Phys.* **52** (1980) 453.

[7] G. 't Hooft, *Nucl. Phys.* **B190** (1981) 455.

[8] D. Mumford, *Tata Lectures on Theta*, vol. I (Birkhauser, 1983).

[9] K. Kikkawa and M. Yamasaki, *Phys.Lett.* **149B**, (1984) 357.

[10] V.P. Nair, A. Shapere, A. Strominger, F. Wilczek, *Nucl. Phys.* **B287** (1987) 414.

[11] K.S. Narain, *Phys. Lett.* **169B** (1986) 41.

[12] K.S. Narain, M.H. Sarmadi, E. Witten, *Nucl. Phys.* **B279** (1987) 369.

[13] R. Dijkgraaf, E. Verlinde and H. Verlinde, *Comm. Math. Phys.* **115** (1988) 649.

[14] P. Ginsparg and C. Vafa, *Nucl. Phys.* **B289** (1987) 414.

[15] M. Dine, P. Huet, and N. Seiberg, City College of New York preprint HEP-88/20.

[16] S. Ferrara, A. Shapere, D. Lüst, and S. Theisen, "Modular Invariance in Supersymmetric Field Theories," to appear in Physics Letters B.

Yang-Baxter Algebras and Quantum Groups

H.J. de Vega

LPTHE, Université Paris VI, Tour 16, 1er etage, 4, place Jussieu,
F-75230 Paris Cedex 05, France

Integrable massive QFT and conformal invariant models follow from lattice integrable models in suitable scaling limits . There are Yang-Baxter algebras (YBA) associated with all these two-dimensional models. These YBA allow one to construct the exact solution (spectrum, S-matrix, form-factors,...) for this class of theories. Braid groups and quantum groups are derived as limiting cases of YBA when θ (spectral parameter) goes to $\pm\infty$.

Let us start by defining a Yang-Baxter algebra (YBA). We consider a set of lines of different types . A vector space $\mathcal{V}_I$ ($I \in \mathcal{J}$) is associated with each type of line. These lines may intersect and a YB operator is associated with each intersection

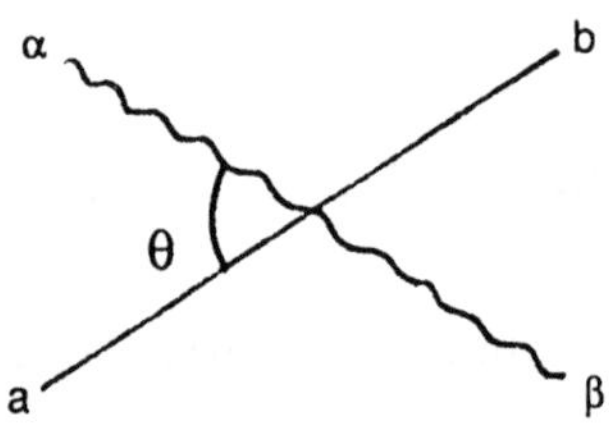

$$= \quad [T_{ab}(\mathcal{A},\mathcal{V})(\theta)]_{\alpha\beta} \qquad (1)$$

Here the lines ——— and ~~~~~~~~ are associated with the vector spaces $\mathcal{A}$ and $\mathcal{V}$ respectively. T(θ) is an operator acting on this couple of spaces , that is on $\mathcal{A} \otimes \mathcal{V}$ The indices a, b ($1 \leq a,b \leq \dim\mathcal{A}$) and α,β ($1 \leq \alpha,\beta \leq \dim\mathcal{V}$) label the basis vectors (states) in $\mathcal{A}$ and $\mathcal{V}$ respectively. The complex variable θ is called the spectral parameter and describes geometrically the angle between the lines.

The operators T(θ) fulfil a YBA when the following relation holds :

$$T^{(K,I)}(\theta-\theta')\; T^{(K,J)}(\theta)\; T^{(I,J)}(\theta') = T^{(I,J)}(\theta')\; T^{(K,J)}(\theta)\; T^{(K,I)}(\theta-\theta') \qquad (2)$$

for any choice of vector spaces $\mathcal{V}_I$, $\mathcal{V}_J$ and $\mathcal{V}_K$ (I, J and K $\in \mathcal{J}$) and any value of θ , $\theta' \in$ **C.** Graphically eq.(2) reads

Springer Proceedings in Physics, Vol. 47 **Number Theory and Physics**
Editors: J.M. Luck · P. Moussa · M. Waldschmidt

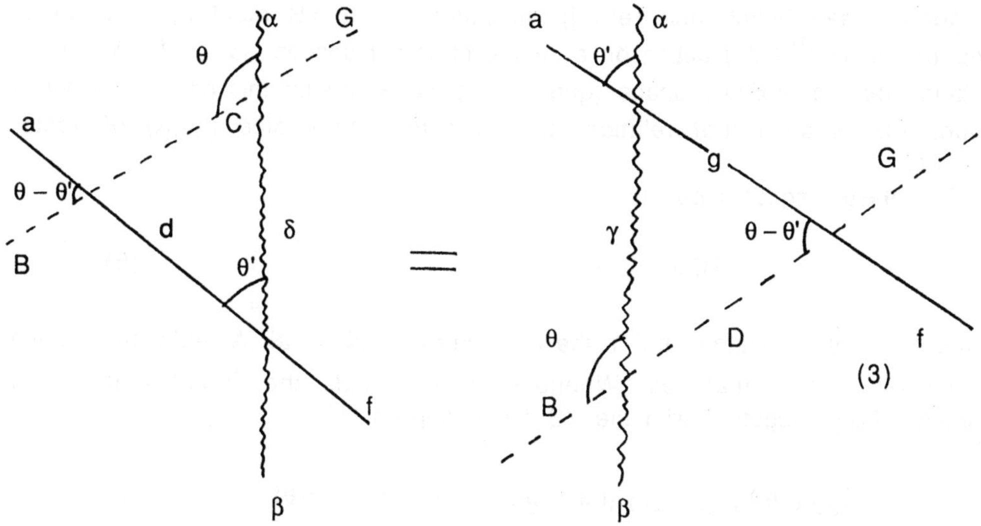

(3)

Here, we use the convention that one must sum over all states in internal lines. That is lines linking two intersections. The graphical meaning of the YBA is clear : one can push any line through the intersection of two others keeping the value of the expression invariant. This invariance holds provided all angles θ are kept fixed under the displacement of the lines. Eqs.(4) or (5) are usually called the Yang-Baxter equations (YBE). They have different physical meanings in different contexts (see below).

The "diagonal" generators $T^{(I,I)}(\theta)$ are called R-matrices. The R-matrix associated with the lowest dimensional V^I is the fundamental one. Usually, the whole set of $T^{(I,J)}(\theta)$ can be constructed once the fundamental R-matrix is found. Therefore this fundamental R-matrix, solution of eq.(2) for the lowest dimensional space I = J = K, defines the YB algebra.

The YBA (4) takes a particularly simple form when two vector spaces are identical. That is $\mathcal{V}_K = \mathcal{V}_I = \mathcal{A}$ and $\mathcal{V}_J = \mathcal{V}$ in eq.(4). One finds

$$R(\theta-\theta')\,[\,T(\theta)\otimes T(\theta')\,] = [\,T(\theta')\otimes T(\theta)\,]\,R(\theta-\theta') \qquad (4)$$

where $R = T^{(\mathcal{A},\mathcal{A})}$ and $T = T^{(\mathcal{A},\mathcal{V})}$ and we have used the tensor product notation $(A\otimes B)_{ab,cd} = A_{ac}\,B_{bd}$. In eq.(4) there is an operational product of the T's in the space $\mathcal{V}$ More explicitly , eq.(4) reads

$$R_{ab}{}^{kl}(\theta-\theta')\;T_{kc}(\theta)\;T_{ld}(\theta') = T_{am}(\theta')\;T_{bn}(\theta)\;R_{cd}{}^{mn}(\theta-\theta'). \quad (5)$$

The R-matrix (with components $R_{ab}{}^{cd}(\Theta-\Theta')$) for the lowest dimensional vector space $\mathcal{V}_0$ defines the YB algebra. These c-numbers play the role of "structure constants" and the $T_{ab}(\Theta)$ that of generators of the YB algebra. It should be noticed that the YB algebras are not in general Lie algebras but Hopf algebras (see below and ref.[1]). In summary, a YB algebra, is a set of operators $T^{(I,J)}(\theta)$ acting on a couple of vector spaces (V^I, V^J). V^I (V^J) stands for the auxiliary space (quantum space = $\mathcal{V}$) for this case. This set of operators is such that relation (2) holds for any choice (I,J,K) of vector spaces.

For $\theta = \theta'$, eq.(5) suggests

$$R(0) = c\, \mathbf{1} \tag{6}$$

where c is a constant and **1** the unit matrix in $\mathcal{A} \otimes \mathcal{A}$. Actually, this is the case for most R-matrices. Whenever eq.(6) holds the R-matrix is called regular. Eq.(6) together with the YB eq.(2) implies [1]

$$[\, T_{ad}(-\theta)\,]_{\alpha\lambda} \quad [T_{\lambda\beta}(\theta)\,]_{db} = \delta_{ab}\,\delta_{\alpha\beta} \quad \rho(\theta)$$

or (7)

$$T(\theta)^{(\mathcal{A},\mathcal{V})\,-1} = (1/\rho(\theta)) \quad T(\theta)^{(\mathcal{V},\mathcal{A})}$$

where $\rho(\theta) = \rho(-\theta)$ is a c-number function.

This expresses unitarity in the S-matrix context.

The necessary and sufficient conditions for a YB algebra to be invariant under a group $\mathcal{G}$ are[2]

$$[\, R(\theta)\,,\, g_{\mathcal{A}} \otimes g_{\mathcal{A}}\,] = 0 \quad , \quad \forall\, g \in \mathcal{G} \tag{8}$$

Here $g_{\mathcal{A}}$ is the representation of $g \in \mathcal{G}$ acting on $\mathcal{A}$. We find from eqs.(5) and (8) that both $T(\theta)$ and $g_{\mathcal{A}}\, T(\theta)$ obey the same YBA. Therefore the YBA is invariant under $\mathcal{G}$.

This implies for the generators S of the Lie algebra $\mathcal{G}$ and the YB generators

$$[\, \mathbf{1}_{\mathcal{A}} \otimes S_{\mathcal{V}} + S_{\mathcal{A}} \otimes \mathbf{1}_{\mathcal{V}} \;,\; T(\theta)\,] = 0 \tag{9}$$

where $S_{\mathcal{A}}$ and $S_{\mathcal{V}}$ are the representations of S acting on the spaces $\mathcal{A}$ and $\mathcal{V}$ respectively. In eq.(9) $T(\theta)$ acts on <u>both</u> the spaces $\mathcal{A}$ and $\mathcal{V}$ as an operator.

The YBA enjoys an invariance under shifts of the spectral parameter. Namely , if $T(\theta)$ fulfils the YBE (4), so does $T(\theta - \alpha)$ for fixed α

$$R(\theta-\theta')\ [\, T(\theta-\alpha) \otimes T(\theta'-\alpha)\,] = [\, T(\theta'-\alpha) \otimes T(\theta-\alpha)\,]\ R(\theta-\theta')\ . \tag{10}$$

The fundamental property of the YBA is the reproduction property. Namely, if $t^{(\mathcal{A},\mathcal{V})}(\theta)$ is a YB operator, one can build more general YB operators by the following rule :

$$T_{ab}^{(\mathcal{A},\mathcal{V}^K)}(\theta) = \sum_{a_1,\ldots,a_{N-1}} t_{a,a_1}^{(\mathcal{A},\mathcal{V})}(\theta) \otimes t_{a_1,a_2}^{(\mathcal{A},\mathcal{V})}(\theta) \otimes \ldots\ldots$$

$$\ldots\ldots t_{a_{N-1},b}^{(\mathcal{A},\mathcal{V})}(\theta) \qquad (11)$$

=

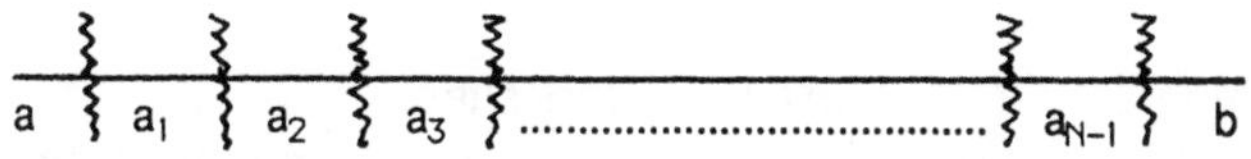

After this short introduction to YB algebras (for details see ref.[1]), let us explain how massive relativistic QFT follow from YB algebras in appropriate scaling limits.

In two dimensions, the light-cone approach is the more general and precise way to construct integrable QFT and conformal invariant theories. One starts from integrable lattice models such as vertex models on a diagonal lattice. This diagonal lattice is a discretization of Minkowski space-time in light-cone coordinates $X_\pm = X \pm T$. The matrix elements

$$[\,T_{ab}(\theta)\,]_\alpha{}^\beta \qquad\qquad (1 \le \alpha, \beta \le \dim\mathcal{V}) \qquad (12)$$

are now interpreted as quantum mechanical transition amplitudes for bare particles propagating to the right or to the left by the bonds at the speed of light. In the simplest case $\mathcal{A} = \mathcal{V}$, $q = 2$ and we interpret these particles as bare fermions without internal degrees of freedom. The allowed microscopical amplitudes, assuming a U(1) charge conservation, correspond to the six-vertex model weights in the statistical mechanical language. We can organize the microscopic amplitudes at a site into a unitary bare scattering matrix:

$R_{ab}{}^{cd}(\theta)$ =

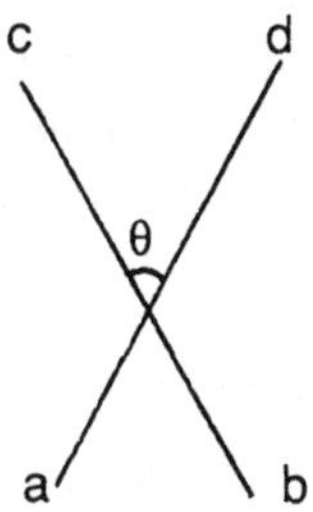

$$= \begin{pmatrix} 1 & 0 & 0 & 0 \\ 0 & c & b & 0 \\ 0 & b & c & 0 \\ 0 & 0 & 0 & \omega \end{pmatrix} . \qquad (13)$$

We can now form the operators describing the evolution by one lattice step in the diagonal directions

$$U_R = \underset{2N\ \ 1}{\overset{1\ \ 2}{X}}\ \underset{2\ \ 3}{\overset{3\ \ 4}{X}}\ \underset{4\ \ 5}{\overset{5\ \ 6}{X}}\ \underset{6\ \ 7}{\overset{7\ \ 8}{X}}\ \ldots\ldots\ldots\ldots\ldots\ \underset{2N\text{-}2\ \ 2N\text{-}1}{\overset{2N\text{-}1\ \ 2N}{X}} \tag{14}$$

$$U_L = \underset{2\ \ 3}{\overset{1\ \ 2}{X}}\ \underset{4\ \ 5}{\overset{3\ \ 4}{X}}\ \underset{6\ \ 7}{\overset{5\ \ 6}{X}}\ \underset{8\ \ 9}{\overset{7\ \ 8}{X}}\ \ldots\ldots\ldots\ldots\ldots\ \underset{2N\ \ 1}{\overset{2N\text{-}1\ \ 2N}{X}} \tag{15}$$

where the numbers 1, 2, 3,...,2N label the sites.

In this way the massive Thirring model is constructed both at the bare and at the renormalized levels [3]. The appropiate scaling limits are defined as $a \to 0$, $i\theta \to \infty$.

$$\begin{aligned} \text{Bare} &: \sin\gamma \exp[-\theta]/a = \text{fixed.} \\ \text{Renormalized} &: \sin\gamma \exp[-K\theta]/a = \text{fixed} \end{aligned} \tag{16}$$

Here a is the lattice spacing , $K = \pi/\gamma$ and γ is the anisotropy parameter. The light-cone transfer matrices rigorously define the Hamiltonian and momentum as

$$H \pm P = (2i/a) \operatorname{Log} U_{R_L}(\theta) \quad . \tag{17}$$

Using U_R and U_L , the Heisenberg equations of motion are derived in the lattice and their scaling limit obtained[3]. They lead to the MTM bare equations, providing a precise identification of the fields, the coupling constant and the mass[3] .The exact spectrum of U_R and U_L follows from the usual row-to-row transfer matrices and provides the particle spectrum of the theory in the renormalized scaling limit (eq. (16)).

We want to recall here that the massless Thirring model follows as the continuum limit of the XXZ Heisenberg chain [10].This Hamiltonian is the logarithmic derivative of the six vertex model transfer matrix at $\theta = 0$. It has been found recently[11] that the exact (Bethe Ansatz) solution of the XXZ chain yields in the continuum limit a Thirring solution which is different from those obtained directly in the continuum[12]. In particular, the Schwinger term is coupling dependent in a way explicitly obtained in close form from the Bethe Ansatz[11].

Let us say two words about the light cone transfer matrices U_R and U_L. The row-to-row inhomogeneous YB generators can be written as [1]

$$T_{ab}(\theta,\alpha) = \sum_{a_1,\ldots,a_{N-1}}^{q} t_{aa_1}(\theta-\alpha_1) \otimes t_{a_1a_2}(\theta-\alpha_2) \otimes \ldots\ t_{a_{N-1}b}(\theta-\alpha_N) \tag{18}$$

where $[t_{ab}(\theta)]_{cd} = R_{ca}^{bd}(\theta)$ fulfils eq. (1). Eq.(18) also verifies the YB equation (2). For N = 2 , eq.(18) defines a coproduct of YB generators providing a Hopf algebra structure. That is, a mapping from the algebra to its tensor product by itself. This is in general a non-commutative and non-cocommutative Hopf algebra having in addition an antipode [1].

The row-to-row transfer matrix $\tau[\theta] = \sum_a T_{aa}[\theta]$ fulfils

$$[\,\tau[\theta,\alpha],\tau[\theta',\alpha]\,] = 0$$

thanks to eq.(1). In particular, if we set $\alpha_j = (-1)^{j+1}\,\theta$ [3],

$$\tau(\theta,\theta) = U_L(\theta) \quad , \quad \tau(-\theta,\theta) = U_R(\theta)^\dagger \quad . \tag{19}$$

Therefore, the eigenvalue problem for U_R and U_L reduces to the one for $\tau(\theta,\alpha)$. This problem is exactly solved by the Bethe Ansatz. That is, calling $B(\theta) \equiv T_{12}(\theta)$, the eigenvectors read [1]

$$\Psi_r(\theta_1,\theta_2,\ldots,\theta_r) = B(\theta_1)\; B(\theta_2)\;\ldots\; B(\theta_r)\;\Omega \tag{20}$$

where Ω is the bare (ferroelectric) ground state and $B(\theta) \equiv T_{12}(\theta)$

$$\Omega = \begin{pmatrix}1\\0\end{pmatrix} \otimes \begin{pmatrix}1\\0\end{pmatrix} \otimes \ldots\ldots \otimes \begin{pmatrix}1\\0\end{pmatrix}$$

and the complex numbers θ_i fulfil the set of equations (Bethe Ansatz eqs.)

$$[\,\mathrm{sh}(\lambda_j + i\gamma/2)\,/\,\mathrm{sh}(\lambda_j - i\gamma/2)\,]^N$$

$$= -\prod_{k=1}^{r}\,[\mathrm{sh}(\lambda_j - \lambda_k + i\gamma)\,/\,\mathrm{sh}(\lambda_j - \lambda_k - i\gamma)]\,. \tag{21}$$

Here $\lambda_j \equiv -i(\theta_j - \gamma/2)$. These Ψ_r are eigenvectors of S_z and $\tau[\theta]$ with eigenvalues N/2 - r and $\Lambda[\theta] = \Lambda_+[\theta] + \Lambda_-[\theta]$ respectively, where

$$\Lambda_+ = a(\theta)^N \prod_{j=1}^{r}\,[a(\theta_j - \theta)\,/\,b(\theta_j - \theta)],\quad \Lambda_- = b(\theta)^N \prod_{j=1}^{r}[a(\theta - \theta_j)/\,b(\theta - \theta_j)]\,. \tag{22}$$

This is the solution for the six vertex model. Eqs.(21) are easily solvable in the N = ∞ limit by Fourier transform. A method to solve them for finite size has been given in refs.[6].

The lattice light-cone approach also works for Chiral fermionic models with Lagrangian

$$\mathcal{L} = i\Psi\partial\Psi - g(\Psi\gamma_\mu t^\alpha\Psi)(\Psi\gamma^\mu t^\beta\Psi)K_{\alpha\beta} \tag{23}$$

where Ψ transforms under an irreducible representation of the symmetry group $\mathfrak{G}$, t_α are the generators of $\mathfrak{G}$ and $K_{\alpha\beta}$ the Killing form.The equations of motion for the vector current (conservation and flatness) were derived for these models as well as their exact particle spectrum in ref. [3]. The renormalized scaling limit is here $\theta \to \infty$, $a \to 0$ with

$$\mu = \exp(-K\theta) / a \quad , \text{ fixed} \tag{24}$$

where K is $2\pi/\gamma$ times the square length of the shortest simple root of $\mathfrak{G}$ in the normalization where $B(E_\alpha, E_{-\alpha}) = -1$. This construction extends to higher dimensional representations of $\mathfrak{G}$ yielding the principal chiral model in the limit of infinite dimension when $\mathfrak{G} = SU(N)$[4].

Relativistic QFT arise from all known gapless integrable models. That is, from YB algebra realizations involving trigonometric functions of θ. Hyperbolic and elliptic YB algebra realizations do not lead to relativistic QFT[14].

A second scaling limit letting the lattice spacing $a \to 0$ with fixed θ yields conformal invariant models provided one starts from a gapless lattice model. That is, a rational or a trigonometric YB algebra. The finite size resolution methods for the Bethe Ansatz (BA) equations of the lattice models [6] give the values of the central charge c and the conformal weights $(\Delta, \bar{\Delta})$ of operators for a large class of models [6,7]. Branching coefficients can be related to one point functions[8].

Let us briefly summarize the results. In all cases , the dominant finite size corrections to the nth eigenvalue of the vertex model transfer matrix $\tau(\theta)$, computed from the BA, have the form

$$\log[\Lambda^n_N / \Lambda^0_\infty] = -(\pi c/6N^2)\sin(\pi\theta/\gamma) + 2\pi[(\Delta+\bar{\Delta})\sin(\pi\theta/\gamma) + i(\Delta-\bar{\Delta})\cos(\pi\theta/\gamma)] \quad . \tag{25}$$

Here c is model dependent and can be identified with the central charge of the Virasoro algebra when periodic boundary conditions have been used. Δ and $\bar{\Delta}$ depend on the eigenvalue considered . Eq.(25) is actually a proof of conformal invariance for these models providing in addition the explicit values of c, Δ and $\bar{\Delta}$. In ref.[3] we propose the formula

$$c = \frac{x \dim\mathfrak{G}}{x+q} \tag{26}$$

where x is the number of fundamental spaces $\mathcal{V} = \mathcal{A}$ fused and q the dual

Coxeter number of $\mathfrak{G}$ It must be stressed that eq.(26) does not follow here from any Sugawara type construction but just from the Bethe Ansatz solution. As particular cases we find $c = \text{rank}\ \mathfrak{G}$ for $x = 1$ and simply laced Lie algebras. For fundamental models the conformal weights always turn out to be [7]

$$\Delta = \frac{1}{2(1-\gamma/\pi)} \sum_{l,l'=1}^{\text{rank}G} (h^+_l - \gamma s_l/2\pi)(M^{-1})_{ll'}(h^+_{l'} - \gamma s_{l'}/2\pi) \tag{27}$$

and a similar formula for $\bar{\Delta}$ with $h^+_l \to h^-_l$. Here M is the Cartan matrix of $\mathfrak{G}$, h^+_l (h^-_l) the number of hole excitations with low energy and positive (negative) speed and s_ℓ the Lie algebra weights for the state . The preceding results correspond to periodic boundary conditions . Twisted boundary conditions mean for the six-vertex model

$$\sigma^\pm_{N+1} = \exp(\pm i\Phi)\,\sigma^\pm_1 \quad , \qquad \sigma^3_{N+1} = \sigma^3_1 \tag{28}$$

The coefficient c and conformal weights result :

$$c = 1 - 6\,\Phi^2 / [\,\pi\,(\pi - \gamma)\,] \quad , \tag{29}$$

$$\Delta = \tfrac{1}{4}(h_+ - \gamma\, s / \pi - \Phi / \pi)^2/(1 - \gamma / \pi)\ ,$$

$$\bar{\Delta} = \tfrac{1}{4}(h_- - \gamma\, s / \pi + \Phi / \pi)^2/(1 - \gamma / \pi) \quad . \tag{30}$$

It must be stressed that c is not the central charge for twisted b.c. .The central charge is not affected by the b.c. but the free energy finite size corrections change their expression. Actually eq. (29) follows by conformal theory arguments[5] .

Unitary conformal models appear for $\gamma = \pi / (m+1)$, $m = 2,3,\ldots$ and $\Phi = \gamma$. For more details, see refs.[1], [7] and [9].

Let us now study the ultrarelativistic limit of hyperbolic/trigonometric YB algebras and its connection with braids, knots and quantum groups. It is useful to introduce the operators

$$X_i(\theta) = \underset{1}{1} \otimes \ldots\ldots\ldots \otimes \underset{(i,\,i+1)}{R(\theta)} \otimes \ldots\ldots\ldots\ldots \otimes \underset{n}{1} \quad . \tag{31}$$

They fulfil the relations

$$[\,X_i(\theta)\ ,\ X_j(\theta')\,] = 0 \qquad \text{if} \quad |\,i - j\,| \geq 2, \quad \forall\ \theta, \theta'$$

$$X_i(\theta)\, X_{i+1}(\theta + \theta')\, X_i(\theta') = X_{i+1}(\theta')\, X_i(\theta + \theta')\, X_{i+1}(\theta) \tag{32}$$

$$X_i(\theta)\ X_i(-\theta) = 1 \quad ,$$

the last two being a consequence of the YB equation (2) and the unitarity eq.(7), respectively. There are two different limits for hyperbolic (trigonometric) YB algebras : $\theta \to \pm\infty$ ($\pm i\infty$) . Upon adequately normalizing $R(\theta)$ we find

$$b_i = \lim_{\theta\to\infty} X_i(\theta) \quad , \quad b_i^{-1} = \lim_{\theta\to-\infty} X_i(\theta) \tag{33}$$

and analogous relations for trigonometric YBA. The operators b_i fulfil:

$$b_i\, b_j = b_j\, b_i \qquad \text{when} \quad |i-j| \geq 2 \ ,$$

$$b_i\, b_{i+1}\, b_i = b_{i+1}\, b_i\, b_{i+1} \quad . \tag{34}$$

This is precisely a braid group [13]. In other words, the ultrarelativistic limit of the YB algebras provide braid group representations. This fact has been recently exploited successfully to compute knot invariants .

Let us study the $\theta = \pm\infty$ limit of the six-vertex YB generators [eq. (18)]. We find [1,2]

$$T_{11}(\theta) \underset{\theta\to\pm\infty}{=} (y_\pm)^N \exp(\pm\gamma S_z)\ [1 + O(y_\pm^{-2})]\ , \tag{35}$$

$$T_{22}(\theta) \underset{\theta\to\pm\infty}{=} (y_\pm)^N \exp(\mp\gamma S_z)\ [1 + O(y_\pm^{-2})] \ , \tag{36}$$

$$T_{12}(\theta) \underset{\theta\to\pm\infty}{=} (y_\pm)^{N-1}\, \mathrm{sh}\,\gamma\, J_-(\mp\gamma)\ [1 + O(y_\pm^{-2})]\ , \tag{37}$$

$$T_{21}(\theta) \underset{\theta\to\pm\infty}{=} (y_\pm)^{N-1}\, \mathrm{sh}\,\gamma\, J_+(\pm\gamma)\ [1 + O(y_\pm^{-2})] \ , \tag{38}$$

where

$$y_\pm = \pm\tfrac{1}{2}\exp[\pm(\theta + \gamma/2)] \quad , \quad S_z = \tfrac{1}{2}\sum_{a=1}^{N} (\sigma_a)_z \tag{39}$$

and

$$J_\pm(\gamma) = \sum_{k=1}^{N} \prod_{j=1}^{k} \exp\{\tfrac{1}{2}\gamma(\sigma_j)_z\}\ (\sigma_\pm)_k \prod_{l=k+1}^{N} \exp\{-\tfrac{1}{2}\gamma(\sigma_l)_z\} \tag{40}$$

The ultrarelativistic limit of the YB algebra relations (4) with the R-matrix (13) yields the algebra of the operators $J_\pm(\gamma)$ and S_z. We find [1]

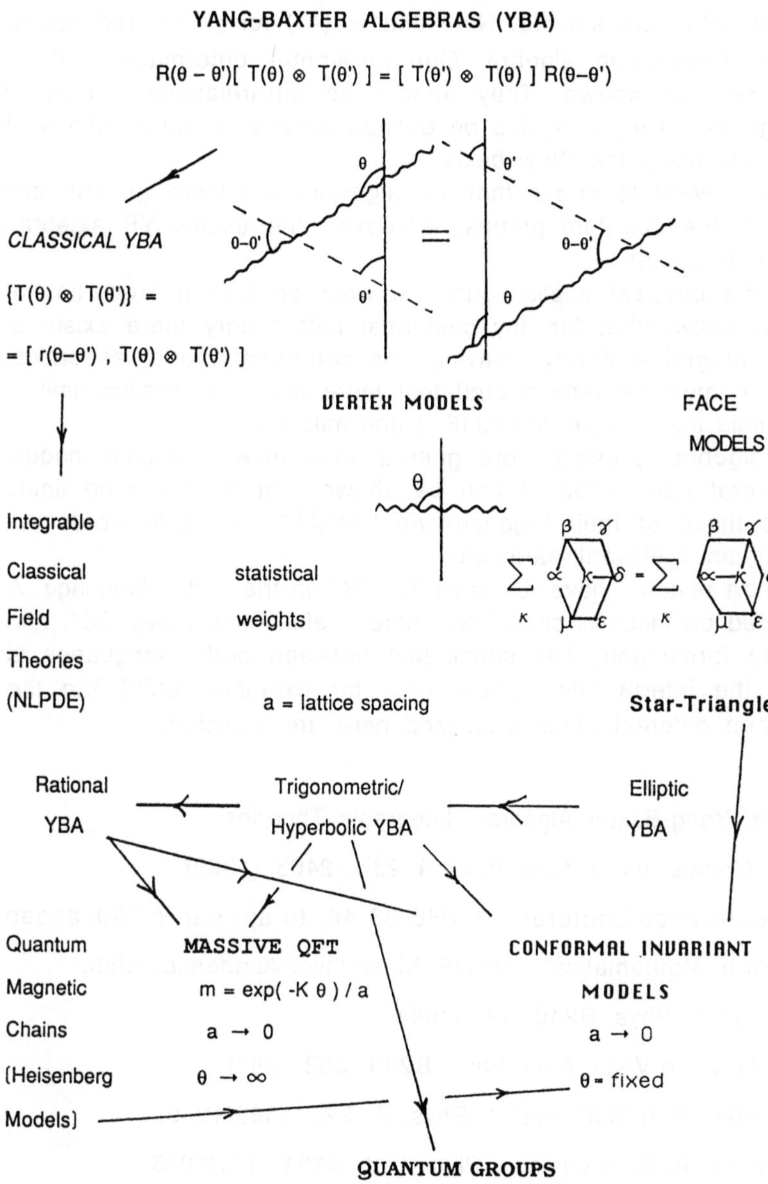

FIGURE : Yang-Baxter algebras and their descendants.

$$[\,J_+(\gamma)\,,\,J_-(\gamma)\,] = \mathrm{sh}(2\gamma S_z)\,/\,\mathrm{sh}\,\gamma$$

$$[\,S_z\,,\,J_\pm\,] = \pm J_\pm \quad . \tag{41}$$

Analogous relations hold with $\gamma \to i\gamma$ in the trigonometric regime. Eqs.(41) define the so-called $SU_\gamma(2)$ quantum group. That is, a one-parameter deformation of the SU(2) Lie algebra. Note that eq.(41) for $\gamma \to 0$ reduces to the usual angular momentum algebra. One parameter deformations of all simple Lie algebras are known. They appear as ultrarelativistic limits of hyperbolic YB algebras. They can also be defined directly as deformations of classical Lie algebras using the Weyl basis.

To conclude, I want to stress that YB algebras are more general and powerful tools than the quantum groups. Moreover, the elliptic YB algebras provide additional structures .

Concerning the physical implications, the work on critical properties of integrable theories shows that for any conformal field theory there exists at least one lattice integrable theory having this conformal model as critical (continuous) limit. It must be remembered that more than one scaling limit is possible. For example massive (as in eq.(16)) and massless.

Elliptic YB algebras provide more general integrable statistical models which are in general non-critical. It can be shown that their scaling limits are the same as those of their trigonometric limits[14]. In other words the elliptic modulus is an irrelevant parameter.

In this very short review I have reviewed the YBA in the vertex language. A dual language based on faces is sometimes more useful. In this way SOS and RSOS models are formulated. The connection between both languages is possible through the intertwining vectors (see for example ref.[9]).In the figure, links between different ideas discussed here are depicted.

1 H. J. de Vega, Yang-Baxter Algebras, Integrable Theories and Quantum Groups. Int. J. Mod. Phys. **4**, 2371-2463 (1989).
Taniguchi Conference Lectures, LPTHE 88-46, to appear in "Advanced Studies in Pure Mathematics" vol.19, Kinokuniya-Academic, 1989.

2 H. J. de Vega, Nucl. Phys. **B240**, 495 (1984).

3 C. Destri and H. J. de Vega, Nucl. Phys. **B290**, 363(1987), Phys. Lett. **B201**, 261(1988) and J. Phys. **A**, **22**, 1329(1989).

4 A.M. Polyakov and P. B. Wiegmann, Phys. Lett. **B131**, 121(1983), **B142**, 173(1984) and **B141**, 217(1984).
L. D. Faddeev and N. Y. Reshetikhin, Ann. Phys. **167**, 227(1986).
C. Destri and H. J. de Vega, Phys. Lett. **B201**, 245 and **B208**, 255 (1988).

5 C. Destri and H. J. de Vega , Physics Lett. **B223,** 365 (1989).

6 H. J. de Vega and F. Woynarovich, Nucl. Phys. **B251**, 439 (1985).
H. P. Eckle and F. Woynarovich, J. Phys. **A20**, L97(1987)
C. J. Hamer et al., J. Phys. **A20**, 5677(1987).

H.J. de Vega and M. Karowski Nucl.Phys. **B285**, 619(1987).

M. Karowski, Nucl. Phys. **B300**, 473(1988).

7 H. J. de Vega, J. Phys. **A20**, 6023(1987) and J. Phys. **A21**, L1089(1988).

8 See for a review, E. Date et al., RIMS-590 (1987) preprint.

9 H. J. de Vega and H. J. Giacomini, J. Phys. **A 22**, 2759 (1989).

10 A. Luther, Phys. Rev. **B14**, 2153 (1976).

M. Lüscher, Nucl. Phys. **B117**, 475 (1976).

11 H.J. de Vega and T.J.M. Simoes, Phys. Lett. **B217**, 142 (1989).

12 B. Klaiber, Boulder Lectures, Vol. **XA**, p.141 (1968).

J. A. Swieca, Fortschr. der Physik, **25**, 303 (1977).

13 See, for example : J. S. Birman Braids et al., Princeton Univ. Press, 1974.

14 C. Destri and H. J. de Vega, LPTHE preprint 89-20, May 1989.

Representations of $U_q\, sl(2)$ for q a Root of Unity

H. Saleur

Service de Physique Théorique de Saclay, Laboratoire de l'Institut de Recherche Fondamentale du Commissariat à l'Energie Atomique, F-91191 Gif-sur-Yvette Cedex, France

Abstract. A review of $U_q s\ell(2)$ representation theory for q a root of unity is given; physical applications to integrable systems or conformal theories are discussed.

1. To each complex simple Lie algebra $\mathcal{G}$, one can associate[1] the "quantum" analog of its enveloping algebra, $U_q\mathcal{G}$, where $q \in \mathbf{C}$ is a deformation parameter. Here we shall deal with the simple case of $U_q s\ell(2)$ which is generated by e, f, h with relations

$$\begin{cases} [h,e] &= 2e \\ [h,f] &= -2f \\ [e,f] &= \dfrac{q^h - q^{-h}}{q - q^{-1}} \end{cases} \tag{1}$$

which reduce to the usual relations among generators (e, f, h) of $s\ell(2)$ in the limit $q \longrightarrow 1$. $U_q s\ell(2)$ can be endorsed with a Hopf algebra[1] structure, the coproduct $\Delta : U_q s\ell(2) \longrightarrow U_q s\ell(2) \times U_q s\ell(2)$ being given in particular by

$$\begin{cases} \qquad \Delta\left(q^{\pm h}\right) = q^{\pm h} \otimes q^{\pm h} \\ \Delta(e) = q^{h/2} \otimes e + e \otimes q^{-h/2} \quad \Delta(f) = q^{h/2} \otimes f + f \otimes q^{-h/2} \end{cases} \tag{2}$$

Note that the other choice in (2) with q replaced by q^{-1} is also possible, leading to a coproduct $\bar{\Delta}$. Relation between Δ and $\bar{\Delta}$ is provided by the universal $\mathcal{R}$ matrix.

The algebra (1) appeared first in the context of integrable systems[1,3] as a powerful tool to build $\mathcal{R}$ matrix solutions of the Yang Baxter equations. Similar structures have been observed in the operator algebras of conformal field(1) theories[4]. Recently, $U_q s\ell(2)$ has also been shown[5] to provide a discrete analog of the Feigin-Fuchs[6] construction of extended Virasoro algebras for $s\ell(2)$ coset[7] theories.

---

(1) For general references on conformal invariance, consult "Conformal invariance and applications to statistical mechanics" C. Itzykson, H. Saleur, J.B. Zuber, World Scientific 1989.

Springer Proceedings in Physics, Vol. 47 — Number Theory and Physics
Editors: J.M. Luck · P. Moussa · M. Waldschmidt

Deformations analogous to (1) can also be defined for other Lie algebras[2,3], with applications of the same kind.

"Quantum groups" appear as unifying algebraic structures to connect conformal field theories and finite integrable systems, and they are presently under active study.

In this context, the most interesting situations arise when q is a root of unity, corresponding to rational conformal theories. For such values of q, the theory of representations has been addressed only recently[5,8,9]. We give in these notes a summary of the results which seem to have physical applications.

2. We consider here for simplicity a vector space $V = \left(\mathbf{C}^2\right)^N$ corresponding to a tensor product of N fundamental representations of $s\ell(2)$. Since

$$\left(q^{\pm 1} - q^{\mp 1}\right) / \left(q - q^{-1}\right) = \pm 1$$

the classical and quantum commutation relations coincide for a single representation, and one finds then, using the coproduct Δ which is an algebra homomorphism, representations of $e, f, q^{\pm h}$ in V

$$\begin{cases} q^{\pm h} = q^{\pm h} \otimes q^{\pm h} \otimes \cdots \otimes q^{\pm h} \\ e = \sum_i e_i = \sum_i q^{h/2} \otimes \cdots \otimes q^{h/2} \otimes e \otimes q^{-h/2} \cdots \otimes q^{-h/2} \\ f = \sum_i f_i = \sum_i q^{h/2} \otimes \cdots \otimes q^{h/2} \otimes f \otimes q^{-h/2} \cdots \otimes q^{-h/2} \end{cases} \tag{3}$$

There are N terms in the tensor product, and an operator in the ith position acts on the ith $\mathbf{C}^2$ component of V. The integrable systems associated to the fundamental representation of $s\ell(2)$ are of the "XXZ spin one half chain" type[10]. For $q = e^{i\pi/x}$ the central charge of their continuum limit conformal theory is $c = 1 - 6/(x-1)x$. The results that follow would hold true (up to minor modifications) for tensor products of N spin j representations $V = \left(\mathbf{C}^{j+1}\right)^N$ as well. If $j = 2$ for instance, the associated integrable systems would have a continuum limit belonging to the superconformal series.

To study representations we need the Casimir element

$$C = fe + \left[\frac{j+1}{2}\right]^2 \tag{4}$$

where $[x] = \left(q^x - q^{-x}\right) / \left(q - q^{-1}\right)$ denotes the q-analog of x.

For generic values of q it is rather straightforward to see that representations ρ_j are in one to one correspondence[11] with those of $s\ell(2)$. For instance the decomposition of $\left(\mathbf{C}^2\right) = \rho_3 \oplus \rho_1 \oplus \rho_1$ is shown (Fig.1). q analogs of the Clebsch

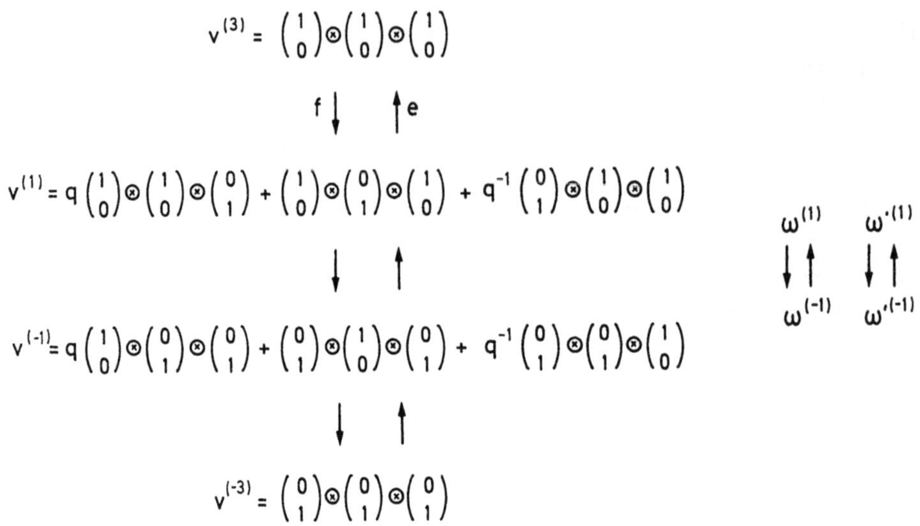

Fig.1. : *Decomposition of* $(\mathbf{C}^2)^3 = \rho_3 + \rho_1 + \rho_1$ *in the generic case.*

Gordan, $6j$ coefficients... can be written[3]. The case q a root of unity ($q^p = \pm 1$ where p is the smallest (odd) integer for which this holds true) is more difficult because the Casimir C can take identical values for highest weight vectors of spins j, j' related by one of the transformations[8]

$$\begin{cases} j' &= j + 2n\ p \\ j' &= 2p - 2 - j + 2n\ p \end{cases} \tag{5}$$

where $n \in \mathbf{Z}$. C is no longer sufficient to label representation ρ_j, the latter being defined here by continuity from the irrational case, and some ρ_j, $\rho_{j'}$ where j, j' satisfy (5) can mix and get connected under the action of e, f.

Another important consequence of q being a root of unity is that e, f become nilpotent[5,8]

$$e^p = 0, \quad f^p = 0 \ . \tag{6}$$

However, $e^p/[p]!$, $f^p/[p]!$ (where $[p]! = [p][p-1]\cdots[2][1]$) are still well defined.

The decomposition of $(\mathbf{C}^2)^3$ for $q^3 = \pm 1$ is shown for instance in Fig.2. We see that ρ_3 and ρ_1 have mixed in a larger representation (ρ_3, ρ_1) which is indecomposable but not irreducible since it contains a subrepresentation ρ_3. ρ_3 is not irreducible either because it contains a subrepresentation formed by fv, f^2v. Hence there is no analog of the Hermann-Weyl complete reducibility theorem here, and accordingly the algebra $U_q s\ell(2)$ can be considered as non-simple. Also it is clear that (ρ_3, ρ_1) is not a highest weight representation.

The analysis of more general situations is similar[8]. In the irrational case $(\mathbf{C}^2)^N$ decomposes as a direct sum of ρ's. For $q^p = \pm 1$ and a given value of $\mathbf{C}^2$,

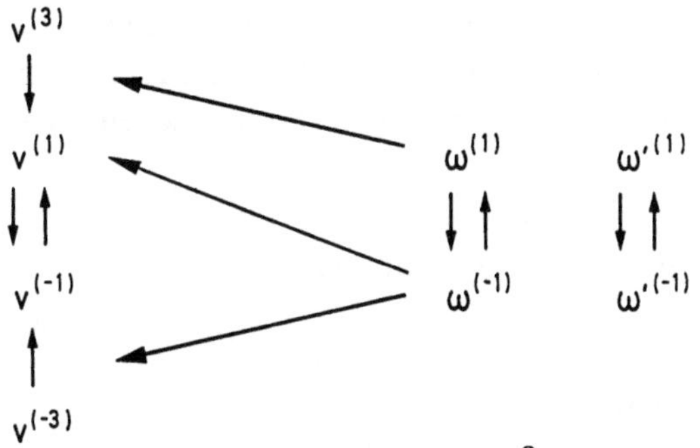

Fig.2 : *Decomposition of* $\left(\mathbf{C}^2\right)^3 = (\rho_3, \rho_1) + \rho_1$ *for* $q^3 = \pm 1$.

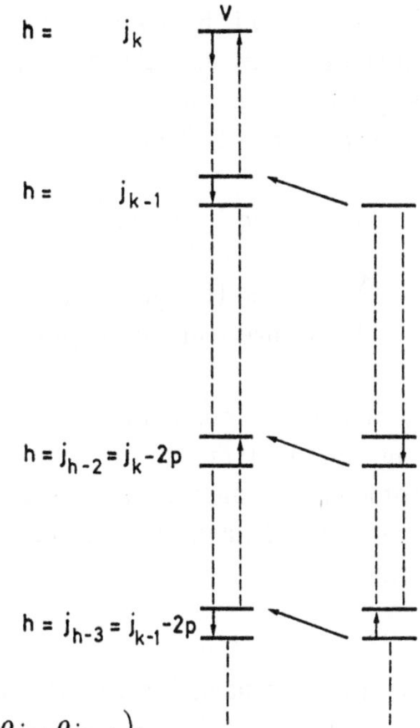

Fig.3 : *Structure of the type I representation* $(\rho_{j_k}, \rho_{j_{k-1}})$.

there are sets of indices like

$$\{j_k > j_{k-1} > \cdots > j_1 \qquad 0 \leq j_1 < p-1\} \tag{7}$$

which are related by the symmetries (5). (If k is odd $j_k = j_1 \bmod 2p$, and if k is even, $j_k = 2p - 2 - j_1 \bmod 2p$). If $k = 1$, ρ_{j_1} cannot get mixed, and is still an irreducible highest weight representation (Type II). If $k > 1$, it can be shown that mixings always occur in pairs - ρ_{j_k} and $\rho_{j_{k-1}}$ on the one hand, $\rho_{j_{k-2}}$ and $\rho_{j_{k-3}}$... mix thus in larger, indecomposable representations (Type I) $(\rho_{j_k}, \rho_{j_{k-1}})$ (Fig.3), $(\rho_{j_{k-2}}, \rho_{j_{k-3}})$... Depending on the number of $\rho's$ we end up with a certain number of ρ_{j_1} that cannot mix and are still irreducible highest weight representations.

A special situation occurs if $j_1 = p - 1$ since $2p - 2 - j_1 = j_1$. In this case the representations ρ_{j_k} are still irreducible and do not pair. By convention we also call them type I.

It is useful to introduce the $q-$dimension of representations

$$\dim_q \rho_j = \sum_{\text{states}} q^h = [j+1] \tag{8}$$

which multiplies as ordinary dimensions

$$\dim_q \rho_{j_1} \times \dim_q \rho_{j_2} = [j_1 + 1]\,[j_2 + 1] = \sum_{|j_1 - j_2|\ ,j = |j_1 - j_2| \bmod 2}^{j_1 + j_2} [j+1] \tag{9}$$

Equation (8) is invariant in the first of the transformations (5) and changes sign in the second case. Accordingly, we see that the representations mix in such a way that their total $q-$dimension is zero. Also (8) vanishes for the degenerate case $j = np - 1$. Thus V splits into[8]

$$\begin{cases} \text{– Type I representaions which have a } q\text{–dimension zero} \\ \text{and are either mixed or of the kind } \rho_{np-1} \\ \text{– Type II representations which have a non – zero } q\text{–dimension} \\ \text{and are still isomorphic to } s\ell(2) \text{ ones} \end{cases} \tag{10}$$

In relation to conformal field theories, only type II representations are interesting. For a given hamiltonian $\mathcal{H}$ which commutes with $U_q s\ell(2)$ (and hence has some eigenvalue associated with each representation), the type I representations are easily thrown away by considering traces of the form (where T denotes the time)

$$Z = \mathrm{Tr}\left[\mathrm{e}^{-T\mathcal{H}} q^h\right] \tag{11}$$

to calculate partition functions. Also, in the product of type II representations one has

$$[j_1 + 1]\,[j_2 + 1] = \sum_{|j_1-j_2|\ ,j=|j_1-j_2| \bmod 2}^{\mathrm{Max}(j_1+j_2,2p-4-j_1-j_2)} [j+1] \tag{12}$$

a formula analogous to those occurring in $s\ell(2)$ level $p-2$ Wess Zumino models[4].

Type II representations are still described by their highest weight vector v_j such that $h\ v_j = j\ v_j$, $0 \leq j < p-1$, and v_j does not belong to a larger $\rho_{j'}$ representation. The latter condition is obtained by restricting to[8]

$$v_j \in \mathrm{Ker}\ e/\mathrm{Im}(e)^{p-1} \tag{13}$$

The number of type II representations of spin j, $\Omega_j^{(N)}$ is immediately deduced from the pairing principle. If we introduce $\Gamma_j^{(N)}$, the number of ρ_j representations in the $s\ell(2)$ case, it reads

$$\Omega_j^{(N)} = \Gamma_j^{(N)} - \Gamma_{2p-2-j}^{(N)} + \Gamma_{j+2p}^{(N)} - \Gamma_{2p-2-j+2p}^{(N)} + \cdots \tag{14}$$

where in the case of $V = \left(\mathbf{C}^2\right)^N$, $\Gamma_j^{(N)} = \binom{N}{(N-j)/2} - \binom{N}{((N-j)/2)+1}$. $\Omega_j^{(N)}$ has a simple interpretation in terms of paths on a Bratteli diagram[12] (Fig.4). The vertical scale is the number N, points correspond to representations of spin j. In the $s\ell(2)$ case, the diagram is not limited on the right side, and $\Gamma_j^{(N)}$ is the number of paths from the point j to the origin. In the $U_q s\ell(2)$, $q^p = \pm 1$ case,

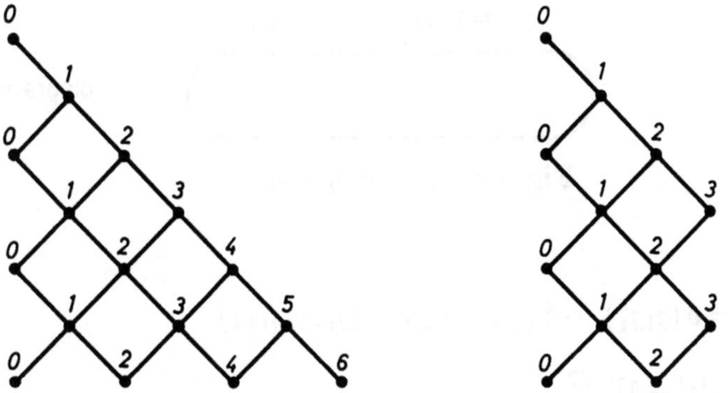

Fig.4 : *Bratteli diagram for $s\ell(2)$ and $U_q s\ell(2)$, $q^5 = \pm 1$.*

the diagram is truncated to exclude spins $j \geq p-1$, and $\Omega_j^{(N)}$ is similarly the number of paths from j to the origin that can be built on the remaining part.

So far, the only property used is $q^p = \pm 1$. The case $q = \exp i\pi/p$ has the further interesting property that all type II representations are then unitary $(e^+ = f)$.

This is related to Jones' result[12] on subfactors as well as to the Friedan Qiu Shenker result[13] on unitarity of Virasoro algebra representations.

3. The commutant of the quantum group (1) in $V = \left(\mathbf{C}^2\right)^N$ is known[2,3] to be the Temperley Lieb algebra with generators e_i $(1 \leq i \leq N-1)$ and relations

$$\begin{cases} e_i e_{i\pm 1} e_i = e_i \\ e_i^2 = \left(q + q^{-1}\right) e_i \end{cases} . \tag{15}$$

A first representation is obtained as

$$e_i = \mathbf{1} \otimes \cdots \otimes \begin{pmatrix} 0 & 0 & 0 & 0 \\ 0 & q^{-1} & -1 & 0 \\ 0 & -1 & q & 0 \\ 0 & 0 & 0 & 0 \end{pmatrix} \otimes \mathbf{1} \cdots \otimes \mathbf{1} \tag{16}$$

where the matrix in the tensor product acts on the ith and $(i+1)^{\text{th}}$ subspaces. It is associated with the unrestricted solid on solid model (6-vertex, XXZ chain).

A basis in V can be chosen in such a way that for any $0 \leq i \leq N$, the first i spins define a state belonging to some ρ_{j_i} representation. The set of ρ'_{j_i}s is in turn associated with a path on the Bratteli diagram, and another representation of e_i is obtained[14] in the space of paths $\{j\}$

Fig.5 : A_{p-1} *diagram.*

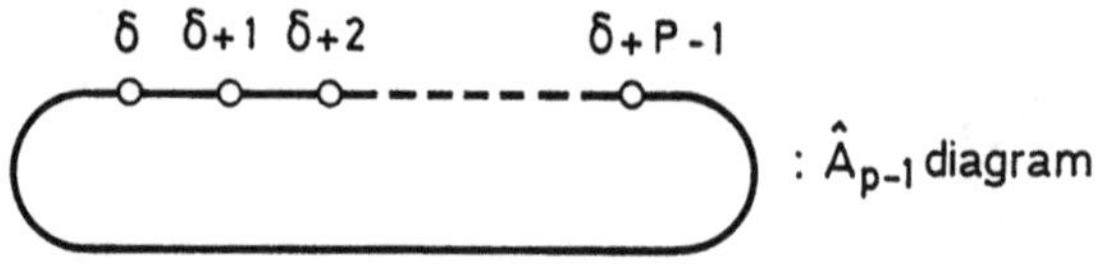

Fig.6 : $\hat{A}_{P-1}$ *diagram.*

$$
\begin{aligned}
(e_i)_{\{j\}\{j'\}} &= \delta\left(j_1 j_1'\right)\cdots\delta\left(j_{i-1}, j_{i-1}'\right)\delta\left(j_{i-1}, j_{i+1}\right) \\
&\times \frac{\left([j_i+1]\,[j_i'+1]\right)^{1/2}}{[j_{i-1}+1]}\delta\left(j_{i+1}, j_{i+1}'\right)\cdots\delta\left(j_N, j_N'\right) \quad . \qquad (17)
\end{aligned}
$$

If $q^p = \pm 1$, $[j+1]$ vanishes for $j = p-1$. Hence the representation (17) is reducible since the e_i's leave the truncated Bratteli diagram invariant. On the latter, for any value of j_N fixed, the e_i's (17) define an irreducible representation of Temperley Lieb algebra[15,16]. It is associated with the restricted solid on solid models of A_{p-1} type[17,18] (Fig.5).

4. For generic q, a basis in ρ_j can be chosen so that

$$
\begin{cases}
e\ v^{(m)} &= \left[\dfrac{\delta - m}{2}\right] v^{(m+2)} \\
f\ v^{(m)} &= \left[\dfrac{\delta + m}{2}\right] v^{(m-2)} \\
q^{\pm h} v^{(m)} &= q^{\pm m} v^{(m)}
\end{cases}
\qquad (18)
$$

with $\delta = j$. Formulas (18) however define a representation of $U_q s\ell(2)$ for arbitrary δ (irrational say), and Casimir $C = \left[\frac{\delta+1}{2}\right]^2$. In the generic case, such a representation is infinite dimensional, and analogous to what is observed in the $s\ell(2)$ case. If $q^P = 1$ however, it is possible to identify $v^{(m)}$ and $v^{(m+2P)}$ in (18), hence giving a finite representation R_δ of dimension P and $q-$dimension 0. R_δ is not of the type previously considered since the action of e, f is here periodic, $e^p = f^p = \left[\frac{\delta}{2}\right]\left[\frac{\delta}{2}+1\right]\cdots\left[\frac{\delta}{2}+P-1\right]\mathbf{1}$. R_δ is irreducible.

Formulas for q analogs of Clebsch Gordan coefficients in tensor products are obtained from the usual ones by replacing j by δ. So for instance $R_\delta \otimes \rho_1$ decomposes as $R_{\delta-1} \otimes R_{\delta+1}$, both having the same dimension P. Considering $R_\delta \otimes (\rho_1)^N$ gives representations $R_\delta, R_{\delta+1} \cdots, R_{\delta+P-1}$ that can be associated with nodes of an $\hat{A}_{P-1}$ diagram (Fig.6). $R_{\delta+P}$ is identified with R_δ since both have the same Casimir value, and differ only by signs in (18).

The number of representations $R_{\delta+j}$ is also the number of paths on the proper periodic Bratteli diagram (Fig.7). It is given by (P even)

Fig.7 : *Periodic Bratteli diagram for the composition of R_δ representations and* $P = 4$.

$$\Omega^{(N)}_{\delta+j} = \binom{N}{(N-j)/2} + \binom{N}{(N-j)/2 \pm (P/2)} + \binom{N}{(N-j)/2 \pm P} + \cdots \qquad (19)$$

Representations of the Temperley Lieb algebra are then obtained by the same formula as (17) where j is replaced by $j+\delta$. They are associated with Boltzmann weights of $\hat{A}_{P-1}$ models[19].

5. The cases considered above exhaust[9] the types of irreducible finite dimensional representations of $U_q s\ell(2)$ for q a root of unity: $q^P = 1$. If P is odd, there are such representations with dimensions $1 \leq d \leq P$, the case $d = P$ corresponding either to the ρ_{P-1} type (10) or R_δ (18). If P is even, there are such representations with dimensions $1 \leq d \leq \frac{P}{2}$ and $d = P$, the case $d = \frac{P}{2}$ ($d = P$) corresponding to $\rho_{(P/2)-1}$ (R_δ).

ACKNOWLEDGMENTS

I thank V. Pasquier with whom I had the pleasure of collaborating on these matters.

REFERENCES

[1] Drinfeld, in Proceedings of the International Congress of Mathematicians, Berkeley (1986).

[2] P.P. Kulish, N. Yu Reshetikhin, J. Sov. Math. 23, 2435 (1983).

[3] Jimbo, Lett. Math. Phys. 10; 63 (1985); Comm. Math. Phys. 102, 537 (1986).

[4] L. Alvarez Gaumé, Gomez, Sierra "Quantum Group Interpretation of some Conformal Field Theories", preprint.

[5] V. Pasquier, H. Saleur, "Common Structures between Finite Sytems and Conformal Field Theories through Quantum Groups", preprint.

[6] B.L. Feigin, D.B. Fuchs, Funct. Anal. and Appl. 16, 114 (1982); 17, 241 (1983).
[7] P. Goddard, A. Kent, D. Olive, Comm. Math. Phys. 103, 105 (1986).
[8] G. Lusztig, "Modular Representaions and Quantum Groups", preprint
[9] P. Roche, D. Arnaudon, " Irreducible Representations of the Quantum Analogue of SU(2)", preprint.
[10] V. Pasquier, H. Saleur, "Symmetry of the XXZ Chain and Quantum SU(2)", preprint.
[11] M. Rosso, Comm. Math. Phys. 117, 581 (1988).
[12] V. Jones, Inv. Math. 72, 1 (1983).
[13] D. Friedan, Z. Qiu, S. Shenker, Phys. Rev. Lett. 52, 1575 (1984).
[14] V. Pasquier, Comm. Math. Phys. 118, 355 (1988).
[15] H. Wenzl, "Representations of the Hecke Algebras and Subfactors", Thesis (1985).
[16] P. Martin, J. Phys. A20, L539 (1987).
[17] G. Andrews, R. Baxter, P.J. Forrester, J. Stat. Phys. 35, 193 (1984).
[18] V. Pasquier, Nucl. Phys. B285, 162 (1987).
[19] V. Pasquier, J. Phys. A20, L1229 (1987).
H. Saleur, unpublished

Part II

Quasicrystals and Related Geometrical Structures

Some Quasiperiodic Tilings as Modulated Lattices

M. Duneau and C. Oguey
Centre de Physique Théorique, Ecole Polytechnique, F-91128 Palaiseau, France

Abstract. We consider quasiperiodic structures obtained by the cut and project method and show that whenever the window satisfies a non-generic tiling property the associated structure can be transformed into a periodic lattice by bounded displacements. The case of the octagonal tiling is examined in detail.

1. Introduction.

Modulated lattices introduced by de Wolf, Janner and Janssen [1] are quasiperiodic structures which are obviously related to average lattices. More general quasiperiodic structures have been considered since the discovery of quasicrystals [2] : 3 D Penrose tilings for instance. Such structures can be obtained as affine cuts of a periodic array of "atomic surfaces" in a higher dimensional space [3-5] — this section method generalizes a well known definition of quasiperiodic functions [6,7].
In quasicrystal models, these atomic surfaces are bounded subsets of the space $E^{\perp}$ complementary to the physical space E (=cutting space) whose orientation with respect to the higher dimensional lattice is known. However the problem of determining the exact profile (shape) of the atomic surfaces is not yet completely solved. Here we ask the question of the existence of an average lattice for those structures and we give a sufficient condition involving the shape of the atomic surfaces. Similar questions were studied by S. Aubry *et al.* (see [8] and included refs.) for 1D chains. We focus on situations where both E and $E^{\perp}$ are of dimension larger than one. The octagonal tiling is examined in detail and we settle the existence of an average lattice for this case.

2. Construction of an average lattice.

Let us briefly recall the general framework of the cut and project method in arbitrary dimensions [3,9]. The "physical" space E is a d-dimensional subspace of $\mathbf{R}^n$ $(0<d<n)$ and E' is a (n–d)-dimensional complement of E. Let p and p' denote the two complementary projectors onto E (with kernel E') and E' (with kernel E). Finally Λ is a n-dimensional lattice generated by the basis $\{a_1,\ldots, a_n\}$:

$$E = p(\mathbf{R}^n) \qquad (1)$$

Springer Proceedings in Physics, Vol. 47 Number Theory and Physics
Editors: J.M. Luck · P. Moussa · M. Waldschmidt

$$E' = p'(\mathbf{R}^n) \tag{2}$$

$$\mathbf{R}^n = E \oplus E'. \tag{3}$$

$$\Lambda = \left\{ \sum_{i=1}^{n} z_i.a_i \mid z_i \in \mathbf{Z} \text{ for } i = 1,\ldots,n \right\} \tag{4}$$

The projections onto E and E' of the lattice Λ are **Z**-modules L and L' given by :

$$L = p(\Lambda) = \left\{ \sum_{i=1}^{n} z_i.e_i \mid z_i \in \mathbf{Z} \text{ for } i = 1,\ldots,n \right\} \tag{5}$$

$$L' = p'(\Lambda) = \left\{ \sum_{i=1}^{n} z_i.e'_i \mid z_i \in \mathbf{Z} \text{ for } i = 1,\ldots,n \right\} \tag{6}$$

with $e_i = p(a_i)$ and $e'_i = p'(a_i)$ for $i = 1,\ldots,n$.

For generic orientations of E and E', the **Z**-modules L and L' are of rank n and are countable dense sets of points. The rank of L may be less than n only if the kernel E' of the projection p contains a rational subspace. In the following we shall assume that E and E' are in generic orientation, which is the case in important applications and in the example of the octagonal tiling studied below.

An important parameter of the CP method is a relative translation between the lattice and the physical space E. We shall assume that E is fixed and that the lattice may undergo global translations. For any given τ in $\mathbf{R}^n$ define $\Lambda_\tau = \Lambda + \tau$, $L_\tau = p(\Lambda_\tau) = L + t$, where $t = p(\tau)$ and $L'_\tau = p'(\Lambda_\tau) = L' + t'$, where $t' = p'(\tau)$.

Once the orientations of E and E' are fixed, the CP method involves a "window" W which is a bounded open subset of E'. Then it is possible to define the corresponding "strip" by

$$S = E \times W = \{ x + x' \mid x \in E \text{ and } x' \in W \} \tag{7}$$

In the case of the "canonical" tilings (see [7,9]) W is a polyhedron obtained as the projection onto E' of the standard unit cell of the lattice Λ (W is therefore a zonohedron). The simple CP method proceeds with the selection of the subset Ξ_τ of points of Λ_τ which belong to the strip; then this subset is projected onto E, yielding a d-dimensional structure X_τ :

$$\Xi_\tau = \Lambda_\tau \cap S \tag{8}$$

$$X_\tau = p(\Xi_\tau) = p(\Lambda_\tau \cap S) \tag{9}$$

If the window W and the translation τ are such that the boundary ∂W of W does not meet L'_τ then X_τ is a regular discrete and quasiperiodic set built out of a finite number of different local patterns [9].

Notice that the set $\Xi_\tau = \Lambda_\tau \cap S$ has no translational invariance; actually if it were invariant with respect to some translation α then α should belong to the lattice Λ and to the physical space E, a situation which has been rejected. So there is no obvious reason for the structure $X_\tau = p(\Xi_\tau)$ to have any translation invariance. As we shall see below X_τ may nevertheless show a weak translation invariance, as a modulated lattice, for certain choices of the window W. This property is defined as follows :

Definition : Given a discrete set X, we say that there is an average lattice for X if there exists a periodic lattice L and a one to one mapping ϕ between X and L such that the displacements $\|\phi(x) - x\|$ are uniformly bounded by some finite constant C.

Theorem.

If W is a unit cell of a certain (n–d)-dimensional lattice L'_0 contained in L' then the above defined structure X_τ has an average lattice.

Proof.

Assume W is a fundamental cell for a certain (n–d)-dimensional lattice L'_0 contained in L'; then L'_0 is the projection of a (n–d)-dimensional sublattice Λ_0 of Λ. Let E_0 be the (n–d)-dimensional subspace spanned by Λ_0 and let Ω denote the intersection of E_0 with the strip S; thus :

$$L'_0 = p'(\Lambda_0) \tag{10}$$
$$W = p'(\Omega) \tag{11}$$
$$\Omega = S \cap E_0 \tag{12}$$

Since W is a fundamental cell for L'_0, Ω is a fundamental cell for Λ_0. The strip S has the equivalent definition

$$S = E \times \Omega = \{ x + y \mid x \in E \text{ and } y \in \Omega \} \tag{13}$$

Obviously, Λ_0 is a sublattice of the intersection lattice $\Lambda \cap E_0$. Now we consider two cases according to whether Λ_0 is equal to $\Lambda \cap E_0$ or not.

1) The simple case : $\Lambda_0 = \Lambda \cap E_0$. It is easily checked that there exists a complementary sublattice Λ_1, spanning a subspace E_1, such that we have the direct sums (see [10]):

$$\Lambda = \Lambda_0 \oplus \Lambda_1 \tag{14}$$
$$\mathbf{R}^n = E_0 \oplus E_1 \tag{15}$$

Define π as the oblique projector with range E and kernel E_0. We prove now that an average lattice of X_τ is given by the oblique projection $\pi(\Lambda_1)$ of Λ_1 on E :

$$L_1 = \pi(\Lambda) = \pi(\Lambda_1) \tag{16}$$

We show first that the oblique projection π gives a one to one mapping between Ξ_τ and $L_1 + t_1$ where $t_1 = \pi(\tau)$:

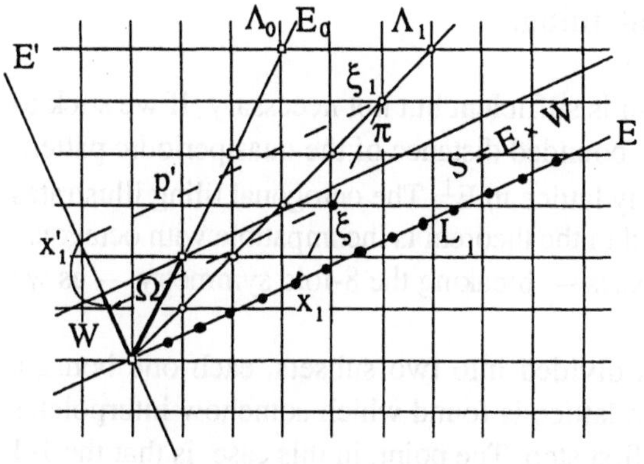

Figure 1. *The window W is the projection of a fundamental cell Ω of a sublattice Λ_0. There is a one to one mapping between lattice points ξ belonging to the strip* S *and the rational projection* L_1 *of Λ in the physical space.*

i) Since $\Xi_\tau \subset \Lambda_\tau$ we have the obvious inclusion $\pi(\Xi_\tau) \subset L_1+t_1$.

ii) Now for any x_1 in $L_1+ t_1$ there exists a point ξ_1 of Λ_τ such that $x_1 = \pi(\xi_1)$; since E_0 is in the kernel of π we also have $x_1 = \pi(E_0+ \xi_1)$. The (n–p)-plane $E_0+ \xi_1$ has an intersection with the strip S which is a translate of the unit cell Ω and therefore $E_0+ \xi_1$ contains a unique point ξ of the lattice $\Lambda_0+ \xi_1$; this point ξ belongs to the set Ξ_τ and satisfies $\pi(\xi) = x_1$.

The one to one mapping between X_τ and L_1 is obtained as follows : for all points ξ in Ξ_τ we consider the projections $x = p(\xi)$ in X_τ and $x_1 = \pi(\xi)$ in L_1+t_1; the mapping ϕ is defined by $\phi(x) = x_1 - t_1$. The existence of a finite upper bound C for the distances $\|x - \phi(x)\|$ follows merely from the assumption that W, and therefore Ω are bounded. See figure 1.

2) Case $\Lambda_0 \neq \Lambda \cap E_0$.

In this case, Λ_0 is a sublattice of $\Lambda \cap E_0$ and there is a finite set of points B such that

$$\Lambda_0 + B = \Lambda \cap E_0 \tag{17}$$

The oblique projection $\pi(\Xi_\tau)$ of Ξ_τ on $L_1+ t_1$ is no longer one to one; actually, for any point x_1 in $L_1+ t_1$ there exists a point ξ in Ξ_τ such that $B+ \xi$ is the intersection of the lattice $\Lambda_0+ \xi$ with the strip S; consequently, $x_1 = \pi(B+ \xi)$.

The construction given above in the simple case can easily be adapted and a one to one mapping ϕ can be defined between the structure X_τ and a lattice $L_1 + B_1$ with the basis B_1 (associated to B by a one to one mapping so that $B+ \xi$ is mapped onto $B_1+ x_1$). By compactness, an upper bound C for the distances $\|x - \phi(x)\|$ obviously exists and can possibly be improved using a global translation s of L_1 and an optimized basis B_1.•

3. Average lattices of the octagonal tiling.

The condition stated in the above theorem is sufficient but not necessary. If we stick to the definitions, there may exist lattices at bounded distance of the quasiperiodic pattern even when the window is not a cell for any lattice in $E^{\perp}$. The octogonal tiling illustrates this fact : a priori the cell property required in the theorem is incompatible with octagonal symmetry. A mean lattice nevertheless exists — breaking the 8-fold symmetry — as we show in this section.

First the set of vertices of the tiling is divided into two subsets, each one being a modulated lattice; next, a common mean lattice is found which somehow interpolates between the two lattices obtained in the first step. The point, in this case, is that the 1-1 map from the quasiperiodic pattern to the mean lattice is not inferred from a manipulation in the torus T^n; so it is not, strictly speaking, a modulation.

The octagonal tiling is obtained by the cut and project method from $\mathbf{R}^4$ in which we consider the hyper-cubic lattice $\Lambda = \mathbf{Z}^4$ with the standard orthonormal basis $\{a_1,..., a_4\}$. The two orthogonal planes E and E' are defined as the ranges of the following orthogonal projectors :

$$p = \frac{1}{2}\begin{pmatrix} 1 & 0 & \alpha & -\alpha \\ 0 & 1 & \alpha & \alpha \\ \alpha & \alpha & 1 & 0 \\ -\alpha & \alpha & 0 & 1 \end{pmatrix} \quad \text{and} \quad p' = \frac{1}{2}\begin{pmatrix} 1 & 0 & -\alpha & \alpha \\ 0 & 1 & -\alpha & -\alpha \\ -\alpha & -\alpha & 1 & 0 \\ \alpha & -\alpha & 0 & 1 \end{pmatrix} \qquad (18)$$

where $\alpha = 1/\sqrt{2}$.

The projections $L = p(\Lambda)$ and $L' = p'(\Lambda)$ are dense **Z**-modules respectively generated by $\{e_1,..., e_4\}$ and $\{e'_1,..., e'_4\}$ where $e_i = p(a_i)$ and $e'_i = p'(a_i)$. In orthonormal coordinates

$$\begin{aligned} e_1 &= \alpha.(1, 0) & e'_1 &= \alpha.(1, 0) \\ e_2 &= \alpha.(0, 1) & e'_2 &= \alpha.(0, 1) \\ e_3 &= \alpha.(\alpha, \alpha) = (1/2, 1/2) & e'_3 &= \alpha.(-\alpha,-\alpha) = (-1/2, -1/2) \\ e_4 &= \alpha.(-\alpha, \alpha) = (-1/2, 1/2) & e'_4 &= \alpha.(\alpha, -\alpha) = (1/2, -1/2) \end{aligned}$$

The lattice Λ is invariant with respect to the transformation γ defined by $\gamma(a_1) = a_3$, $\gamma(a_2) = a_4$, $\gamma(a_3) = a_2$ and $\gamma(a_4) = -a_1$. One easily checks that γ is an element of the hyper-cubic group and is of order 8. The corresponding representation of the 8-fold cyclic group generated by γ is reducible and the decomposition into irreducible invariant subspaces precisely yields the planes E and E'. We have two inequivalent representations : γ induces a rotation of $\pi/4$ in E and a rotation of $5\pi/4$ in E'.

The octagonal tilings correspond to a strip $S = E \times W$ where W is the octagon obtained as the projection $p'(\Omega)$ of the unit cube of $\mathbf{R}^4$. W is the zonohedron generated by the four vectors $\{e'_1,..., e'_4\}$ (see figure 2).

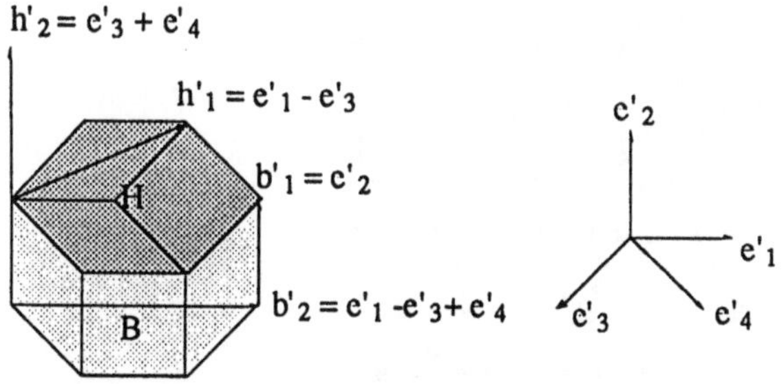

Figure 2. *The projection of* $\mathbf{Z}^4$ *in* E' *is a* **Z***-module generated by the four vectors* $\{e'_1, e'_2, e'_3, e'_4\}$. *The projection of the unit cube of* $\mathbf{R}^4$ *is a regular octagon which can be decomposed in two pieces :* B *which is a unit cell for the lattice generated by* $\{b'_1, b'_2\}$ *and* H *which is a unit cell for the lattice generated by* $\{h'_1, h'_2\}$.

W is not a prototile for any tiling of E' and consequently, the construction presented in the above section cannot apply directly. However the octagon can be cut in two pieces, an hexagon W_H and a "boat" W_B, shown in figure 2, which are prototiles. More precisely, W_H is a unit cell for the 2-dimensional lattice L'_H generated by $h'_1 = e'_1 - e'_3$ and $h'_2 = e'_3 + e'_4$ whereas W_B is a unit cell for the 2-dimensional lattice L'_B generated by $b'_1 = e'_1 - e'_3 + e'_4$ and $b'_2 = e'_2$.

Thus, if τ is a generic translation, the vertices $X_\tau = p(\Lambda_\tau \cap S)$ of the corresponding octagonal tiling can be split into two complementary subsets $X_{H,\tau}$ and $X_{B,\tau}$ respectively associated to the sub-strips S_H and S_B defined by $S_H = E \times W_H$ and $S_B = E \times W_B$:

$$X_{H,\tau} = p(\Lambda_\tau \cap S_H) \tag{19}$$
$$X_{B,\tau} = p(\Lambda_\tau \cap S_B) \tag{20}$$
$$X_\tau = X_{H,\tau} \cup X_{B,\tau} \tag{21}$$

Since W_H tiles E', the general method presented in the above section can be applied to build a average lattice for the structure $X_{H,\tau}$:

Define Λ_H as the 2-dimensional sublattice of Λ generated by $\eta_1 = a_1 - a_3$ and $\eta_2 = a_3 + a_4$ (in such a way that $L'_H = p'(\Lambda_H)$) and let E_H be the 2-dimensional lattice plane containing Λ_H. Notice that $\{\eta_1, \eta_2, a_1, a_2\}$ is a basis of Λ and consequently $\{a_1, a_2\}$ generates a 2-dimensional sublattice which is a complement of Λ_H. If π_H denotes the oblique projector with range E and kernel E_H then the average lattice L_H of $X_{H,\tau}$ is the oblique projection $\pi_H(\Lambda)$ on E. Elementary linear algebra shows that L_H is generated by

$$h_1 = \pi_H(a_1) = (2(\sqrt{2} - 1), 0) \tag{22}$$
$$h_2 = \pi_H(a_2) = (0, 2) \tag{23}$$

Thus, this average lattice is rectangular and the one to one mapping between $X_{H,\tau}$ and L_H is shown in figure 3a.

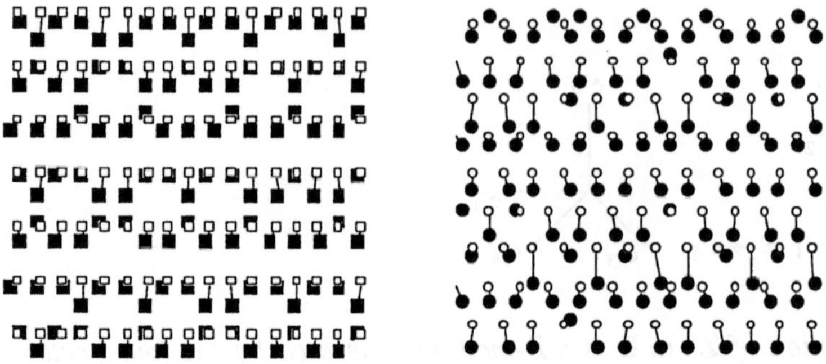

a b

Figure 3.a) *Black squares represent the* H-*part of the octagonal tiling; the white squares correspond to their average lattice.* b) *Black dots represent the* B-*part of the octagonal tiling and the white dots correspond to their average lattice.*

Similarly, define Λ_B as the 2-dimensional sublattice of Λ generated by $\beta_1 = a_1 - a_3 + a_4$ and $\beta_2 = a_2$ (so that $L'_B = p'(\Lambda_B)$) and let E_B be the 2-dimensional plane containing Λ_B. Notice that $\{\beta_1, \beta_2, a_3, a_4\}$ is a basis of Λ. Define π_B as the projector with range E and kernel E_B. The average lattice L_B of $X_{B,\tau}$ is the oblique projection $\pi_B(\Lambda)$ on E and is generated by

$$b_1 = \pi_B(a_3) = (2(\sqrt{2} - 1), 0) \qquad (24)$$

$$b_2 = \pi_B(a_4) = (0, 2) \qquad (25)$$

The one to one mapping between $X_{B,\tau}$ and L_B is shown in figure 3b.

Finally the octagonal tiling X_τ can be viewed as the union of two modulated structures $X_{H,\tau}$ and $X_{B,\tau}$, each one being associated with an average lattice.

However these two lattices L_H and L_B are different and their union is not even a Delaunay set. Nevertheless, it can be proved that the whole structure admits a global average lattice; the construction is based on a particular property that was not used before, namely L_H and L_B have the same density. This property readily follows from the remark that the two windows W_H and W_B have the same area $(1+\sqrt{2})/2$ (corresponding to a square and two rhombi). In such a situation there is a bounded one to one mapping ϕ between L_B and L_H [11] (see figure 4). The final average lattice is obtained as the union of L_H with a convenient translate of $\phi(L_B)$ in order to obtain, say, a centered rectangular lattice $L_H \cup [L_H + (h_1+h_2)/2]$.

To conclude, we mention an application of the average lattice related to a quasiperiodic tiling. By means of the method explained above, a deterministic algorithm can be devised to construct the octagonal tiling. We give here the main lines of the construction for the $X_{H,\tau}$ structure (the $X_{B,\tau}$ structure is obtained in a similar way).

The main loop of the algorithm consists in scanning a finite subset of the square lattice of E' generated by $\{e'_1, e'_2\}$ (which is isomorphic to the average lattice L_H).

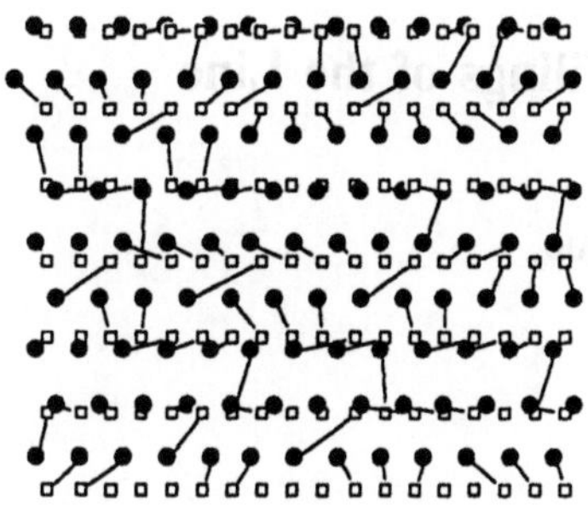

Figure 4. *The one-to one mapping between the lattices* L_H *(white squares) and* L_B *(black dots).*

For each $x'_1 = p_1e'_1 + p_2e'_2$ find the translation $y' = q_1h'_1 + q_2h'_2$ of the lattice L'_H such that $x' = x'_1 + y'$ belongs to the hexagonal unit cell W_H (possibly translated by a fixed translation t'). This step is the generalization of the modulo operation used in one-dimensional algorithms such as the circle map. Then $x' = p_1e'_1 + p_2e'_2 + q_1h'_1 + q_1h'_1 = (p_1 + q_1)e'_1 + p_2\, e'_2 + (q_2 - q_1)e'_3 + (q_1 + q_2)e'_4$; the corresponding point of the structure is then $x = (p_1 + q_1)e_1 + p_2\, e_2 + (q_2 - q_1)e_3 + (q_1 + q_2)e_4$.

4. References.

[1] de Wolf P. M., Janssen T., Janner A. : *Acta Cryst.* A**37** (1981) 625 .

[2] Shechtman D., Blech I., Gratias D. and Cahn J. 1984 : Phys. Rev. Lett. **53**, 1951.

[3] Kalugin P. A., Kitaev A. Yu., Levitov L. C. : *JETP Lett.***41** (1985) 145; *J. Physique Lett.* **46** (1985) L–601.
Elser V. : *Phys. Rev. Lett.* **54** (1985) 1730 ; *Phys. Rev.* B**32** (1985) 4982.
Duneau M., Katz A. : *Phys. Rev. Lett.* **54** (1985) 2477 .

[4] Janssen T. : *Acta Cryst.* A**42** (1986) 261 .

[5] Bak P. : Scr.Metal. **20** 1199 (1986).

[6] Besicovitch A. S. : *Almost Periodic Functions.* Cambridge (1932).
Bohr H. : *Almost Periodic Functions.*Chelsea Publi. Co. New-York (1951).

[7] Oguey C., Duneau M., Katz A. 1988 : Commun. Math. Phys. **118**, 99-118.

[8] S. Aubry, C. Godrèche, J.-M. Luck : J. Stat. Phys. **51**, 1033 (1988).

[9] Katz A., Duneau M. : J. de Physique **47** (1986) 181 .

[10] Cohn H. 1980 : *Advanced number theory*, Dover.

[11] Duneau M. and Oguey C. : to be published.

Types of Order and Diffraction Spectra for Tilings of the Line

C. Godrèche

Service de Physique du Solide et de Résonance Magnétique,
CEA, Saclay, F-91191 Gif-sur-Yvette Cedex, France

A. Introduction.

1. I take the opportunity of this talk to summarize several papers written by S. Aubry, J.M. Luck, F. Vallet and myself, which consider aspects of the general problem of types of order in one dimension [1-4]. I will try to set out the ideas contained in these works in a coherent fashion, without giving all the details of the computations. A review of the two-dimensional case will appear elsewhere [5].

2. The area in which this work falls may be described by the key words and concepts it uses :
- Sequences and structures generated by a circle map.
- Renormalisation of these sequences and structures.
- Existence of an average lattice and Kesten theorem.
- Substitution properties, Pisot theorem and Bombieri-Taylor method.
- Nature of the Fourier spectrum : discrete, absolutely continuous, singular continuous.
- Multifractal analysis of the Fourier spectrum of structures generated by a substitution.

3. Let me first describe one building block of many aspects of this work : a tiling model generated by a circle map. Consider a uniform rotation of angle ω on the unit circle. Define a window $(0, \Delta)$ with characteristic function

$$\chi_\Delta(x) = \mathrm{Int}(x) - \mathrm{Int}(x - \Delta); \qquad (A.1)$$

$\chi_\Delta(x)$ is a 1-periodic function. The sequence $\{\chi_\Delta(n\omega)\}$, consisting of two symbols 0 and 1, is central in this study. From it, one builds a tiling model as follows. With 1, associate a bond length ℓ_1, with 0, associate a bond length ℓ_2, i.e. tile the line with two types of tiles (segments).

Springer Proceedings in Physics, Vol. 47 Number Theory and Physics
Editors: J.M. Luck · P. Moussa · M. Waldschmidt

We can associate with each tile a distribution of matter. From now on let us put atomic masses at the end of each bond. The case of a general distribution of matter will be considered later on. More notations : u_n is the position of the nth atom, and the atomic density reads

$$\rho(x) = \sum_n \delta(x - u_n). \tag{A.2}$$

Hence, the Fourier transform of this density is

$$G(q) = \int \rho(x) e^{-iqx}\, dx = \sum_n e^{-iqu_n}. \tag{A.3}$$

4. One of the questions one may ask about this structure is the nature of its Fourier spectrum for generical values of the parameters $\omega, \Delta, \ell_1, \ell_2$ (in fact only ℓ_1/ℓ_2 is a relevant parameter, since one may rescale the lengths). A guess would be that the structure is quasiperiodic, since the sequence $\{\chi_\Delta(n\omega)\}$ is quasiperiodic itself. By this is meant that this sequence is obtained by chopping a 1-periodic function $\chi_\Delta(x)$ at periodic intervals, with a period ω incommensurate with 1. This guess is wrong as will be shown below .

5. In part B we describe the sequence $\{\chi_\Delta(n\omega)\}$ and the atomic structure, in direct space. In part C, we comment on another model, which can be derived from the same sequence $\{\chi_\Delta(n\omega)\}$ and which was the starting point of this long study. The Fourier spectra are described in part D.

B. Description of the sequence $\{\chi_\Delta(n\omega)\}$ and of the tiling model.

Three aspects will be considered:
- Kesten theorem and existence of an average lattice.
- Renormalisation of these sequences and substitution properties.
- Relationships between the circle map tilings and the cut and project $2D \longrightarrow 1D$ tilings.

1. Kesten theorem and existence of an average lattice.

1.1. From section A3, we have

$$u_n - u_{n-1} = \ell_2 + (\ell_1 - \ell_2)\chi_\Delta(n\omega) \tag{B.1}$$

Hence

$$u_n = n\ell_2 + (\ell_1 - \ell_2) \sum_1^n \chi_\Delta(m\omega) \tag{B.2}$$

(we take $u_0 = 0$).

The mean interatomic distance (or inverse density) is

$$a = \lim_{n\to\infty} \frac{u_n}{n} = \ell_2 + (\ell_1 - \ell_2)\Delta. \qquad (B.3)$$

Indeed, since the points $n\omega$ mod 1 are equidistributed on the unit circle, the proportion of points falling in the window is equal to Δ. Let us write u_n as

$$u_n = na + (\ell_1 - \ell_2)\left[\sum_1^n \chi_\Delta(m\omega) - n\Delta\right] \qquad (B.4)$$

or

$$u_n = na + (\ell_1 - \ell_2)\delta_n \qquad (B.5)$$

with

$$\delta_n = \sum_1^n \chi_\Delta(m\omega) - n\Delta \qquad (B.6)$$

which represents the fluctuation of the number of points falling in the window (i.e. of symbols 1 in the sequence $\{\chi_\Delta(n\omega)\}$) with respect to its average value $n\Delta$. Let us comment on eq.(B.5) : na represents the nth point of the "average lattice" and $(\ell_1 - \ell_2)\delta_n$, also denoted g_n, represents the modulation of the position of the nth atom around the corresponding point of the average lattice.

1.2. This separation of u_n as

$$u_n = na + g_n \qquad (B.7)$$

comes from the field of one-dimensional incommensurate structures, where it appears as a natural description. Indeed in a so-called "displacive" incommensurate structure, the diffraction spectrum shows a pattern of bright spots, organized in a lattice, the dual of the "average lattice", around which one finds less intense spots, the satellites. So the separation into average lattice and modulation is natural. This may also be understood by the description of the structure in direct space. The prototype model for an incommensurate structure is the Frenkel-Kontorova model, which grasps the essence of the structure of an incommensurate material, namely the presence of two lengths in competition. A chain of atoms bonded by harmonic springs is subjected to a b-periodic substrate potential, and to the constraint of a fixed interatomic distance a. The two lengths in competition are a and b. The energy of the chain of atoms reads

$$E = \sum \frac{1}{2}(u_{n+1} - u_n)^2 + \lambda \sin \frac{2\pi}{b} u_n. \qquad (B.8)$$

Aubry [6] showed that, in the ground state, obtained by minimizing the energy of the chain of atoms subjected to the above mentioned constraint, the position of the nth atom is given by

$$u_n = na + g(na) \quad , \tag{B.9}$$

where $g(x)$, the "hull" function, is b-periodic, as is the substrate potential.

Though the discussion of Fourier spectra is postponed to section D, it is worth noting that eq.(B.9) implies a quasiperiodic Fourier spectrum for the structure, with wave vectors $2\pi/a$ and $2\pi/b$. To show this, expand $e^{-iqg(x)}$ in a Fourier series, then use this expansion in

$$G(q) = \sum e^{-iqna}\, e^{-iqg(na)}. \qquad (B.10).$$

More generally, any structure of the type given by eq.(B.9), with a periodic function $g(x)$, possibly not smooth, has a quasiperiodic Fourier spectrum.

1.3. The circle map tiling, for generic values of its parameters ω and Δ, cannot be described by an average lattice with a bounded modulation $g_n \equiv u_n - na$ of the positions of atoms. This is a consequence of a theorem due to Kesten [7]. This theorem asserts that δ_n is bounded with increasing n if, and only if,

$$\Delta = r\omega \bmod 1 \;\equiv\; \mathrm{Frac}(r\omega) \tag{B.11}$$

for some integer r. In other words, after r steps of the uniform rotation of angle ω on the unit circle, the point should coincide with Δ. Cases fulfilling this condition are called "integrable". It is possible to gain understanding of this theorem by several arguments. One approach is as follows. Another one will be met later on in this article. In order to understand the divergence of δ_n with n for non-integrable cases, we compute δ_n for an integrable case. This yields

$$\delta_n = -\gamma_r(n\omega) \tag{B.12}$$

where $\gamma_r(x)$ is a 1-periodic function (see Refs. 2,4). As seen in Fig.1, $\gamma_r(x)$ is a piece-wise linear function. The linear parts have slope r, and there are $r-1$ discontinuities between 0 and 1. Each discontinuity has magnitude 1. When $r \to \infty$, i.e. for a non-integrable case, this function becomes wild. In particular the variation of the function, i.e. the difference between its highest and lowest values, diverges when $r \to \infty$.

Fig. 1 : The function $\gamma_r(x)$ for $r = 4$.

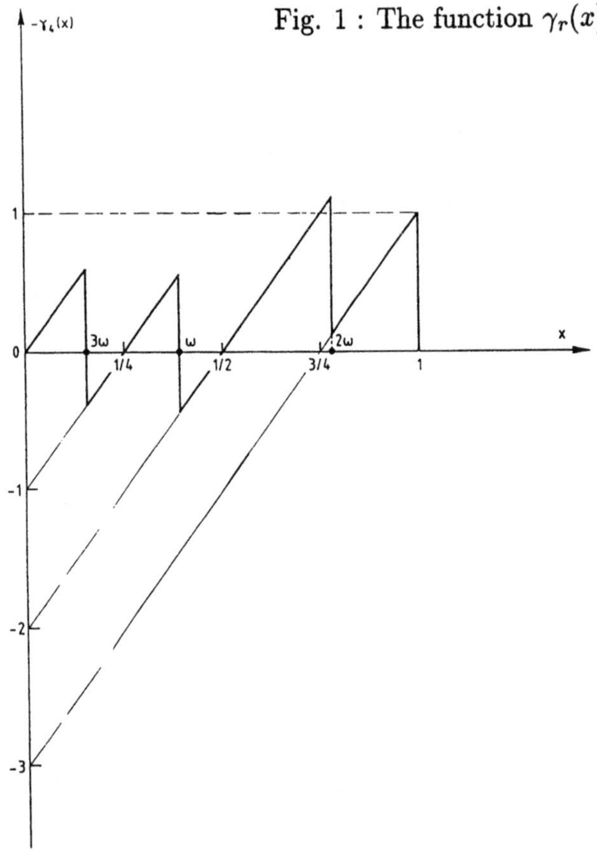

2. Renormalisation of $\{\chi_\Delta(n\omega)\}$ and substitution properties.

Let us define an intermediate labelling sequence on the unit circle. A sequence of letters L_n is associated with the sequence of numbers $\alpha_n = \text{Frac}(n\omega)$ by the rule

$$L_n = \begin{cases} A & \text{if} \quad 0 \le \alpha_n < \inf(\Delta, \omega) \\ B & \text{if} \quad \inf(\Delta, \omega) \le \alpha_n < \sup(\Delta, \omega) \\ C & \text{otherwise.} \end{cases} \tag{B.13}$$

Then by choosing

$$A \equiv 1; \qquad B \equiv 0 \ \text{ if } \ \Delta < \omega \quad \text{or} \quad B \equiv 1 \ \text{ if } \ \omega < \Delta; \qquad C \equiv 0 \tag{B.14}$$

the sequence $\{\chi_\Delta(n\omega)\}$ is recovered. It is possible to renormalize this sequence. By this we mean that there exists a renormalisation operation $\mathcal{R}$ that changes both the values of (ω, Δ) and the sequence $\{L_n\}$:

$$\begin{aligned} (\omega, \Delta) \quad &\longrightarrow (\omega', \Delta') \equiv \mathcal{R}(\omega, \Delta) \\ L_n \quad &\longrightarrow L'_n \equiv \mathcal{R} L_n \end{aligned} \tag{B.15}$$

In the last equation, $\mathcal{R}$ is a condensed notation for a substitution $\sigma_{\mathcal{R}}$ acting on the letters

$$\begin{aligned} A &\longrightarrow A' &&\equiv \sigma_{\mathcal{R}}(A) \\ B &\longrightarrow B' &&\equiv \sigma_{\mathcal{R}}(B) \\ C &\longrightarrow C' &&\equiv \sigma_{\mathcal{R}}(C). \end{aligned} \tag{B.16}$$

If, for particular values of (ω, Δ), one has

$$\mathcal{R}(\omega, \Delta) = (\omega, \Delta) \tag{B.17}$$

(ω, Δ) is called a fixed point of the renormalization transform. In these cases the infinite sequence $\{L_n\}$ is invariant under a substitution $\sigma_{\mathcal{R}}$, or in other words, it may be built by "inflation rules". Two examples are

- $(\omega = \tau^{-2}, \Delta = \omega)$, $\tau = \frac{1+\sqrt{5}}{2}$. This gives the Fibonacci sequence. The substitution is

$$\sigma \left| \begin{aligned} A' &= B \\ B' &= BA \end{aligned} \right. \quad \text{or} \quad \left| \begin{aligned} 0 &\longrightarrow 1 \\ 1 &\longrightarrow 10 \end{aligned} \right. \tag{B.18}$$

- $(\omega = \tau^{-2}, \Delta = 1/2)$. The substitution is

$$\sigma \left| \begin{aligned} A' &= CAC \\ B' &= ACCAC \\ C' &= ABCAC \end{aligned} \right. \tag{B.19}$$

To summarize, if one agrees to restrict the study of the circle map to particular cases, one gains the possibility of generating the sequence $\{\chi_\Delta(n\omega)\}$ by a substitution.

3. Relationships between the circle map tilings and the cut and project 2 ⟶ 1 tilings.

We already noticed above that the case $\Delta = \omega = \tau^{-2}$ generates the Fibonacci sequence, which may also be built by a cut and project method. More generally whenever $\omega = \Delta$, the sequence $\{\chi_\Delta(n\omega)\}$ and the resultant atomic structure on the line may also be generated by the canonical cut and project method. This is illustrated by Fig.2. The slope of the strip, $t = \tan\theta$ is related to Δ by

$$t = \frac{1-\Delta}{\Delta} \tag{B.20}$$

In fact, all integrable cases $(\omega, \Delta = r\omega \bmod 1)$ may be constructed from 2D by section. This is explained in Ref.[8].

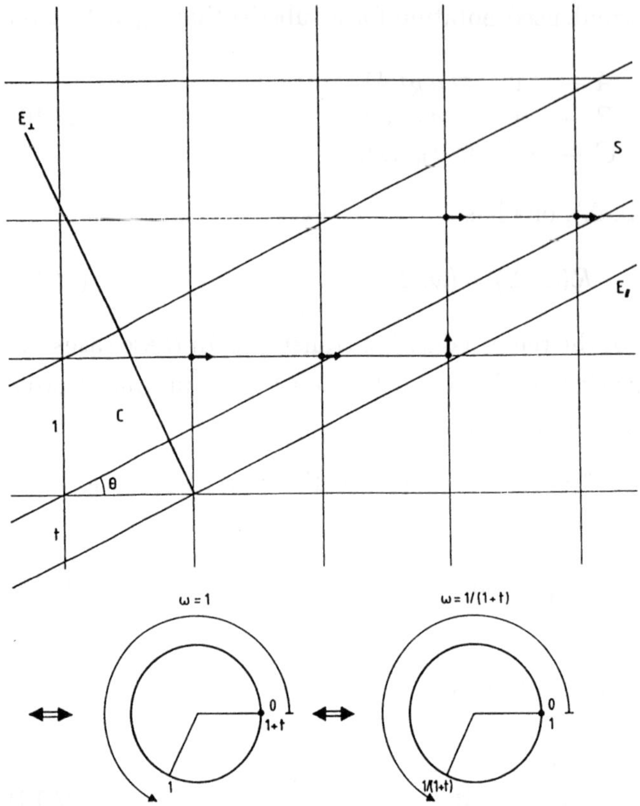

Fig. 2 : Equivalence between the canonical cut and project tiling and the circle map tiling.

C. Atoms and vacancies on a line.

The sequence $\{\chi_\Delta(n\omega)\}$, appears in another model [1]. Consider a chain of atoms subject to a periodic potential, e.g. a sinusoidal one. The atoms may be at a distance ω or 2ω and the concentration of vacancies on the chain is fixed. The ground state of the chain is obtained by minimizing the following energy:

$$E = \sum_n \sin 2\pi u_n \qquad (C.1)$$

with the constraint

$$u_{n+1} - u_n = \omega \quad \text{or} \quad 2\omega. \qquad (C.2)$$

It is easy to realize that the vacancies should be put near the top of the sinusoidal potential, where the cost in energy is maximum. More precisely, for a given concentration of vacancies, there is a forbidden region of size Δ near the top of the potential. Since we imposed only two lengths between atoms, that implies some relationships between ω and Δ. In particular Δ should be smaller than ω.

Folding the sinusoidal potential, with a forbidden region, on a unit circle and knowing that the sites of the atoms and vacancies are on a lattice $\{n\omega\}$, we realize that the order of atoms and vacancies on the line is given by the circle map of sections A and B. In particular, from $\{\chi_\Delta(n\omega)\}$ one gets the structure of the ground state by assigning a vacancy to 1, and an atom to 0. Let us call this model the atoms and vacancies model, or model 1, since it was the first that we considered. Hence, let us call the circle map tiling model considered at the beginning model 2. It is easy to see two facts about model 1.

(i) It has no average lattice with bounded modulation since the theorem of Kesten plays a similar role here as for model 2.

(ii) However, the structure is quasiperiodic, since its atomic density

$$\rho(x) = \chi_\Delta(x) \sum \delta(x - n\omega) \qquad (C.3)$$

is the product of two periodic "functions", with period 1 and ω, respectively. Hence its Fourier transform is discrete: there are Dirac peaks for values of the wave vector such that

$$q = j2\pi + k\frac{2\pi}{\omega}. \qquad (C.4)$$

The Fourier spectrum of model 2 is richer, as will be seen in Section D. There are several other interesting aspects of model 1 that I now comment on.

(a) Using this model, one gains some insight into the meaning of Kesten's theorem by the following argument : suppose that there is an average lattice in the structure. This means that one may write u_n as $u_n = na + g_n$, where g_n is bounded and periodic. As a consequence, by a remark already made in section B.1.2., $2\pi/a$ should be in the spectrum. In other words, $2\pi/a$ must belong to the module spanned by 2π and $2\pi/\omega$. Knowing that

$$a = \frac{\omega}{1 - \Delta} \qquad (C.5)$$

one gets

$$\frac{1 - \Delta}{\omega} = m + \frac{n}{\omega} \qquad (C.6)$$

or $\Delta = r\omega - s$, which is the condition for an integrable case.

(b) Model 1 has a natural application in the cut and project method. Let us consider a canonical strip in 2D which selects points that are projected on a

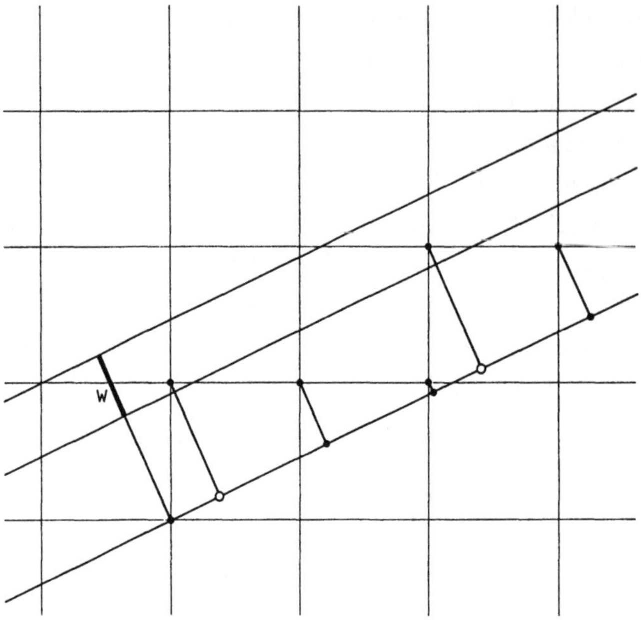

Fig. 3 : Vacancies generated by a window.

line. Then consider a smaller strip (see Fig. 3). Those points that fall in the sub-band denoted W (as window) on the figure will be withdrawn from the structure, thus creating a vacancy at this location. This mechanism is similar to that used in model 1. Hence the resulting structure, though quasiperiodic (as is the structure before withdrawing points) has no average lattice.

(c) There is in fact a connection between model 1 and model 2, which we explain now. Let us come back to the general case mentioned in section A.3, namely model 2 with any distribution of matter $f_1(x)$ and $f_2(x)$ on the bonds ℓ_1 and ℓ_2 respectively. Let us consider several cases :

(i) $\ell_1 = \ell_2$; $f_1(x) = 0$; $f_2(x)$ is a delta peak, say at one end of bond ℓ_2, Fig. 4.a shows the structure thus obtained : one has thus recovered model 1 ! In fact we obtain a model of atoms and vacancies with possibly more than two distances, e.g. two vacancies may be consecutive. As seen in section C, the Fourier spectrum of the structure is quasiperiodic. Note that for model 2, the condition $\ell_1 = \ell_2$ implies a trivial periodic structure. Quasiperiodicity arises from the condition $f_1(x) = 0$.

(ii) Same as case (i) except that $\ell_1 \neq \ell_2$ (see Fig. 4.b). The model obtained that way corresponds to model 1 after relaxation of the atoms. Since the resulting structure comes from model 2 it has the same Fourier spectrum, which we describe now.

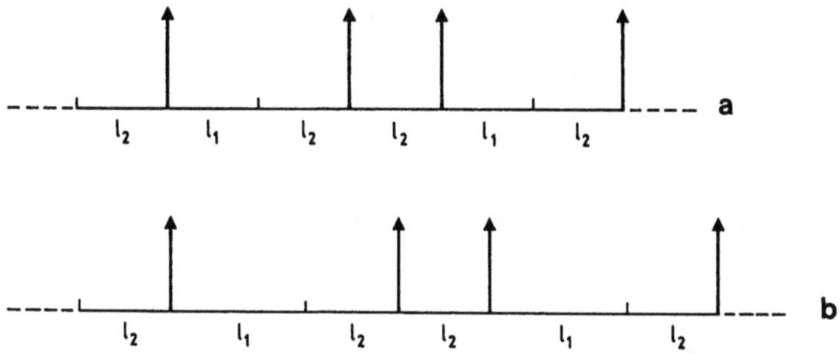

Fig. 4 : Connections between model 1 and model 2 :
a) $\ell_1 = \ell_2$, $f_1(x) = 0$.
b) $\ell_1 \neq \ell_2$, $f_1(x) = 0$.

D. Fourier spectrum of the circle map tiling.

Let us extract from Refs. (3,4) the main ideas and give the line of reasoning rather than the details.

1. Some definitions.

For a finite number N of atoms, we define the partial Fourier sums

$$G_N(q) = \sum_{n=1}^{N} \exp(-iqu_n) \tag{D.1}$$

where u_n is the abscissa of the nth atom, the associated structure factor

$$S_N(q) = \frac{1}{N} |G_N(q)|^2 , \tag{D.2}$$

and the intensity measure

$$dH_N(q) = S_N(q)dq. \tag{D.3}$$

In the thermodynamic limit

$$dH(q) = \lim_{N\to\infty} dH_N(q) \tag{D.4}$$

is the spectral (intensity) measure of the structure, and

$$H(q) = \lim_{N\to\infty} \int_0^q S_N(q')dq' \tag{D.5}$$

is the distribution function (or integrated density) of the intensity measure. The structure factor of the infinite structure is

$$dH(q) = S(q)dq, \tag{D.6}$$

which does not lead to a function $S(q)$ in general, but rather to a generalized function.

2. Characterization of the Fourier spectrum.

Sections 2.1 and 2.2 give the general framework.

2.1. Study of the local behavior of the Fourier spectrum.

Three cases of spectra may be considered:

(1) In a periodic or quasiperiodic structure, there are values q_0 of q such that

$$G_N(q_0) \approx C(q_0)N \tag{D.7}$$

$C(q_0)$ being some complex amplitude. $H(q)$ has then a discontinuity of strength $|C(q_0)|^2$ at $q = q_0$, and the structure factor $S(q)$ contains a Dirac peak : $|C(q_0)|^2 \delta(q - q_0)$.

(2) In amorphous systems, the averaged structure factor $S(q)$ is usually a smooth function : $G_N(q)$ grows typically as $N^{1/2}$.

(3) In the case of a singular continuous spectrum, there is usually a dense set of values q_0 of q, for which $G_N(q_0)$ grows as $N^{\gamma(q_0)}$, with $1/2 < \gamma(q_0) < 1$. More precisely, let q_0 be such a value of q. We then have $S_N(q_0) \approx N^{2\gamma-1}$. A standard finite size scaling argument leads to

$$|H(q) - H(q_0)| \approx |q - q_0|^{\alpha} \tag{D.8}$$

with $\alpha = 2(1-\gamma)$. Hence $0 < \alpha < 1$, and the intensity $S(q_0)$, which is formally equal to the derivative of $H(q)$ at q_0, is divergent, but "less infinite" than in the presence of a Dirac peak, which corresponds to $\gamma = 1$ $(\alpha = 0)$.

2.2. Study of the global behavior of the Fourier spectrum.

Consider the quantity

$$J_N = \int_0^{q_{max}} dq\, |S_N(q)|^{1/2} \tag{D.9}$$

(1) If the spectrum is discrete, this integral should behave as $1/\sqrt{N}$ with increasing N.
(2) If the spectrum is absolutely continuous, i.e. for a random structure, J_N should behave as a constant with increasing N.
(3) If the spectrum is singular continuous, one expects an intermediate dependence of J_N in N : $J_N \sim 1/N^{\beta}$ with $\beta < 1/2$.

2.3. We now apply these considerations to the spectrum of the circle map tiling.

(a) Local behavior of the Fourier spectrum.

There are two ways of studying this behavior.

• The first one consists in approaching a particular value of Δ incommensurate with ω, i.e. a non-integrable case, by a sequence

$$\Delta_{r_p} = \text{Frac}(r_p\omega) \qquad (D.10)$$

of integrable cases, such that $\Delta_{r_p} \longrightarrow \Delta$ in some optimal way when $r_p \longrightarrow \infty$. This method may, in principle, be used for any values of ω and Δ. In practice, we have chosen to study the case ($\omega = \tau^{-2}$, $\Delta = 1/2$), which plays a simple and exemplary role in this problem. Since any integrable case corresponds to a quasiperiodic structure, it is easy to compute the Fourier structure factor for the successive values of r_p, which are in our case

$$r_0 = 1, \quad r_1 = 4, \quad r_2 = 17, \quad r_3 = 72, ..., r_p = \frac{1}{2}F_{3p+3}. \qquad (D.11)$$

From the comparison of the corresponding spectra, one observes the following. Select a particular peak q_p (this may be defined in a proper way, cf. Ref.3). As $r_p \to \infty$, q_p goes to q_∞. A self-similar structure can be observed in the vicinity of q_∞ : the spectra corresponding to different values of p are identical, after an appropriate rescaling. Moreover, the heights of the peaks at q_p behave as λ^p for large p, where $\lambda < 1$. In simple words, the Dirac peaks vanish.

This method has the advantage of being usable, as already mentioned, for any values of ω and Δ. But it is mainly a heuristic and numerical method. So we shift to the second one.

• The second method relies upon the existence of a substitution for the case under study, namely $\omega = \tau^{-2}$, $\Delta = 1/2$. (see eq.B.19). It uses the line of thought drawn by Bombieri and Taylor [9]. Recursive relations on the Fourier amplitudes of successive sizes of the structure may be obtained from the substitution. The presence, for a particular value q_0 of the wavevector, of a Dirac peak is signaled by a maximal growth of the Fourier amplitude, proportional to the size of the sample. Bombieri and Taylor showed that if there exists a discrete part in the spectrum, the leading eigenvalue of the substitution should be a Pisot number.

For the case under study, the substitution matrix

$$M = \begin{pmatrix} 1 & 2 & 2 \\ 0 & 0 & 1 \\ 2 & 3 & 2 \end{pmatrix} \qquad (D.12)$$

has eigenvalues $\tau^3, -\tau^{-3}, -1$. Although τ^3 is a Pisot number, the presence of the subleading eigenvalue (-1) mimics a non-Pisot case. In Ref.4 the consequence of this fact is studied and the existence of a singular continuous part in the spectrum is shown.

(b) Global behavior of the Fourier spectrum.
The absence of a continuous part is then ruled out by the method of section 2.2. It is found numerically that $J_N \sim 1/N^{\beta}$ with $\beta = .16$.

E. Discussion and conclusion.

1. A more systematic way of studying the global behavior of the Fourier spectrum is provided by the multifractal analysis of the intensity measure [10]. For instance, the exponent β appearing above is related to a generalized dimension.
2. It may be shown that there is a deep connection between the disappearance of the average lattice for the structure and the singular continuous nature of its Fourier spectrum.
3. The results found in the preceding sections imply for the circle map tiling an intermediate kind of behavior between quasiperiodic and random. From a remark made above, we expect the atoms and vacancies model to lead to a similar order after relaxation of the positions of the atoms. Hence we believe that such a type of order should exist in Nature. The Fourier spectrum of such a structure would, at first sight, look like a quasiperiodic one. However, the experimental scattering factor would exhibit a scaling behavior and some finer structure if observed with a better resolution.

References.

[1] S. Aubry and C. Godrèche, J. Phys. (Paris) Coll. C3 (1986)187; S. Aubry, C. Godrèche and F. Vallet, J. Phys. (Paris) 48, 327 (1987).

[2] C. Godrèche, J. M. Luck and F. Vallet, J. Phys. A 20, 4483 (1987).
[3] S. Aubry, C. Godrèche and J. M. Luck, Europhys. Lett. 4, 639 (1987).
[4] S. Aubry, C. Godrèche and J. M. Luck, J. Stat. Phys. 51 (1988) 1033.
[5] C. Godrèche and J. M. Luck, to appear.
[6] S. Aubry, J. Phys. 44, 147 (1983).
[7] H. Kesten, Acta Arith. 12, 193 (1966/67).
[8] C. Godrèche and C. Oguey, preprint.
[9] E. Bombieri and J. E. Taylor, J. Phys. (Paris) Coll. C3 :19 (1986); Contemp. Mat. 64 :241 (1987).
[10] C. Godrèche and J. M. Luck, to appear.

A Topological Constraint on the Atomic Structure of Quasicrystals

A. Katz

Centre de Physique Théorique, Ecole Polytechnique, F-91128 Palaiseau, France

Abstract. This paper gives a short account of some mathematical problems related to the structure determination of icosahedral quasicrystals. It is shown that a natural requirement on the geometry of the so-called "atomic surfaces" leads in a straightforward way to the study of their homology classes, and that the corresponding constraints on the density of quasicrystals are compatible with experimental results.

1. Introduction

Since the discovery of quasicrystals [1], a lot of work has been devoted to the study of their atomic structure. Their quasiperiodic character was soon recognized [2] and the general framework for the description of such quasiperiodic sets of points was defined. See [3] and [4] for general information on quasicrystals.

However, the atomic structure of quasicrystals is still not known, and no systematic approach to this problem is known. As will be explained later on, the main reason for this is that the single assumption of quasiperiodicity is not sufficient to allow the direct construction of a model, because a general quasiperiodic set of points depends on infinitely many parameters. As a consequence of this situation, it is important to try and find physically plausible constraints on the structure in order to reduce the number of parameters on

In this paper, we shall explain how such constraints arise in a natural way when one tries to understand the propagation of ordering in quasicrystals.

In the next section, we recall the natural description of quasiperiodic sets of points, which is a straightforward generalization [5] of the classical theory of incommensurate crystals. Then we describe the 3-dimensional Penrose tilings, which are the best known and simplest examples of icosahedrally symmetric quasiperiodic tilings.

Finally we show how a special property of Penrose tilings (the so-called "worm flips") is related to the topology of the corresponding atomic surface and in the last section we explain the physical meaning of this geometrical property and how it can be expressed in the general case, resulting in strong restrictions on the possible atomic models for quasicrystals.

Springer Proceedings in Physics, Vol. 47 Number Theory and Physics
Editors: J.M. Luck · P. Moussa · M. Waldschmidt

2. Construction of Icosahedral Quasiperiodic Sets of Points

Quasicrystals were first discovered in aluminium-based alloys. Their most striking feature is that their diffraction patterns (which are roughly speaking, the squared modulus of the Fourier transform of the set of atoms) exhibit both sharp Bragg peaks and icosahedral symmetry.

As is well known, the sharpness of the Bragg peaks means long range ordering of the atoms, while the icosahedral symmetry is not compatible with any periodic order. This is why this discovery was so surprising: a (solid) phase with long range positional order and without periodicity had never been observed. Although several tentative explanations (such as "icosahedral glasses") were given at first, it now seems clear that the correct approach refers to quasiperiodic ordering, and in the following we shall only consider this point of view.

By its very definition, a quasiperiodic function on an affine space E is the restriction to E (embedded as an affine subspace) of a periodic function defined in a higher dimensional space. Of course, if the embedded space E (which will be referred to as the "cut") is rationally oriented with respect to the lattice of periods of the periodic function, then the restriction of this function to E is also periodic. But if the orientation of the cut is irrational, that is, if the vector subspace parallel to the cut contains no point of the lattice besides the origin, then the restriction is not a periodic function.

To describe a quasiperiodic set of points rather than a quasiperiodic function, the natural generalization is to consider the restriction of a periodic measure instead of the restriction of a periodic (smooth) function. This measure is to be carried by a periodic set of submanifolds (with boundary) with a dimension equal to the codimension of the cut in the high dimensional space, in such a way that the quasiperiodic set of points appears as the set of intersections of the cut with the periodic set of submanifolds, which are called "atomic surfaces". These intersections are well defined each time the atomic surface is transverse to the cut, and we shall only consider situations in which this condition is generic.

The next step is to introduce symmetry considerations to this construction: In order for the diffraction patterns to present icosahedral symmetry, the elements of the construction must fulfill three independent symmetry requirements. We have to define in the high dimensional space an action of the icosahedral group such that the set of atomic surfaces is invariant, and this is a two-fold condition: on one hand, the lattice of translations of the set of surfaces must be invariant and on the other hand the "shape" of each surface must also be invariant. Finally, the orientation of the cut has to be invariant under the same action of the icosahedral group.

These requirements are readily achieved in $\mathbb{R}^6$ equipped with the following action of the icosahedral group: choose six vertices of an icosahedron (no two of them being opposite) and consider the representation of the group in $\mathbb{R}^6$ which permutes the six vectors of the canonical orthonormal basis in the same way as the ordinary 3-dimensional action permutes the six given vertices of the icosahedron. Then it is clear that the 6-dimensional representation thus defined is a symmetry group for the simple cubic lattice $\mathbb{Z}^6$, since it acts by signed permutations of the basis vectors, and our first requirement is fulfilled if we choose this lattice as the translation group of the set of atomic surfaces. On the other hand, the 6-dimensional representation is reducible, and upon reduction we get the two well-known non-equivalent 3-dimensional representations of the icosahedral group. Thus we have to choose as the direction of the cut one of the two corresponding invariant 3-dimensional vector subspaces.

More generally, one can show [6] that in order to describe icosahedral quasicrystals it is sufficient to consider periodic lattices in $\mathbb{R}^6$ and that there exist only three "Bravais lattices" in $\mathbb{R}^6$ for which the icosahedral group is a symmetry group. They are the simple cubic lattice, just defined, the face-centered cubic lattice and the body-centered cubic lattice. See the references [6] for more details. The important point here is to observe that the analysis of the diffraction patterns is sufficient to assess which Bravais lattice (and even which space group) corresponds to a given quasicrystal, as in ordinary crystallography. Up to now, two of the previous three cases have been experimentally observed: the simple cubic in several quasicrystals such as Al-Mn-Si and the face-centered cubic in the newly discovered [7] Al-Fe-Cu quasicrystals.

As one can see, two of the three elements of the construction, namely the high dimension space with its translation lattice and the orientation of the cut with respect to this lattice, can be determined through a "qualitative" analysis of the diffraction patterns. Thus the main question in the structure determination of quasicrystals is the determination of the atomic surfaces, which can be considered either as a lattice of 3-dimensional manifolds with boundary in $\mathbb{R}^6$ or as a (single) manifold with boundary in the quotient of $\mathbb{R}^6$ by the translation lattice, which is the 6-dimensional torus $\mathbb{T}^6$.

In some sense, this situation is a natural generalization of the structure determination problem in classical crystallography: in that case too, the first step is to find the space group of the crystal, and the second is to determine the atomic positions in the unit cell, that is in the 3-dimensional torus. We can say that this second step is to determine the embedding of a 0-dimensional manifold (a finite set of points...) in $\mathbb{T}^3$ just as it is for quasicrystals to determine the embedding of the 3-dimensional atomic surface in $\mathbb{T}^6$.

Now, the main difficulty of "quasicrystallography" becomes clear: loosely speaking, the position of a 3-dimensional manifold in $\mathbb{T}^6$ depends on "infinitely more" parameters than the position of a finite set of points in $\mathbb{T}^3$. On the other hand, we get from diffraction experiments essentially the same amount of data: a finite set of numbers, measured of course with a finite precision. It seems that the only way to solve this difficulty is to find enough *a priori* constraints on the atomic surface to make its position in $\mathbb{T}^6$ depend on a finite number of real parameters only.

Before proceeding to a discussion of this point, it may be useful to give a brief description of the Penrose tilings, which are the best known quasiperiodic structures, in order to provide the reader with a better idea of what a quasiperiodic system looks like.

3. The 3-Dimensional Penrose Tilings

Within the previous framework, the definition of the Penrose tilings reduces to the definition of the corresponding atomic surface, which is the following: Let us call the two 3-dimensional vector subspaces of $\mathbb{R}^6$ which are invariant under the action of the icosahedral group $\mathbf{E}^{//}$ and $\mathbf{E}^{\perp}$, and let us decide that the cut will be parallel to $\mathbf{E}^{//}$. Then we define the atomic surface as the orthogonal projection of the unit hypercube of the canonical simple cubic lattice $\mathbb{Z}^6$ on the subspace $\mathbf{E}^{\perp}$. We get in this way a polyhedron (named a rhombic triacontahedron, Fig. 1) contained in $\mathbf{E}^{\perp}$ and we attach a copy of this triacontahedron at each vertex of $\mathbb{Z}^6$. The vertices of the Penrose tilings are the intersections of this set of atomic surfaces with any plane cut parallel to $\mathbf{E}^{//}$ which does not intersect the 2-dimensional boundary of the triacontahedra. It is clear that almost all positions of the cut will fulfill this transversality condition.

Alternatively, we may prefer to describe the construction in the quotient space $\mathbb{T}^6$: because of their irrational orientations, the two subspaces $\mathbf{E}^{//}$ and $\mathbf{E}^{\perp}$ project in $\mathbb{T}^6$ onto two immersed non-compact manifolds which "wind" infinitely around $\mathbb{T}^6$ and which are everywhere dense. The set of atomic surfaces projects onto a single atomic surface, which is a compact manifold with boundary embedded in $\mathbb{T}^6$ and contained in the projection of $\mathbf{E}^{\perp}$. Finally, we get the vertices of a Penrose tiling in $\mathbb{R}^3$ through the following process: embed $\mathbb{R}^3$ in $\mathbb{R}^6$ as a subspace parallel to $\mathbf{E}^{//}$, project it in $\mathbb{T}^6$, and select all the intersection points of this projection with the single atomic surface.

As is well known (see for instance [8]) the resulting structure is the set of vertices of a tiling of $\mathbb{R}^3$ by means of the two Ammann rhombohedra (Fig. 1). Since there is

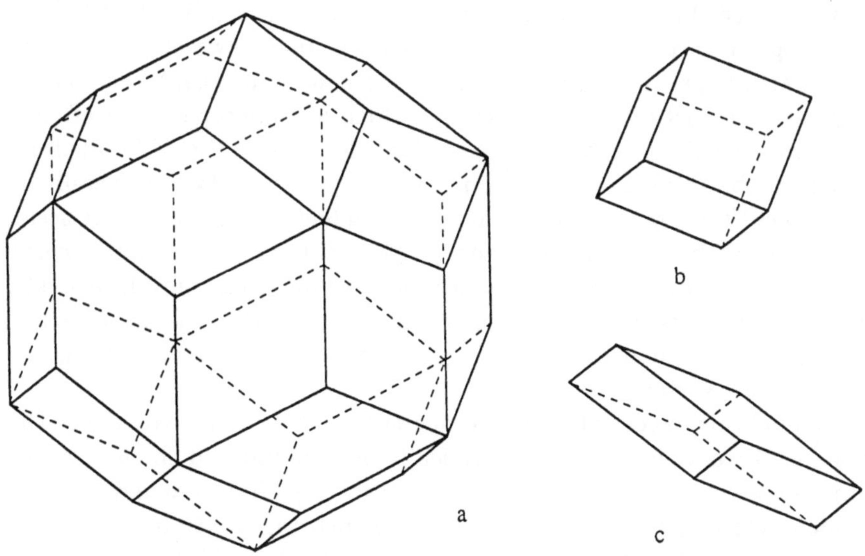

Fig. 1: The polyhedra referred to in the text. (a): the rhombic triacontahedron; (b): the prolate rhombohedron; (c): the oblate rhombohedron.

no translational symmetry in the tiling, any two different vertices have different global environments, but one can study the local neighborhoods of each vertex [8,9] and show that there are in any Penrose tiling only 24 different ways (up to the icosahedral symmetry) of packing tiles around a vertex.

Another important property is that when one moves the cut (while keeping it parallel to $\mathbf{E}^{//}$), one generates infinitely many different tilings. We are especially interested in the elementary changes in the tilings which take place when one shifts the cut. To explain what happens, let us first consider the low dimensional model of Fig. 2, which is self-explanatory: the orientation of the line $\mathbf{E}^{//}$ is irrational and the atomic surface is the projection of the unit square of the lattice onto the orthogonal line $\mathbf{E}^{\perp}$. The intersections of the cut with the atomic surfaces are easily seen to be the vertices of a tiling of $\mathbb{R}$ constructed from the projections on $\mathbf{E}^{//}$ of the two edges (horizontal and vertical) of the unit square.

Then one sees clearly what happens when one moves the cut (say, upward). Each time the cut reaches the upper tip of an atomic surface, it begins at the same time to cut a neighboring atomic surface, because the projection on $\mathbf{E}^{\perp}$ of the upper tip of any atomic surface coincides with the projection of the lower tip of a neighboring one. This depends of course on the choice of the atomic surface (the length of the segment in this low-dimensional model) and can be interpreted as the "jump" of a vertex from one site to a neighboring one.

This is the only elementary change in the tiling upon a shift of the cut but, since the orientation of the cut is irrational, such elementary changes occur infinitely many times for any finite translation of the cut, resulting in an uncountable infinity of different tilings.

The same phenomenon exists for the 3-dimensional Penrose tilings: when one shifts the cut, some vertices of the tiling undergo the same kind of jump. More precisely, one can see that there is only one elementary change in the tiling: it involves the simplest vertex neighborhood, which is made of two oblate and two prolate rhombohedra, packed together to form a rhombic dodecahedron. There are two ways of constructing such a packing, which are mirror images of each other in the symmetry plane orthogonal to the long diagonal of the dodecahedron, and the "jump" of the internal vertex simply exchanges these two tilings of the dodecahedron. Such "flips" of the tiles are not independent of each other, but occur along whole planes at the same time during the shift of the cut (the name "worm flip" comes from the 2-dimensional case, where the flips occur simultaneously along whole lines named "worms"). It is important to observe that the length of the jump is short, and in particular that during the jump the vertex does not come very close to any other vertex.

The main claim of this paper is that this property, which depends strongly on the shape and size of the atomic surface, is physically meaningful and should be preserved in any realistic atomic model of quasicrystals. We shall now develop the underlying idea, keeping in mind that we are not proving anything, but more modestly trying to argue a working hypothesis about difficult problems.

4. The Closeness Condition for the Atomic Surface

In fact, structure determination is only one of the problems raised by the discovery of quasicrystals. The main one is to understand the simple possibility of a non-periodic long range order. Recall that in such a perfect quasiperiodic structure the positions of any two atoms are correlated, whatever the distance between them, and that the number of different neighborhoods of atoms grows with the size of the neighborhood. Moreover, the effective range of the interatomic forces is short, so that the big question is: "How can an atom know where it has to sit ?".

This is the difficult question of *propagation of order*. It seems that it splits into two sub-problems. The first one deals with the existence of local constraints which could force the quasiperiodic order. It has received a partial answer with the theory of "matching rules", which gives a solution for the 3-dimensional Penrose tiling [10], considered as a "toy-model" for quasicrystals. The second sub-problem concerns the possibility of atomic rearrangement, which could allow the relaxation of defects or of strain during the growth of the quasicrystal, and would thus permit the long range manifestation of a locally-governed propensity for icosahedral symmetry.

It is this second question that we address here. Observe first that we can describe a lot of non-quasiperiodic structures within the same framework simply by considering non-plane cuts. Now one can imagine the growth of a quasicrystal from the melt as the following process: atoms in the melt stick on the boundary of the growing quasicrystal at positions governed by short range interactions and which are not necessarily compatible with the quasiperiodic order, but which nevertheless could perhaps be described by an "undulating" cut in our framework. Then, as new layers of atoms stick to the solid phase, the propensity to form a quasicrystal manifests itself through a *bulk reconstruction*, a process through which atoms jump to their final positions and which geometrically corresponds to the progressive flattening of the cut behind the "quasicrystallization front".

This (imaginary !) description of the growth of a quasicrystal is the reason for splitting the problem of propagation of order into two parts. On one hand, the atomic surface should be such that the set of local configurations defined by cuts parallel to $\mathbf{E}^{//}$ is energetically favored with respect to the set associated with any other orientation of the cut (the special orientation of $\mathbf{E}^{//}$ makes it possible, see [10]). This is the "local rules" part of the problem. On the other hand, the deformation of the cut and its flattening should be energetically possible at the quasicrystallization temperature, i.e. should be a low activation energy process.

The "jumps" which occur in Penrose tilings are good candidates for this low activation energy process, for two reasons. First, they correspond to a symmetry operation of the first neighbors' configuration, so that if the effective range of the interatomic interaction is short, the energies of the two configurations related by the jump should be close to each other (they may differ only by the second and further shells, whose contribution to the energy is small). Second, the energy barrier between the two equilibriums of the jumping atom should be low, simply for

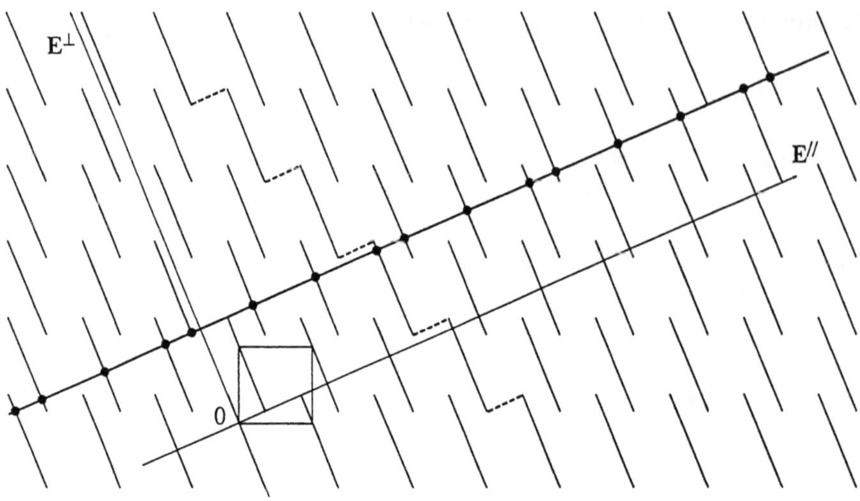

Fig. 2: The low dimensional construction of a quasiperiodic set. The dashed segments correspond to the completion of the atomic surface.

geometrical reasons: the distance between the two sites is small and during the jump the atom does not come too close to its neighbors.

Thus, we are looking for a characterization of all atomic surfaces with the same property, and this happens to be very simple. Let us first consider the low dimensional model of Fig. 2, and let us add the jump trajectories to connect the atomic surfaces together. This does not change anything for the quasiperiodic system, since the added segments are parallel to $\mathbf{E}^{//}$ and thus do not generically intersect the cuts parallel to $\mathbf{E}^{//}$. Moreover, we have explicitly excluded those cuts which go through an end point of an atomic surface and entirely contain one of the new segments. Now observe that this enlarged atomic surface projects in the 2-dimensional torus onto a manifold *without boundary,* which is of course homeomorphic to a circle.

Conversely, let us examine what happens if we alter the length of the atomic surfaces, in which case it is no longer possible to connect them with line segments parallel to $\mathbf{E}^{//}$. Suppose for instance that we decrease their length a little, and let us move the cut upward. When the cut reaches the upper end point of an atomic surface and leaves it, the corresponding atom has no neighboring site to jump to. The prescription is that there should no longer exist an atom on this site, but of course atoms cannot vanish and we have to see it as a point defect in the structure (an insertion). Similarly, when the cut reaches the lower end point of an atomic surface, resulting in a new intersection, no atom can magically appear and we find a vacancy at this site. Such defects are of relatively high energy and the system will resist the production of them: the end points of the atomic surfaces will "anchor" the cut and the flattening of the cut will not occur. We see that the cut will be "glissile" only when the appearance of an insertion atom coincides with the appearance of a neighboring vacancy, so that the insertion atom can fall in the vacancy and both defects cancel: this is the previous "jump".

Although they are only intuitive, these arguments suggest that it is natural to add the following constraint to a realistic model of a quasicrystal: it is possible to "complete" the corresponding atomic surface embedded in $\mathbb{T}^6$ to form a manifold *without boundary*, by gluing on its boundary some 3-dimensional cells nowhere transverse to the projection of $\mathbf{E}^{//}$. This last condition implies that a generic cut will intersect the original atomic surface and its completion at exactly the same points, and the added cells can be considered as the union of all possible jump trajectories.

Thus our constraint is to make the atomic surface a manifold without boundary when projected in $\mathbb{T}^6$. Observe that in the low dimensional model this manifold is a circle, so that each time the completion is possible, a 1-dimensional quasicrystal is nothing other than a modulated crystal. In higher dimensions, the geometry is far richer. It is easy to see that (3-dimensional) modulated crystals correspond to complete atomic surfaces which are topologically 3-dimensional tori (these questions were first raised in the stimulating paper [11]), and that the completion of the triacontahedron is not of this type. Actually, the topology of the complete atomic surface seems to embody the very difference between modulated crystals and quasicrystals, which appears to be of a topological nature.

The physical meaning of our constraint is clear: the previous point defects appear in the general case each time the cut crosses the boundary of an atomic surface, and we suppress them simply by eliminating the boundary.

5. The Homology Class of the Atomic Surface

As an immediate consequence, observe that a compact manifold without boundary embedded in $\mathbb{T}^6$ is homologically a cycle and thus defines a homology class. Although its homology class does not provide a lot of information about an embedded manifold, it is worthwhile considering it because it is directly related to the density (or more precisely the atomic concentration) of the quasicrystal.

The simplest way to handle such homology classes is to make use of the intersection index. Recall that on the n-th homology group of any compact manifold of even dimension 2n, there is a natural pairing which is the homological version of the Poincaré duality usually defined in cohomology. Concretely, the pairing between two classes (their intersection index) is defined as follows: Consider any representatives of the two classes; it is always possible to choose them so that they are transverse and thus intersect at a finite number of points. Orientation considerations allow us to ascribe a sign ± 1 to each of these points and the intersection index is simply the sum of these signs. As is well known, the intersection index depends only on the homology classes and is a non-degenerate pairing (scalar product). In other words, an embedded manifold is homologous to zero if and only if its intersection indices with the elements of a basis of the n-th homology group are all zero.

In our case, n=3 and the 2n-dimensional manifold is $\mathbb{T}^6$. Using the six canonical angles θ_ι as coordinates on $\mathbb{T}^6$, a convenient basis of $H_3(\mathbb{T}^6)$ is provided by the classes of the twenty 3-dimensional tori T_{ijk} defined by the equations $\theta_l = \theta_m = \theta_n = 0$, where (i, j, k, l, m, n) is a permutation of (1, 2,..., 6). Thus the homology class of a (complete) atomic surface Σ is characterized by the 20 integers $(\Sigma \mid T_{ijk})$. This is only a very crude characterization of Σ, since quite different Σ's may define the same homology class. Nevertheless, this is sufficient to compute the atomic concentration of the quasicrystal, as we will now show.

Let us first note that our basis is invariant (up to a sign) under the action of the icosahedral group, and that there are only two orbits under this action, associated with the two types of Ammann rhombohedra. It is clear that since Σ is itself invariant, the intersection indices of Σ with two tori related by a symmetry operation have to be equal, so that on account of symmetry the number of independent indices reduces to 2.

To compute the atomic concentration, let us consider a large volume V in a plane cut parallel to $\mathbf{E}^{//}$ and identified with the physical space. Let N be the number of intersection points in V between the cut and the set of atomic surfaces. Then the atomic concentration is equal to the ratio N/V and we are looking for an expression of this ratio in terms of intersection indices. For this, let us construct on the cut the Penrose tiling corresponding to the given simple cubic lattice of $\mathbb{R}^6$ and consider the projection of a tile in $\mathbb{T}^6$. The key point is that this rhombohedron wound round $\mathbb{T}^6$ along the projection of $\mathbf{E}^{//}$ can be completed to yield a manifold isotopic to one of the 3-dimensional tori T_{ijk} by gluing on its boundary a set of cells which are not transverse to $\mathbf{E}^{\perp}$: the construction is exactly the same as for the completion of the atomic surface, with the exchange of $\mathbf{E}^{\perp}$ and $\mathbf{E}^{//}$. The first step is to connect opposite facets of the rhombohedron with three cells homeomorphic to 3-dimensional cubes, contained in the projections of 3-dimensional planes which intersect $\mathbf{E}^{//}$ along a plane and $\mathbf{E}^{\perp}$ along a line. This eliminates the original boundary but since the new cells are glued along only 2 of their 6 facets, new boundaries appear. Then we glue along them three other cells, whose carriers intersect $\mathbf{E}^{\perp}$ along a plane and $\mathbf{E}^{//}$ along a line. This introduces again new boundaries but it is easy to observe that these remaining boundaries are topologically the set of facets of a cube so that we can eliminate them by adding a last cell completely contained in the projection of $\mathbf{E}^{\perp}$ and we get a closed manifold isotopic to a 3-dimensional torus.

This shows that the mean number of atoms falling in that direction of tile is equal to the intersection index of the atomic surface with the corresponding torus T_{ijk}. If the ideas explained above are correct, then this mean number is an integer and we recover in some sense the usual notion of integral stoichiometric coefficients.

On the other hand, we cannot measure this mean number of atoms per tile for each kind of tile, but only compute the mean over the set of 20 directions of tiles, using the experimental density of the quasicrystal to assess the atomic concentration and the diffractional data to assess the unit length in $\mathbb{R}^6$ or the length of the edges of the corresponding Ammann rhombohedra.

Let v_o and v_p be the volumes of the oblate and prolate rhombohedra, n_o and n_p their respective numbers (for each of the 10 orientations) inside the volume V, and let us denote by τ the golden ratio $(\sqrt{5}+1)/2$. As already mentioned, the intersection indices $(\Sigma \mid T_{ijk})$ depend only on the type (oblate or prolate) of the rhombohedron associated with the torus T_{ijk}. Let ι_o and ι_p denote these two numbers.

We have:

$V = 10\,(n_o.v_o + n_p.v_p)$ and $N = 10\,(n_o.\iota_o + n_p.\iota_p)$, so that

$N/V = (n_o.\iota_o + n_p.\iota_p)/(n_o.v_o + n_p.v_p).$

As is well known, for a Penrose tiling and if V is large enough:

$n_p \approx \tau.n_o$,

and from the study of Ammann rhombohedra we know that:

$v_p = \tau.v_o$

Thus:

$N/V = (1/v_o).(\iota_o + \tau.\iota_p)/(2 + \tau).$

Finally, we see that the experimentally measurable quantity:

$(2 + \tau).v_o.N/V,$

should be of the form $\iota_o + \tau.\iota_p$. Of course, one can trivially remark that these linear combinations are dense in $\mathbb{R}$, but we are expecting reasonably small intersection indices. In fact the experimental value for the quasicrystal Al-Mn-Si is equal to 10.815 and this is close to either $3\tau + 6 = 10.854$ or $5\tau + 3 = 11.090$. This last expression is specially attractive since $5\tau + 3 = \tau^5$ corresponds to a triacontahedron τ times larger than the triacontahedron which defines the Penrose tilings, and could lead to interesting geometrical structures.

To sum up, the ideas are: first, the propagation of order throughout the quasicrystal is difficult to understand and may require a kind of "bulk reconstruction" i.e. a rearrangement of atoms behind the solidification front. Second, the simplest way to express the possibility for such processes is to ask that the atomic surface can be completed to a manifold without boundary in $\mathbb{T}^6$. Third, this condition yields constraints on the possible atomic concentrations.

Finally, we conclude that the rather abstract scheme sketched in this paper leads to results which are directly comparable to and appear compatible with experimental data. We hope that these ideas will be helpful for solving the difficult problem of the structure determination of quasicrystals.

References

1 Shechtman, D. Blech, I., Gratias, D. and Cahn, J. W., *Phys. Rev. Lett.* **53** 1951(1984).

2 Duneau, M and Katz, A., *Phys. Rev. Lett.* **54**, 2688 (1985)
Kalugin, P. A., Kitayev, A. Yu and Levitov, L. S., *J. Physique Lett.* **46** 601-607 (1985).
Elser, V., *Acta Crys.*, **A42**,36 (1986).

3 Les Houches, Workshop on Aperiodic Crystals, *J. Phys. France* **47** C3 (1986).

4 The Physics of Quasicrystals (P. Steinhardt and S. Ostlund eds) World Scientific, Singapore (1987)
5 Bak, P., *Phys. Rev. Lett.* **56**, (1986) 861
6 Levitov, L. S. and Rhyner, J., *J. Phys. France* **49** (1988) 1835-1849
Martinais, D., *C. R. Acad. Sci. Paris* **305 I** (1987) 509-512
7 Gratias, D., *A.P.S. Meeting Abstracts*, Saint Louis (1989), and private communications
8 Katz, A. and Duneau, M., *J. Phys. France* **47** (1986) 181-196
9 Katz, A., *in* Aperiodicity and Order Vol. 2 (ed. M. Jaric) Academic Press Boston (1988)
10 Katz, A., *Comm. Math. Phys.* **118** (1988) 263-288.
11 Frankel, D. M., Henley, C. L. and Siggia, E. D., *Phys. Rev. B* **34**, 3649 (1986)

From Approximants to Quasicrystals: A Non-standard Approach

R. Mosseri and F. Bailly

Laboratoire de Physique des Solides de Bellevue, CNRS,
F-92195 Meudon Cedex, France

Quasicrystal approximants are periodic structures, with usually large unit cells, such that the local and medium range order is close to that present in the 'true' quasicrystal. We study the transition from an infinite series of approximants to the asymptotic quasicrystal. This is done in the framework of non-standard analysis which provides simple tools for the manipulation of the infinitely large and small quantities in the problem.

1. Introduction

There has been a growing interest since 1985 in the study of quasiperiodic space tilings, due to the experimental discovery of quasicrystalline alloys [1]. These materials display, under diffraction experiments (electrons or X-rays) , a point-like pattern (related to the atom distribution Fourier spectrum) with an icosahedral symmetry, a kind of symmetry which is incompatible with space periodicity. A relation with Penrose-like tilings was soon invoked to explain the experimental results [2]. Within a few months many geometrical models had been investigated [3], shedding light on their fascinating properties, with subtle interplay between self-similarity, quasiperiodicity and (non-crystallographic) orientational order. There are two main methods of generating n-dimensional Penrose patterns, the grid method [4] and the cut and projection method [5]. With on one hand the Penrose tiling playing a paradigmatic role, a lot of work remains to be done. At the experimental level, the main collective result was to find that quasicrystallinity was observed in many different alloys, being therefore a candidate for a new kind of state in solids. At the theoretical level, many results concerning the quasicrystal geometry are available, but very little is known (except in 1 dimension) about the expected properties.

One limitation is probably due to the fact that the tiling vertex positions are not a priori known but result from an algorithmic procedure : a selection of points in an n-dimensional space for the cut and projection method or a precise ordering of plane intersection for the grid method. This is not an important point for numerical studies but it makes it difficult to construct appropriate theoretical tools. These problems can be by-passed in the case of approximant structures , which are periodic structures with larger and larger unit cells inside which the geometry asymptotically approaches the quasi-periodic one. In many respects these approximants are the n-dimensional analogues of the rational approximants to irrational numbers. One of us (R.M. [6]) has shown that the coordinates can be very simply obtained for codimension 1 (mapping from n to n-1 dimensions) quasicrystal approximants. The generalization for any codimension was then investigated [7] and is now much better understood in the case of self-similar tilings [8]. Here again the coordinates are a priori calculated in an explicit form for any approximant.

For numerical purposes, this knowledge is sufficient. After all, computers do not calculate with irrational numbers but with their rational approximants. Nevertheless the next important step consists in dealing with the asymptotic transition to the "real" quasicrystal. In doing this, one faces conceptual difficulties like handling such quantities as infinite numbers modulo infinite numbers, points at infinity and, at a geometrical level, quadrilaterals with unit area and two vanishing angles. In this paper we solve this problem within the framework of non-standard analysis, which is the natural tool when dealing with infinitely small and large numbers.

In a first step , we recall the method of generation for quasicrystal approximants. For the sake of simplicity we focus on a 1D example but this construction applies to higher dimensions, for codimension 1 tilings [6]. One gets a simple expression for vertex coordinates. The particularity here is that the sites are not numbered according to their natural order along the line but according to their distance to the irrational direction before the projection. The structure factor $S(q)$ is then very simply calculated. In the second step we introduce non-standard numbers to treat the asymptotic quasicrystalline case. This is especially simple when there exists a recurrent sequence of numbers, like the Fibonacci sequence, such that the ratio of successive terms gives the best rational approximants of the irrational number. The sequence is extended to non-standard terms with infinitely large indices. The set of all successive ratios for infinite indices forms the "halo" of the irrational number. One can proceed to calculate the site coordinates. One finds as expected that all the sites with a finite index lie at infinite distance from the origin. But those sites at finite distance are recovered as sites with an infinite index and can be treated without problem. The structure factor is also calculated.

2. Approximants of 1-Dimensional Quasiperiodic Structures

Let α be an irrational number. In the cut and projection method framework the quasicrystalline tiling is obtained by selecting points of $\mathbf{Z}^2$ which belong to the semi-open strip obtained by translating the unit square along the line E of slope $1/\alpha$. These points are then orthogonally mapped onto E using the projector Π_α:

$$\Pi_\alpha = \frac{1}{\beta^2}\begin{pmatrix} \alpha^2 & \alpha \\ \alpha & 1 \end{pmatrix}, \qquad \beta = (1+\alpha^2)^{1/2}. \tag{2-1}$$

A quasiperiodic sequence is thus generated, with two distinct distances in the ratio α.

APPROXIMANTS:.

Let F_n be a sequence of integers such that $\alpha_n = F_{n+1}/F_n$ approximates α better and better as n increases and such that

$$F_{n+1}F_{n-1} - F_n^2 = (-1)^n. \tag{2-2}$$

The approximants are generated by translating the unit square along the 'approximate direction' E_p generated by $\mathbf{u}_p = (F_{p+1}, F_p)$. The vector $\mathbf{u}_p$ norm is the unit cell length, which contains $M_p = F_{p+1} + F_p$ sites. To generate the sites in the unit cell, one needs to define a new vector $\mathbf{r}_p$ which connects the origin to the point closest to the line E_p. There are infinitely many such points , equivalent under translations. One can choose for example $\mathbf{r}_p$ with smallest norm:

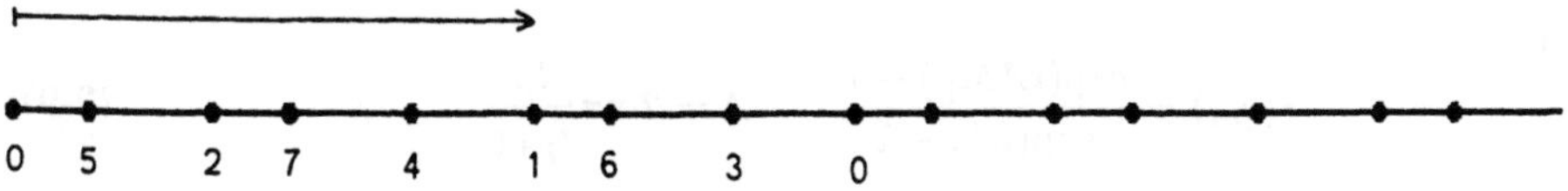

Figure 1 : The p=4 Fibonacci approximant repeated along 2 cells

$$\mathbf{r}_p = (F_p, F_{p-1}). \tag{2-3}$$

Now it is easy to show [6] that the unit cell sites are obtained by translation along $\mathbf{r}_p$:

$$\mathbf{x}_j^p = j\Pi_\alpha(\mathbf{r}_p) \quad mod\left(\Pi_\alpha(\mathbf{u}_p)\right) \qquad j = 0, \ldots, M_p - 1, \tag{2-4}$$

where all the sites are in the same unit cell under the 'modulo' operator. Let

$$t_p = \alpha F_p + F_{p-1}, \tag{2-5}$$

then, using (2-1) and (2-5) one gets

$$x_j^p = \frac{1}{\beta}\left(jt_p \, mod\,(t_{p+1})\right). \tag{2-6}$$

The p = 4 approximant of a Fibonacci quasicrystal is shown on figure 1 (then F_p is the Fibonacci sequence).

Observe the unusual ordering of sites. It is easy to show that the site nearest to the origin is F_{p+1} for p even and F_p for p odd. Furthermore the site nearest to site j is indexed by $(j + F_{p+1})\, mod\, M_p$ (replace F_{p+1} by F_p for p odd). It is possible to generalize this approach to any irrational number, when there is no such sequence F_n with the above property. Indeed one can always find a sequence of best approximants. A given rational approximant defines $\mathbf{u}_p$. The vector $\mathbf{r}_p$ is found by taking in an appropriate Farey sequence the rational which is the closest to the given approximant [6].

Note that the quadrilateral generated by $\mathbf{u}_p$ and $\mathbf{r}_p$ has unit area. It forms a $\mathbf{Z}^2$ unit cell. There exists a unimodular matrix with integer entries and α and $-\alpha^{-1}$ as eigenvalues which transforms upon repeated action the unit square into this quadrilateral. This property is at the heart of the general method proposed in ref [8] to generate approximants.

Structure Factor.

The structure factor $S(q)$ is easily calculated. One can use the above expression for the site coordinates without the "modulo" operation since the sites do not need to be all in the same unit cell. The sites thus form a limited periodic chain and the calculation of $S(q)$ follows immediately:

$$S(q_m) = \sum_{j=0}^{M_p-1} \exp\left(iq_m x_j^p\right) \tag{2-7}$$

$\mathbf{q}_m$ belong to the approximant reciprocal lattice

$$q_m = 2m\pi \frac{\beta}{t_{p+1}} \quad . \tag{2-8}$$

One gets

$$S(q_m) = \frac{\exp(iAM_p) - 1}{\exp(iA) - 1} \qquad A = 2m\pi\frac{t_p}{t_{p+1}} \quad . \tag{2-9}$$

3. The Asymptotic Quasicrystal : Non-standard Approach

Non-Standard Analysis (NSA) was introduced by Robinson [9] to give a sound basis to the use of infinitesimals in differential calculus. We do not intend to give here a detailed description of NSA. It can be found in the book of Robinson [9] or in the article by Nelson [10] who gave an axiomatic presentation of NSA, the 'Internal Set Theory' (IST). See also Ref.[11]. The real number system **R** is extended to a totally ordered non-archimedean field ***R** which contains **R** as a subset, but contains also infinitely small and large numbers. It is also possible to define the set ***N** containing infinitely large integers. Each usual real number r is surrounded by a "halo" in ***R** (the set of numbers which differ from r by only an infinitesimal number). Let a be in this halo; then r is called the standard part of a, denoted $r = {}^{st}(a)$. In addition to the strict equality between two numbers, one introduces the notation $x \simeq y$,which means that x and y only differ by an infinitesimal, or ${}^{st}(x-y) = 0$. Suppose L is an infinitely large number in ***N**; one can also define F_L and F_{L+1} which belong to ***N** as well. Then

$$\alpha_L = \frac{F_{L+1}}{F_L} \quad , \alpha_L \in {}^*R \quad \text{and } \alpha =^{st} (\alpha_L). \tag{3-1}$$

Such an approach is legitimated in particular by the fact that it is possible to construct a model of **R** from a non-standard enlargement of the rational ordered field **Q** [13][14]. Indeed, let ***Q** be such an enlargement, with the finite subring $^*\mathbf{Q}_f$ (the numbers in ***Q** whose integral part is finite in the non-standard sense). If we denote by $\mathcal{M}$ the maximum ideal of $^*\mathbf{Q}_f$ (the halo of 0, the infinitesimals), then it is known that the quotient $^*\mathbf{Q}_f/\mathcal{M}$ is equivalent to the ordered field **R**. This equivalence is central in our identification of irrational numbers with non-standard rationals. In the following , we shall omit as often as possible the usual '*' on the upper-left part of a character to denote its non-standard nature. For simplicity we suppose at first that L is even. Equation (2-2) still holds for L infinite in ***N**. We can also extend the definition of t_L which becomes a non-standard (infinite) real number $t_L = \alpha F_L + F_{L-1}$. The sites' coordinates are now

$$X_j = \frac{1}{\beta}(jt_L \quad mod\, t_{L+1}) = \frac{1}{\beta}\left[jt_L - t_{L+1}\, Int\left(\frac{jt_L}{t_{L+1}}\right)\right] \qquad j = 0,\ldots, M_L - 1. \tag{3-2}$$

When j is a 'limited' integer , these sites lie at infinity. In accordance with the approximant case, we expect that the site nearest to the origin is such that $j = F_{L+1}$. Let us verify this:

$$\begin{aligned} X_{F_{L+1}} &= \frac{1}{\beta}\left[F_{L+1}t_L - t_{L+1}\, Int\left(\frac{t_L F_{L+1}}{t_{L+1}}\right)\right] \\ &= \frac{1}{\beta}\left[F_{L+1}t_L - t_{L+1}\, Int\left(\frac{(\alpha\alpha_L + 1)F_L^2 + 1}{(\alpha\alpha_L + 1)F_L}\right)\right] \end{aligned} \tag{3-3}$$

$$
\begin{aligned}
&= \frac{1}{\beta}\left[F_{L+1}t_L - t_{L+1}F_L\right] \\
&= \frac{1}{\beta}\left[F_{L+1}(\alpha F_L + F_{L-1}) - F_L(\alpha F_{L+1} + F_L)\right] \\
&= \frac{1}{\beta} \qquad \text{Q.E.D.}
\end{aligned}
$$

We have used equation (3-2) and the fact that whatever α is (being "limited"), the quantity $((\alpha\alpha_L + 1)F_L)^{-1}$ is infinitely small and so its integer part is zero. Taking the analogy one step further we expect the k first sites from the origin to be labelled by

$$j = kF_{L+1} \quad mod(F_L + F_{L+1}) \qquad \text{L even,} \tag{3-4-a}$$

$$j = kF_L \quad mod(F_L + F_{L+1}) \qquad \text{L odd.} \tag{3-4-b}$$

Inserting these values in the expression for X_j ,we get after some manipulations

$$X_j = \frac{1}{\beta}\left[k + (\alpha - 1)\, Int\left(\frac{k\alpha}{\alpha + 1}\right)\right] \qquad \text{L even,} \tag{3-5-a}$$

$$X_j = \frac{1}{\beta}\left[k\alpha + (1 - \alpha)\, Int\left(\frac{k}{\alpha + 1}\right)\right] \qquad \text{L odd,} \tag{3-5-b}$$

with in both cases $j = 0, \dots, M_L - 1$.

This approach can again be extended to any irrational number by using Farey series with infinite index belonging to $^*\mathbf{N}$. We can then find a (non standard) rational number $\alpha_L = p_L/q_L$ such that α is its standard part. Extending to the non-standard terms the intrinsic properties of Farey series [12] we have

$$p_{L+1}q_L - p_L q_{L+1} = 1, \tag{3-6}$$

which will play a role analogous to equation (2-2). If t_L is defined as

$$t_L = \alpha q_L + p_L \tag{3-7}$$

the site coordinates are still given by equation (3-2). The first k sites from the origin, given by

$$j = kp_{L+1} - (p_{L+1} + q_{L+1})\; Int\left(\frac{kp_{L+1}}{p_{L+1} + q_{L+1}}\right) \tag{3-8}$$

are now located at

$$X_j = \frac{1}{\beta}\left[k\alpha + (1 - \alpha)\, Int\left(\frac{k\alpha}{\alpha + 1}\right)\right]. \tag{3-9}$$

Here again we can recognize a familiar expression for 1-dimensional quasicrystals.

Structure Factor.

The calculation done for approximants can here also been extended to the non-standard case. The real space unit cell has become of infinite length, so the reciprocal space unit cell becomes infinitely small. To get information about the structure factor at finite non-zero reciprocal space vector, one understands that in equation (2-8) m should become infinite. This is very similar to the real space coordinates where the site index j was taken infinite to obtain the first k sites from the origin.

We start from the equation (2-9) and focus on the first case , described by equations (3-5). Peaks are found in the Fourier spectrum whenever the denominator vanishes in equation (2-9) while the numerator remains finite. This translates to the conditions

$$A \simeq 2\pi \times \text{ an integer} \tag{3-10-a ,}$$
$$AM_L \neq 2\pi \times \text{ an integer.} \tag{3-10-b}$$

Let us try the following values for m:

$$m = kF_L + lF_{L+1} \tag{3-11}$$

where k and l are limited integers. This gives

$$\frac{A}{2\pi} = kF_{L-1} + lF_L + \frac{(-1)^L(l-\alpha k)}{F_L\,(\alpha\alpha_L + 1)}, \tag{3-12-a}$$
$$\frac{AM_L}{2\pi} = M_L\,(kF_{L-1} + lF_L) + (-1)^L\,(1+\alpha_L)\,(l-\alpha k). \tag{3-12-b}$$

As long as l and k remain limited the quantity $(l - \alpha k)$ never becomes infinitely small. So the R.H.S of eq.(3-12-a) is infinitely close to an integer (recall that F_L is infinitely large) while the R.H.S of eq.(3-12-b) remains distinct from an integer. This was the condition for having peaks in S(q). These peaks are located at

$$\begin{aligned} q_m &= 2\pi\beta\,\frac{k + l\alpha_L}{\alpha\alpha_L + 1} \\ &\simeq \frac{2\pi}{\beta}(k + l\alpha). \end{aligned} \tag{3-13}$$

Finally if we look to the Farey case (equation 3-9) we find peaks for

$$m = lp_{L+1} + kq_{L+1}, \tag{3-14}$$

which are located at

$$\begin{aligned} q_m &= 2\pi\beta\,\frac{k + l\alpha_{L+1}}{\alpha + \alpha_{L+1}} \\ &\simeq \pi\beta\,\frac{k + l\alpha}{\alpha}. \end{aligned} \tag{3-15}$$

Conclusions

In this paper we have shown that non-standard analysis is a good candidate to formulate the transition between approximant structures and the asymptotic quasicrystal. The quasicrystal is explicitly treated as a " non-standard " crystal with an infinite unit cell length and the real quasicrystal is recovered as its "standard part". The reciprocal space formalism can also be adopted with an reciprocal lattice having an infinitesimal unit cell length. Classical results for 1-dimensional quasicrystals are easily recovered. The application to d-dimensional quasicrystals , in the general framework for approximants introduced in refs. [6],[7] and [8], is presently under study.

References

[1] Shechtman,D.,Blech,I.,Gratias,D.,Cahn,J.W.,Phys.Rev.Lett. **53**,1951(1984).

[2] Levine,D.,Steinhardt,P.J.,Phys.Rev.Lett. **53**,2477(1984).

[3] see Proc. of 'International Workshop on Aperiodic Crystals ' Les Houches ,March 1986, J. de Physique,**C3** (86).

[4] Bruijn,N.G. de, Proc. K. Ned. Akad. Wet. Ser.A, **43**,39-52,53-66 (1981)

[5] Duneau,M., Katz,A.,Phys.Rev.Lett.**54**,2688(1985);
Elser,V., Acta Cryst.,**A42**,36(1988);
Kalugin,P.A., Kitaev,A.Y.,Levitov,L.C.,J. Physique Lett. **46**, L601(1985)

[6] Mosseri,R., in "Universality in condensed matter", Les Houches March 88 (ed. Jullien, R., Peliti, L., Rammal, R. and Boccara,N., Springer 1988) page 9

[7] Mosseri,R. Oguey,C.,Duneau,M., in "Quasicrystalline materials" ,ILL-CODEST Workshop (ed. Janot,C. and Dubois,J.M., World Scientific 1988)page 224

[8] Duneau,M.,Mosseri,R.,Oguey,C.,preprint ,submitted for publication.

[9] Robinson,A.,"Non Standard Analysis",(North Holland 1980)

[10] Nelson,E.,Bull. Amer. Math. Soc.,**83**,1165(1977)

[11] 'La Mathematique non standard' ,(edited by Barreau,H. and Harthong,J., CNRS 1989)

[12] Hardy and Wright, "An introduction to the theory of numbers " (Oxford,1938)

[13] Petitot, J. in Reference 11.

[14] Cartier, P., in 'Journées X-UPS, 1985' , reprint from Ecole Polytechnique.

The Topological Structure of Grain Boundaries

*N. Rivier**

Materials Science Division, Argonne National Laboratory, Argonne, IL 60439, USA

Abstract. It is shown that grain boundaries are quasiperodic tilings made of inflatable sequences of structural units. A general grain boundary is a quasiperodic grid of dislocations with the same inflation multiplier (a quadratic irrational) in every direction.

1. Introduction

Grain boundaries are interfaces between two perfect crystalline lattices. They occur naturally as the result of nucleation and growth of small crystalline grains from the liquid phase, which stops when the grains meet. The final product is a space-filling mosaic of polyhedral crystalline grains, chemically and structurally identical, but of arbitrary orientations. Grain boundaries can also be manufactured over wide ranges of orientations [1]. A crystal lattice can be viewed as three periodic sets of parallel planes, intersecting at atomic positions. Because they involve atoms in externally determined or manufactured arrangements, grain boundaries are geometrical manifestations of number theory.

High resolution electron micrographs [2] and computer simulations [3] confirm the following geometrical idealization:

- the grain boundary is planar (there is no roughening)
- the two incident grains impose from "infinity" the geometrical parameters of the boundary, which are the grains orientations, the orientation of the boundary plane, and the interatomic or interplane spacing a_o as length scale.

The topological structure of the boundary is described therefore by five geometrical parameters (for example, Euler angles for the relative orientations of the two grains and of the boundary plane) [4] and one length scale. On the other hand, the technologically important (diffusion, embrittlement, corrosion, etc.) distance d between the two grains is a microscopic parameter, solution of a many body problem involving secondary diffusion and relaxation of atoms within a few atomic layers of the boundary, the primary geometrical substrate. After heat treatment, atomic relaxation may spread out, and particles of diameters up to 200 nm grow at the boundary [5]. Their orientation, and their quasicrystalline structure are still controlled by the grain geometry.

There are two special types of grain boundaries, twist and tilt, corresponding to rotation of the grains about a principal axis of the lattice. In twist

*Work supported by the U.S. Department of Energy, BES-Materials Sciences, under contract #W-31-109-ENG-38.

Springer Proceedings in Physics, Vol. 47 Number Theory and Physics
Editors: J.M. Luck · P. Moussa · M. Waldschmidt

boundaries, the axis of rotation is perpendicular to the boundary plane; it is parallel in tilt boundaries, which retain the translation symmetry of the crystal along the rotation axis.

2. **Topological Structural Units**

The numbers of lattice planes incident on the boundary from either grain are, in general, different (Fig. 1), so that additional planes must be introduced at the boundary. This is done through edge dislocations, topological objects of codimension 2 (points in 2D, lines in 3D).

Dislocations are defined independently of a metric by drawing the fundamental domain or Voronoi polyhedron of each atom. The material is represented by a froth of space-filling Voronoi cells. In two dimensions, the only structurally stable, regular froth is hexagonal (corresponding to a triangular lattice). A pentagon-heptagon dipole constitutes a dislocation (Fig. 2.) It glides under shear and inserts an additional row of cells upon climbing [6].

In three dimensions, the dislocation line is also represented by a double row of non-crystalline Voronoi polyhedra. One distinguishes between edge dislocations with Burgers vector - which measures the lack of closure (torsion) under parallel transport around the dislocation - perpendicular to the line, and screw dislocation lines with parallel Burgers vector. Since dislocations are structural defects, they must lie on the boundary separating the two perfect grains. Edge dislocations, by adding lattice planes from one grain to the other, characterize tilt boundaries. A grid of screw dislocations rotates one grain with respect to the other about an axis perpendicular to the boundary, and characterizes twist boundaries. General grain boundaries are grids of screw and edge dislocations.

There are only a finite number of topologically distinct Voronoi polyhedra (polygons in 2D) filling space as a structurally stable, tetracoordinated (tricoordinated) froth, and each plays a specific structural role as the source of some Burgers vector, measured in number of interatomic spacings. (E.g. in 2D, hexagons are neutral and characterize a perfect grain, a dipole of contiguous pentagon and heptagon is an edge dislocation with unit Burgers vector, and inserts one lattice plane. Square-octagon dipoles or pentagon-heptagon separated by one hexagon insert two planes, etc. In 3D, truncated octahedra 4.6^2 are the Voronoi cells of bcc crystals).

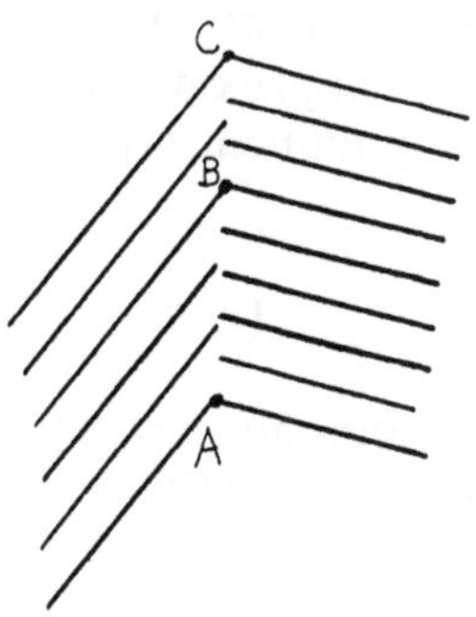

Fig. 1 Grain boundary (schematic).

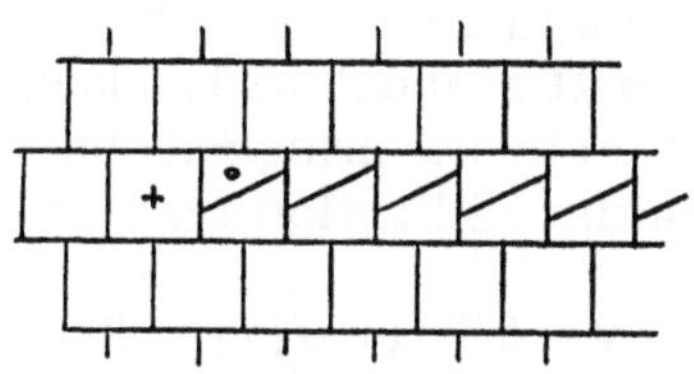

Fig. 2 Topological edge dislocation (pentagon-heptagon dipole). It introduces an extra layer of material, and is generated by cell division.

Operationally, a small edge dislocation loop can be created by division of a Voronoi cell. Subsequent divisions of neighboring cells make the dislocation glide and the loop expand. Another local elementary topological transformation of the Voronoi froth, which switches edges on the polyhedra (T1 of [6]), generates a screw dislocation.

So the boundary consists of Voronoi polyhedra, with each type playing a specific, integral structural role defined by an elementary topological transformation. One can group them into structural units, isolated crystalline Voronoi cells, and rows of non-crystalline polyhedra forming dislocation. Topological stability of the polyhedra grants structural stability to the boundary in the presence of atomic relaxation and diffusion.

3. Inflation Symmetry

The simplest grain boundary involves one family of lattice planes in each (two-dimensional) grain (Fig. 1). The other families may go through the boundary uninterrupted as in compound flowers such as daisies [7]. There are two different structural units, dislocations and crystalline cells. A finite segment AB of boundary involves, say, a=3 planes on the left and b=5 on the right grain. The two extra planes originate from dislocations, which lie on the boundary between the two perfect grains. Dislocations strain the lattice and repel each other, so they must be as equally spaced as possible along the boundary segment. The lowest energy solution for segment AB is the smoothest alternance between three planes going through and two dislocations.

Consider now a longer segment AC of the boundary. If the ratio between planes on the left (a) and dislocations (b-a) were equal to that for AB, the solution for AB could simply be repeated and the grain boundary would have translation symmetry. This is not so in general; the ratio a/(b-a) fluctuates about its value in the limit of an infinite boundary, which is an irrational ν. The structures for AC and AB are then different rational approximants of ν, a macroscopic parameter determined by the grains orientations.

The irrational ν (<1, say) can be represented as the slope of a line E in a lattice Λ with dislocations d=(b-a) in abscissa and through planes a in ordinate. The strip method [8] gives the geometrical structure of the boundary, an irregular staircase in Λ which is indeed the smoothest alternance of through planes and dislocations in proportion approximating ν. Planes are vertical and dislocations horizontal steps of the staircase. Dislocations are kept apart as much as is allowed by ν. Numerically, the best approximations are given by the principal convergents A_i/B_i to ν, obtained by truncation of its continued fraction representation,

$$\nu = 1/\{[q_1+1/[q_2+\ldots]\} \approx 1/\{q_1+1/[\ldots+1/q_i]\} = A_i/B_i. \qquad (1)$$

A_i and B_i are the numbers of through planes and of dislocations on the finite segment, respectively. They are related to the natural integers q_i by the recursion relation

$$B_i = q_i B_{i-1} + B_{i-2} \quad , \quad B_0 = 1, B_{-1} = 0,$$
$$A_i = q_i A_{i-1} + A_{i-2} \quad , \quad A_0 = 0, A_{-1} = 1. \qquad (2)$$

Inflation of a finite sequence generates a longer one with the same structural properties. It can be done either geometrically (by a different choice of the unit cell of the lattice whose translation along E defines the strip), algebraically (by matrix multiplication) or by automatic substitution (replacing by "words" the letters d, a in the sequence). All these methods are equivalent. They are also iterative.

Equation (2) is a particular inflation operation, replacing staircase w_{i-1} from 0 to (B_{i-1}, A_{i-1}) by a longer one, w_i, to (B_i, A_i). Let us approximate line E by a broken, continuous line E_B made of segments with rational slopes joining principal (and intermediate) convergents. E can be moved to E_B without crossing any lattice point, and the strip on E and E_B yields identical staircases. We obtain the substitution rules for constructing a sequence of B_i dislocations and A_i through planes,

$$w_i = (w_{i-1})^{q_i}\, w_{i-2} \qquad ; \qquad w_0 = d\,, w_{-1} = a. \tag{3}$$

(The order of the words in (3) follows from the fact that the last two steps of a segment are da or ad if its end point is above or below E. Successive principal convergents lie alternatively above (i odd) and below (i even) E, so w_i, a sequence of q_i w_{i-1}'s and one w_{i-2} must end with w_{i-2}).

It is the integer q_i which sets the inflation rule (2) or (3) at every stage i. Inflation symmetry generates longer structures by repeated application of the same rule, and thus requires the sequence of integers $\{q_i\}$ to be periodic after some stage and ν to be quadratic irrational [9]. Conversely, if the sequence $q_1, q_2, \dots q_i, \dots$ were not periodic, the rules would depend on the stage of inflation or, through (3), on the position in the structure, and inflation would not be context-free [10]. Quasicrystalline structures with inflation symmetry are characterized by quadratic irrationals ν. (Finite sequences $\{q_i\}$ have rational ν and periodic (crystalline) structures).

One constructs an inflation matrix Σ with integral coefficients Σ_{ij}, $\mathrm{Tr}\Sigma>0$ and $|\det \Sigma| = 1$. Σ sends lattice point $\mathbf{w}=(d,a)$ into $\mathbf{w}'=(d',a')$, $\mathbf{w}' = \Sigma\,\mathbf{w}$. (Alternatively, Σ^{-1} transforms the unit cell of Λ which defines the strip [8]). Σ is determined by the requirement that the eigenvector corresponding to eigenvalue $\lambda_+>1$, $\mathbf{e}_+=(1,\nu)$, is along E. Then,

$$\lambda_+ = \Sigma_{12}\,\nu + \Sigma_{11}. \tag{4}$$

Σ maps any point $\mathbf{w}$ of Λ within the strip into another, closer to E ($|\lambda_-|<1$) but farther from the origin, and inflates structure $0\mathbf{w}$ into $0\mathbf{w}'$, with inflation multiplier λ_+. Quadratic irrationals ν and λ_+ belong to the same field $Q(\lambda_+)$.

The physical structure is the staircase $0\mathbf{w}$ projected on E. Its length w_+ is the component of $\mathbf{w}$ along $\mathbf{e}_+$,

$$w_+ = d + a\,(\Sigma_{12}/\Sigma_{21})\,\nu. \tag{5}$$

It is inflated consistently, i.e., $w'_+ = \lambda_+ w_+ = d' + a'\,(\Sigma_{12}/\Sigma_{21})\,\nu$.

4. Fourier transform of inflatable structures [8]

Experimental information on the structure of grain boundaries comes chiefly from x-ray or electron diffraction [11], supplemented only recently by high-resolution electron microscopy [2] and computer modelling [3]. The observation of sharp diffraction peaks suggested a periodic structure (the coincidence site lattice) [11]. We now know that the former is not a necessary condition for the latter, and argue that the structure is quasiperiodic.

The intensity of a diffraction pattern is the modulus square of the Fourier transform (of the characteristic function) of the diffracting object. Here, the object is the irregular staircase generated by the strip method, that is a lattice Λ, cut and projected on line E, $\Pi_E(S_E . \Lambda)$ (suitably generalized to higher dimensions if necessary). Each operation has its Fourier image. Cut (multiplication (.) by the strip's characteristic function S_E) becomes convolution (*) by its form factor (S_E"), and projection (Π_E) becomes intersection ($\cap_E$). Here, x" denotes Fourier transform of x. Λ"= Λ (up to scale) are both square lattices, and the line E remains the same in Fourier space. Thus,

$$\{\Pi_E(S_E . \Lambda)\}'' = \cap_E(S_E'' * \Lambda''). \qquad (6)$$

Diffraction peaks of intensity above a threshold form a discrete set, and the most intense originate from the points in Λ" closest to E, which are principal convergents to the slope of E.

5. General boundary between three-dimensional grains

5.1 Topological description

Since the boundary separates two perfect grains, the structural defects which define the mismatch topologically must lie on the grain boundary itself. The mismatch involves, in general, both rotation and numbers of incident lattice planes within a given interval. The corresponding structural defects are screw, and edge dislocations, respectively. The Burgers vector (conserved quantity which measures the mismatch under parallel transport around the dislocation line) is parallel to the screw, and perpendicular to the edge dislocation lines.

The grids of screw and edge dislocations are independent. They can go through the same Voronoi polyhedron without interference, because the elementary transformations of its faces (division and edge-switching) are performed independently. The two grids must, however, have the same inflation multiplier λ_+ for the whole boundary to be inflatable.

Rotation (the twist in the boundary) is effected by a grid of two orthogonal arrays of screw dislocations. The two-dimensional structure is a direct product of individual arrays. It is automatically inflatable if the array is inflatable. So, rotation is a one-dimensional problem. This is not the case for numerical mismatch, for which a two-dimensional tiling, product of inflatable arrays of edge dislocations in different directions, made with a finite number of types of tiles with edges of finite sizes, must be constructed.

The necessary and sufficient condition for the existence of such a grid has been given by Pleasants [10].

But let us see first where the dislocations should be located.

5.2 Geometrical description

On the plane of the grain boundary, the incident lattice planes of each grain form a tri-grid of 3 regular arrays of straight lines. The grid is not periodic in general (i.e., there is only one point, A, say, at which three lines intersect). Three orthogonal lattice planes, concurrent at some point O_1 within grain 1, define on the boundary a triangle T_1 of edges a_1, b_1, c_1, with A as one of its vertices. The lines of other planes are parallel to the edges of T_1.

A similar triangle T_2 can be constructed for grain 2. We shall assume for definition that A is also the point where the three lines of grain 2 intersect. A is therefore a point of exact coincidence (the only one, in general) on the boundary. It is common to both triangles. Note that, in order for O_2 to be within grain 2, T_2 and T_1 have opposite orientation (abc) (Fig. 3). For definition, select edge a_2 so that triangles T_1 and T_2 have the same area. (Then a_2 is parallel to, but not exactly on the intersection of a lattice plane of grain 2).

If the two grains have identical orientation, there is no grain boundary, the two triangles T_1 and T_2 form a parallelogram, with matching lines from planes of both grains (Fig. 4).

This suggests a strategy: An affine collineation (matrix A_{ij}, with det $A_{ij}= 1$ to conserve areas) transforms T_2 into T_2', the parallelogram complement to T_1. A_{ij} consists of dilatations and rotation by an angle ϕ given by $\tan \phi = (A_{12}-A_{21})/(A_{11}+A_{22})$, regardless of the order of these operations. Affine transformations preserve parallelism, so the lattice planes still define lines parallel to the edges of the triangle, but with different spacings in different grains (Fig.5). The problem reduces to that of matching two sets of lines in a given interval. (It was solved in § 3 for one set of lines.)

Let each triangle T_i' contain p_i, q_i, and r_i lines parallel to its edges a_i, b_i and c_i, respectively (i=1,2). Along edge c, one must match p_1 with p_2 and q_1 with q_2 in an inflatable fashion, that is

$$p_1/p_2,\ q_1/q_2 \in \mathbb{Q}(\lambda_+). \qquad (7)$$

Ditto along edges a and b, so that $r_1/r_2 \in \mathbb{Q}(\lambda_+)$ as well. Fig. 6 shows that this is possible, as long as $\mathbb{Q}(\lambda_+)$ contains the cosine of the effective tilt angle θ about edge c. Indeed, the lines intersecting every edge can be lifted back to their respective grain planes, with $c_1 = p_1^2 + q_1^2 = c_2' = p_2^2 + q_2^2$.

5.3 Particular cases

a) Pure tilt. Grains 1 and 2 are tilted from the boundary by θ_1 and θ_2, respectively. Triangles T_1 and T_2 have edges b and a parallel to each other, and one common edge $c_1 = c_2$. Point C is at ∞. The structure is periodic along b (and a), and quasiperiodic along direction c only. Edge dislocations are perpendi-

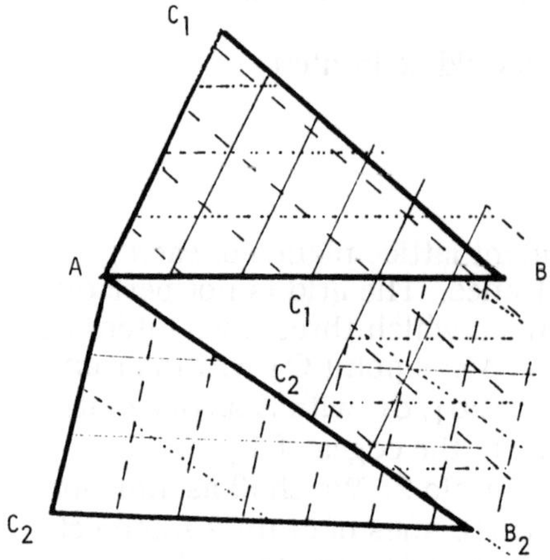

Fig. 3 The triangles formed by the intersection of three orthogonal lattice planes in each grain with the boundary plane.

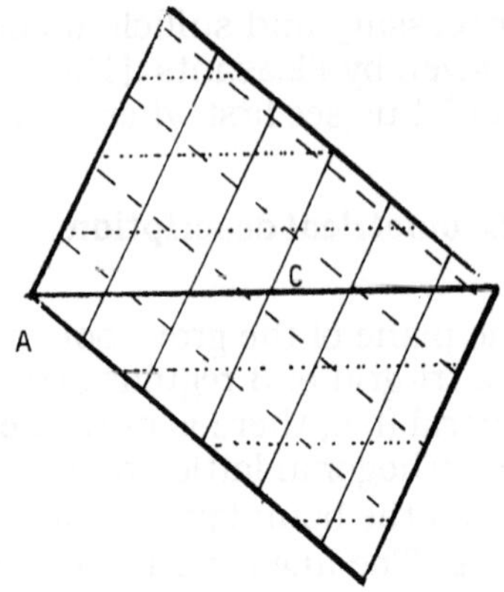

Fig. 4 Identical grains, trivial boundary.

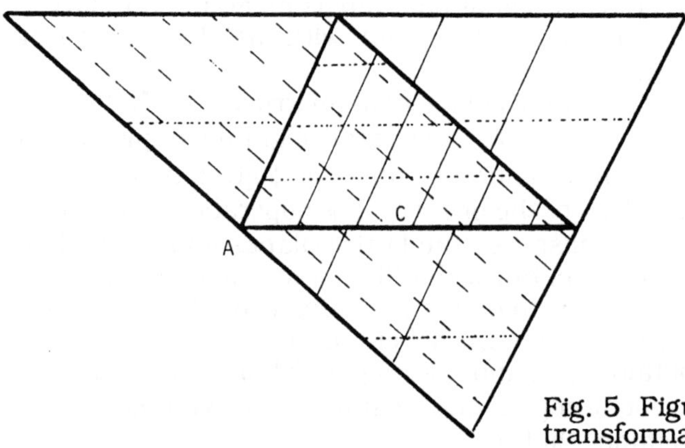

Fig. 5 Figure 3, after affine transformation.

cular to c. The geometry is pictured in Fig. 6. One obtains, by trigonometry, the cosine of the tilt angle $\theta = \theta_2 - \theta_1$,

$$\cos\theta = [(p_1/p_2)(q_1/q_2) + 1] \,/\, [p_1/p_2 + q_1/q_2]. \tag{8}$$

$\cos\theta \in \mathbb{Q}(\lambda_+)$ if p_1/p_2 and $q_1/q_2 \in \mathbb{Q}(\lambda_+)$, and cyclically so, because $\mathbb{Q}(\lambda_+)$ is a field. Conversely, if any two have different inflation multipliers, the third is not context-free inflatable.

A fully asymmetric tilt has $\theta_1=0=q_1$, with one family of planes parallel to the boundary. Then $\cos\theta = p_2/p_1$, regardless of the q_2 incident planes of the other grain. The corresponding edge dislocations are periodically spaced.

In daisies, $q_1=q_2$ and these parastichies (lattice planes) go through the grain boundaries. $\theta=0$, and one has one-dimensional inflation [7].

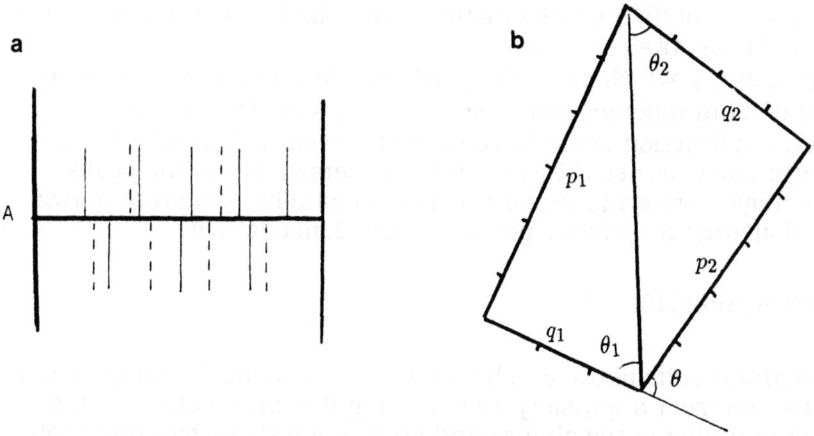

Fig. 6 Pure tilt boundary. a) The triangles of Fig. 3 have two parallel edges. b) Geometry perpendicular to the boundary plane.

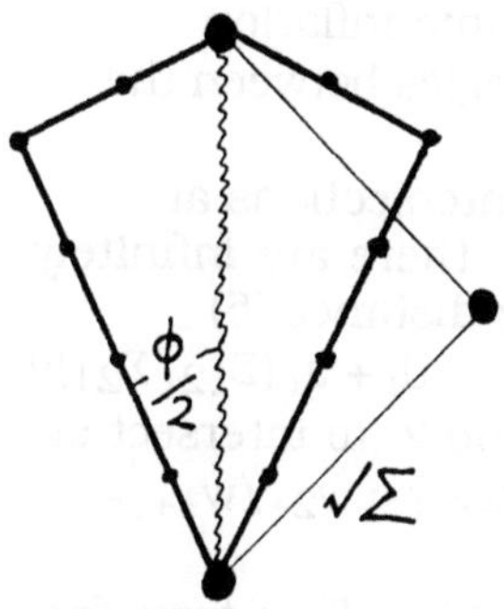

Fig. 7 Twist boundary. Grain 2 has been peeled off on the left to reveal grain 1.

b) Pure twist. Here, the affine transformation is a rotation by ϕ, which is effected by a rectangular grid of screw dislocations. The numerical matching is represented in Fig. 7. If r and s were (coprime) integers, then atom B, as well as A, would be at a coincidence site in both grains, and the arrays of dislocations would be periodic, with period of the "coincidence lattice" $\sqrt{\Sigma}$ given by

$$\Sigma = n \text{ (n odd)}, \quad \Sigma = n/2 \text{ (n even)}; \qquad n = r^2 + s^2. \tag{9}$$

Exact coincidence is insignificantly rare, but approximate coincidence (which requires no or little atomic relaxation) is not There is near coincidence for all integers r, s in (10) such that s/r are principal convergents to tan ($\phi/2$), and $\sqrt{\Sigma}$ indexes the corresponding strong diffraction peaks. A substantial range of tan ($\phi/2$)'s share the same principal convergents, and observed diffraction peaks persist indeed over some range of twist angles ϕ. (By contrast, the exact period of the coincidence lattice is a non-monotonic function of ϕ).

Which integers n are the sum of two squares and in how many distinct ways, is a classic problem in Number Theory [9]. There are 2^{t-1} essentially distinct solutions r, s, of (9) if n has t distinct prime factors of the form 4k+1; none if n is divisible by 4k+3 or by 4.

Twist angles ϕ and ϕ' which are orthogonal complements, $\phi+\phi'=\pi/2$, have the same value of Σ but different convergents, $r'=s-r$, $s'=s+r$ and $n'=2n$.

Stable, intense diffraction peaks have been observed [11] for all small Σ, but not any degeneracy, as yet. The stability and persistence of the peaks over a range of angles are evidence for the quasicrystalline structure of twist boundaries and against a perfectly periodic coincidence lattice.

5.4 Pleasants' Theorem [10]

The discussion and results above can be summarized in one formal question: Is it possible to construct a quasicrystalline tiling (i.e. an inflatable tiling with only a finite number of tile shapes and sizes, such that every finite configuration occurs throughout the tiling with uniform frequency and there are only finitely many such configurations) as a direct product of inflatable arrays along the symmetry directions of a finite symmetry group? The answer is yes, provided the inflatable arrays have the same inflation multiplier λ_+, and $Q(\lambda_+)$ contains the cosines of all the angles between the symmetry directions [10].

It is sufficient to show that the arrays have multiple intersections at points besides the (arbitrary) origin. Then, by inflation, there are infinitely many such points and the tiling is quasicrystalline. The distance (5) between any two lines of the array along direction i is $w_{i+} = d_i + a_i (\Sigma_{12}/\Sigma_{21})\nu$. γ is the angle between any two directions. For arrays 1 and 2 to intersect on direction 1, say, we require that $w_{1+} \cos\gamma = w_{2+}$, so that $\cos\gamma = w_{2+}/w_{1+} = w_{2+}'/w_{1+}' \in Q(\lambda_+)$, and is invariant by inflation.

Remarks. a) Interaction and parallelism are conserved by affine transformations, so that the symmetry group generating the various directions is only a topological framework. To know which topological symmetry generates the hexagrid of a given boundary, one must measure the inflation multiplier λ_+. b) Crystallographic patterns are restricted by the compatibility of translation and rotation symmetries. Quasicrystallographic patterns are similarly restricted, with inflation replacing translation. c) For context-free inflation, λ_+, hence $\cos\gamma$ must be quadratic irrational. Quasiperiodic patterns are generated by symmetry of regular s-gon, $\gamma=2\pi/s$, with s=1, 2, 3, 4, 5, 6, 8, 10 and 12. Out of these, s=5, 8, 10 and 12 are non-crystallographic, and have been found in real quasicrystalline materials. Only s=5 corresponds to a truly three-dimensional finite subgroup E_8 of SO(3) [12]. The other quasicrystalline patterns s=8, 10 or 12 are periodic in the third dimension.

In grain boundaries, conflicting "boundary" conditions are imposed from both grains, and the quasicrystalline structure is the best compromise between them. It has the lowest elastic energy. The grain boundary can act as a substrate for the growth of icosahedral particles [5]. (Whether isolated quasicrystals are structures of lowest free energy remains an open question.)

References

[1] T. Schober and R. W. Balluffi, Phil. Mag 20, 511 (1969)
F. Cosandey, Y. Konem and C. L. Bauer, Phys. Stat. Sol.(a) 48, 555 (1978)
[2] A. Bourret and J. M. Penisson, JEOL News 25E, 2 (1987)
[3] A. P. Sutton and V. Vitek, Phil. Trans. A309, 1 (1983)
V. Vitek, in Dislocations, P. Veyssiere, L . Kubin and J. Castaing, eds. (Editions du CNRS 1984), p. 434
[4] F. C. Frank, in Plastic Deformation of Crystalline Solids (Washington: U.S. Government Printing Office 1950), p.150
D. Wolf, J. Physique Coll. 46, C4-197 (1985)
[5] W. A. Cassada, G. J. Shiflet, and S. J. Poon, Phys. Rev. Lett. 56, 2276 (1986)
[6] D. Weaire and N. Rivier, Contemp. Phys. 25, 59 (1984)
[7] N. Rivier, J. Physique Colloq. 47, C3-299 (1986)
[8] A. Katz and M. Duneau, J. Physique 47, 181 (1986) (independently discovered by V. Elser and by P. A. Kalugin, A. Yu. Kitayev and L. S. Levitov, following N. G. de Bruijn)
[9] H. Davenport, The Higher Arithmetic (London:Hutchinson 1952)
[10] P. A. B. Pleasants, Banach Center Publ. 17, 493 (1985)
W. F. Lunnon and P. A. B. Pleasants, J. Math. pures et appl. 66, 217 (1987)
[11] S. L. Sass, T. Y. Tan and R. V. Balluffi, Phil.Mag. 31, 559 (1975)
W. Gaudig and S. L. Sass, Phil. Mag. 39, 725 (1979)
M. R. Fitzsimmons and S.L. Sass, Acta Met. 36, 3103 (1988)
[12] B. Doubrovnine, S. Novikov and A. Fomenko, Geometrie Contemporaine (Moscow: Mir 1982), I, §20.

Calculation of 6D Atomic Surfaces from a Given Approximant Crystalline Structure Using Approximate Icosahedral Periodic Tilings

J.-L. Verger-Gaugry

Institut National Polytechnique de Grenoble, LTPCM/ENSEEG, BP 75,
CNRS UA no. 29, Domaine Universitaire, F-38402 Saint Martin d'Hères, France

Abstract. Approximant structures of quasicrystals such as α-AlMnSi, R-AlLiCu and $c(Ti_2Ni)$ are well described while atomic organization laws in quasicrystals of similar composition are still unknown. We suggest a method based on the projection method from $\mathbb{R}^6$ to calculate, in a general way, the 6D atomic surfaces associated with a given approximant structure, of known atomic decoration, of any Bravais lattice, with respect to the geometrical features inside the unit cell and the distortions in Fourier space. From these atomic surfaces, we deduce three decorated 3D quasilattices of possibly different icosahedral Bravais lattice type (P, F, I) and lattice parameters, which in general present aperiodic density and stoichiometry fluctuations.

1. Introduction

Structure determination of quasicrystals is now one of the main goals in the field of quasicrystals. In order to find the atomic decoration of the icosahedral phase, at first in AlMn and AlMnSi since Shechtmann et al. /1/ discovered the i-phase in the AlMn system in 1984, and subsequently in many other systems, several methods were followed /2/.

One experimental method consists in doing a Patterson Fourier analysis in 3 and 6 dimensions of the neutron or X-ray diffraction data on the quasicrystalline phase investigated. Millimetric-sized monoquasicrystalline particles or powders composed only (or essentially) of the quasicrystalline phase are required. Therefore, only a few systems can be tested, such as AlMnSi /3,4/, AlLiCu, AlFeCu /5,6/, many other systems exhibiting too small a volume fraction of quasicrystals and presenting often quasicrystalline particles of too small dimensions, typically micron-sized, in a multi-phased matrix. The resulting Patterson function is represented in 2D sections of $\mathbb{R}^6$ and shows "continuous" distributions of atoms which reflects the atomic organization in the i-phase.

Another experimental method, which necessitates the same requirements as for the feasibility of the quasicrystalline phase in large volume fractions as above, consists in calculating partial pair distribution functions using neutron or X-ray diffusion data, as in the case of disordered systems /7,8/. By this method, atomic surfaces can sometimes be determined but within a certain accuracy, assuming confidence in the phase reconstruction procedure.

The Patterson approach leads to results which have to be completed by a 6D modelization, that is an explicit allocation of atoms to sets of sites if we consider here that atoms in 6 dimensions are ponctual entities, or to elongated continuous distributions in 3D subspaces in $\mathbb{R}^6$ in the section framework /9,10/. Other experimental methods, such as electron microscopy, provide useful insights into the decoration of the i-phase, and a useful basis for modelization as well /11-15/.

Another class of methods, "global" methods, assumes that icosahedral units, such as Mackay's icosahedra, are located at each vertex of a perfect 3D quasilattice /2, 16/. Icosahedral units are glued in order to form a compact assembly of atoms presenting globally the icosahedral symmetry. Local problems are encountered in the use of such methods: density, resemblance of local environments in approximant structures and i-phase, as in AlMnSi,..., and such formal constructions are not yet satisfactory.

Springer Proceedings in Physics, Vol. 47 **Number Theory and Physics**
Editors: J.M. Luck · P. Moussa · M. Waldschmidt

Other methods, "local" methods, start from the principle that an approximant structure has a unit cell which can be partitioned with the same tiles as those encountered in the quasilattice, the non-distorted prolate and oblate rhombohedra, such that atoms are positioned always at equivalent places in each kind of tile, irrespective of the orientation of the tiles in the unit cell. The quasicrystal is then formed by the tiling with both rhombohedra now decorated and placed quasiperiodically /17/. This approach has the advantage of keeping the same local clusters in both the crystalline approximant phase and the icosahedral phase. However, some constraints arise : matching rules /18/ have to be satisfied, and it is clear that many structures have no reason a priori to be tiled by such non-distorted polyhedra in the way described above. This led some authors to slightly distort local clusters for the matching rules be satisfied /19/ to generate quasicrystals.

Because of the existence of a large number of systems in which the i-phase forms, under the form of micron-sized particles which cannot be investigated by the methods mentioned above, but only by electron microscopy to obtain reliable diffraction data, there is therefore the need to develop methods to generate quasicrystals based on what we know about approximant structures. This approach is in some sense followed experimentally by Kuo /20/ since Frank-Kasper and related phases are considered to be approximant phases, and the feasibility of quasicrystals in systems in which such phases form is an attractive challenge /21/.

Here, we suggest a method making use of rational cuts in R^6 to obtain directly atomic surfaces calculated from the known atomic decoration of a given approximant structure. 6D atoms are considered here as ponctual entities. The present method is new, though the general framework is the projection method /22/, and is not based on a rigid geometry model. The general purpose of this work is to go from approximant structures towards quasicrystals via approximant tiling properties.

2. Embedding of physical space E in $\mathbb{R}^6$

2.1 Real and integral (over $\mathbb{Z}$) linear representations of the icosahedral group

Let G the icosahedral group, $|G| = 120$. Recalling /22/, we take a real linear representation of G, $\rho : G \to GL(\mathbb{R}^6)$, $g \to \rho(g)$ such that it is faithful, leaves invariant a symmetric definite positive bilinear form, and that 1) it is realizable over $\mathbb{Q}$, which is equivalent to the existence of a G-invariant lattice, 2) $\mathbb{R}^6$ is the orthogonal sum of two three-dimensional subspaces endowed with non equivalent actions of G, that is non isomorphic simple $\mathbb{R}(G)$-modules, E and $E_\perp$. E represents physical space and is the image of the irrational projector $P_{\parallel}$ given by the following matrix in the canonical basis of $\mathbb{R}^6$:

$$P_{\parallel} = 1/2\sqrt{5}\begin{pmatrix} \sqrt{5} & 1 & 1 & 1 & 1 & 1 \\ 1 & \sqrt{5} & 1 & -1 & -1 & 1 \\ 1 & 1 & \sqrt{5} & 1 & -1 & -1 \\ 1 & -1 & 1 & \sqrt{5} & 1 & -1 \\ 1 & -1 & -1 & 1 & \sqrt{5} & 1 \\ 1 & 1 & -1 & -1 & 1 & \sqrt{5} \end{pmatrix} \qquad (1)$$

Let us call $P_\perp = Id - P_{\parallel}$, $\rho_{\parallel} : G \to GL(E)$ the real subrepresentation of ρ, and L the generic notation for a G-invariant lattice in $\mathbb{R}^6$. Following /23/, we note P the primitive lattice $\sqrt{2}\mathbb{Z}^6$, where the normalization factor is just used to normalize the images of the vectors e_i, with $\sqrt{2}\mathbb{Z}^6 = \Sigma\,\mathbb{Z}e_i$, ie $\|P_{\parallel}(e_i)\| = 1$ for i = 1 to 6. $P_{\parallel}$ and $P_\perp$ are orthogonal projectors. For a generic vector x in $\mathbb{R}^6$, we call $x_{\parallel}$ or x'' its projection on E by $P_{\parallel}$ (resp. $x_\perp$ or $x^\perp$ on $E_\perp$ by $P_\perp$). When restricted to $\Sigma\,\mathbb{Q}e_i$, for i = 1 to 6, $P_{\parallel}$ and $P_\perp$ are $\mathbb{Q}$ - isomorphisms onto their images, $\Sigma\,\mathbb{Q}e_i''$, resp. $\Sigma\,\mathbb{Q}e_i^\perp$. The 3-faces of the 6-cubes of L=P are projected onto both rhombohedra known as the prolate and oblate rhombohedra. Whatever L is, the sets $P_{\parallel}(L)$ and $P_\perp(L)$ are dense and uniformly dense in E, resp. $E_\perp$. $\{e_i/\sqrt{2}\ /\ i = 1 \text{ to } 6\}$ is the canonical basis of $\mathbb{R}^6$.

2.2 Strip algorithm

Assume at first L=P. In order to tile aperiodically E with only two types of tiles, an open strip E+TR is placed around E in $\mathbb{R}^6$ to select nodes of L. TR denotes here the image by $P_{\parallel}$ of the open elementary 6-cube of the lattice L=P at the origin. The quasilattice formed by L and TR is the set of points $\{x_{\parallel} \ / \ x \in L, \ x_{\perp} \in TR\}$. This quasilattice can be seen more geometrically by introducing links between points: $n_{\parallel}$ and $m_{\parallel}$ are linked by an edge in the quasilattice iff n - m is one of the 12 vectors $\pm e_i$. This provides the edges of rhombohedra and the quasilattice appears as a true tiling with both types of rhombohedra. TR is called "cut volume, or window, or profile" as usual. We call it now TR(P). When L is a G-invariant lattice in $\mathbb{R}^6$, we call TR(L) the image by $P_{\perp}$ of an open elementary unit cell associated with the origin. The terminology TR comes from the fact that any elementary 6-cube of L=P projects on a triacontahedron in $E_{\perp}$. With L and TR(L), we can form quasilattices as in the case of L=P. More generally, we can take any bounded G-invariant convex domain K in $E_{\perp}$ to generate a quasilattice by setting the strip E+K around E, selecting nodes in a G-invariant lattice L and projecting them on E by $P_{\parallel}$.

The Fourier transform of a quasilattice has the property that it is G-invariant; its support is dense and, in any bounded domain in reciprocal space, the number of peaks of intensity above a given strictly positive value is finite /23/.

The present method is based on the fact that any radiation interacting with matter will only see diffusive center positions and the nature of the diffusive species at these positions but not the rhombohedra edges, which remain, for the incident beam, an abstraction. Therefore, in the following, we shall mainly use vertices and not the underlying geometrical description of the tilings with polyhedra arising from projections of some faces of a G-invariant lattice in $\mathbb{R}^6$. We shall adopt a similar attitude for periodic tilings generated by the projection method.

2.3 G - invariant lattices in $\mathbb{R}^6$

Let us define as usual the lattices /23,24,25/: $P = \sqrt{2}\mathbb{Z}^6 = \Sigma\, \mathbb{Z} e_i$, for i = 1 to 6, $I = P \cup (P + \Delta/2)$, $2F = \{\Sigma x_i e_i \in P / \Sigma x_i$ is even, with i running from 1 to 6$\}$, $D6 = 2F \cup (2F + \Delta/2)$, $D'6 = 2F \cup (2F + \Delta'/2)$, with $\Delta = e_1 + e_2 + e_3 + e_4 + e_5 + e_6$ and $\Delta' = \Delta - 2e_6$, built on the same basis, the e_i's. They are G-invariant and we have the following sequence of inclusions:

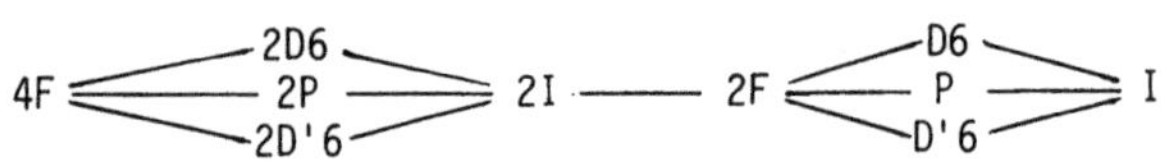

(2)

where any G-invariant lattice included in L_o and containing $2L_o$ lies in this presentation, where L_o is one of the lattices I, P or 2F. Since the present method is actually an algorithm which has to be handled by the computer (§ 4.2), we shall consider only G-invariant lattices included in I, and, more precisely, the partially ordered (by inclusion) set (*) = $\{mL/m \in N - \{0\}$, L is any of the lattices P, I, 2F, D6, D'6$\}$. The concept of quasi-homothety classifies these lattices /p 95 in /24//, in particular in (*): any G-invariant lattice L in $\mathbb{R}^6$, satisfying $2I \subset L \subset I$, is quasi-homothetic to I, P or 2F. Because of such equivalence relations, we shall not consider D6 and D'6 in the following. The set (*) will be used to find a G-invariant host lattice for the atomic decoration of a given approximant structure, lifted up in $\mathbb{R}^6$. The set (*) gives rise to a sequence of quasilattices in E of alternate icosahedral Bravais lattice, which are of the three types P, I or F, of larger and larger lattice parameters.

The i-phase in AlMnSi is of P-type. The first example of F-type icosahedral phase was recently found in the AlFeCu system /26/, where diffraction patterns show additional reflections by comparison with those of the i-phase in AlMnSi. A possible reordering of atoms in the i-phase may occur to explain these diffraction data /6 /.

3. Rational cuts and periodic approximant tilings of physical space with pseudo-icosahedral symmetry /27/

3.1 Preliminary remarks

As mentioned in 2.2, we start from the need to take more into account vertices in periodic and aperiodic tilings than polyhedra for simple arguments based on the nature of radiation-matter interactions.

We start also with another principle which is of metallurgical nature: in melt-spun ribbons, both icosahedral and approximant phases coexist, and particles of both phases often stick together. Orientation relationships between them can be determined by electron microscopy by setting a selected area diaphragm on the phase boundary. The existence of the i-phase in physical space means that our physical space is embedded in $\mathbb{R}^6$ in a "rigid way" since it is then stable under the action of G. Indeed, in the will to interpret the i-phase by a decorated quasilattice, E has to be located in $\mathbb{R}^6$ as described in § 2. The existence of approximant phases also in E indicate that approximant tilings have to be constructed in E, in view of interpreting approximant structures by approximant tilings. Though in some extent harmless, this remark enables the calculation of the atomic surfaces associated with an approximant structure, whatever its Bravais lattice and space group.

3.2 Some definitions and basic ideas

We use the same notations as in /27/. A rational cut E' is by definition a 3D subspace of $\mathbb{R}^6$ such that $E' \cap I$ is a rank 3 lattice in E'. We take I instead of P in this definition because it is the densest lattice in the set (*), but any lattice in (*) would be suitable for this definition. We consider 3-tuples of $\mathbb{R}$-linearly independant vectors $\{w_1, w_2, w_3\}$ in $E' \cap I$. When projected on E by $P_\parallel$, they form a basis of E; let us call $u_i = P_\parallel(w_i)$, for i = 1 to 3. We denote by $P'_\parallel$ the "rational" projector whose image is E' and $P'_\perp = Id - P'_\parallel$. Expressed in the e_i's basis, the matrices of $P'_\parallel$ and $P'_\perp$ have coefficients in $\mathbb{Q}$ /27/. Consequently, the image of any G-invariant lattice of the set (*) by $P'_\parallel$ on E' is discrete and included in $\Sigma\, \mathbb{Q}e_i = \sqrt{2}\mathbb{Q}^6$. Because the restriction of $P_\parallel$ to $\Sigma\, \mathbb{Q}e_i$, with i = 1 to 6, is a $\mathbb{Q}$-isomorphism onto $\Sigma\, \mathbb{Q}e_i^\parallel$, with i = 1 to 6, the lattice $P'_\parallel(L)$ is in one to one correspondence with its image $P_\parallel P'_\parallel(L)$ by $P_\parallel$. By this means, we form two periodic lattices in E: the first one is generated by the u_i's, i = 1 to 3, and will fit the crystal system of the approximant structure (cubic, quadratic,...), the other one $P_\parallel P'_\parallel(L)$ will serve as a host lattice for the atomic sites, and is of higher density than the first one. The rational cut E' together with the w_i's and the lattice L will not be chosen independently but according to criteria, at first in Fourier space and then in real space, attached to the characteristics of the approximant structure.

<u>Periodic tilings with non distorted rhombohedra</u>. We consider here L = P and E' a rational cut. TR' denotes the convex set generated by the vectors $P'_\perp(e_i)$ in $E'_\perp$, slightly translated in $E'_\perp$ if necessary to be sure of having no point of $P'_\perp(P)$ on the boundary. As in /23,27/, we set the strip E'+TR' around E' to select nodes in L and project them on E by $P_\parallel$. This tiling is periodic and constituted by non distorted prolate and oblate rhombohedra. When the volume of TR' increases, the density of selected points in L increases which implies, by projection by $P_\parallel$, an enhancement of the density of vertices inside the unit cell (assuming for instance that $\{w_i / i = 1 \text{ to } 3\}$ is a basis of $E' \cap P$) $\{u_1, u_2, u_3\}$. Rhombohedra are partitioned into smaller and smaller parts. In the present approach, we are looking for coincidence between the true atom positions in the unit cell of the approximant structure and these sets of vertices in order to lift up the decoration on L. But, such a method based on "rational selection and irrational projection" cannot be used with reliability. Indeed, atom positions are determined experimentally within a certain accuracy in the unit cell of the approximant structure and $P_\parallel(L)$ is dense. Thus, there are infinitely many

ways to adjust a well-defined cut volume in $E_\perp'$ to obtain a covering of these atom positions, and, among them, no canonical way exists. A choice would be arbitrary and would consist in the use of an arbitrary section of the projector $P_\parallel$.

Covering by coincidence of an approximant crystalline structure. The idea consists in comparing the crystallographic data of an approximant structure, that is atom positions and their affectation, to both lattices $\Sigma \mathbb{Z}u_i$ and $P_\parallel P_\parallel'(L)$ where L is G-invariant and contains the three vectors w_i. When E' is close to E, that is, when, for each i = 1 to 3, $\|w_i - u_i\| \ll 1$, the lattice $P_\parallel P_\parallel'(L)$ can be seen as an infinite periodic set of vertices of slightly distorted icosahedral symmetry polyhedra which overlap. As an example, with L = P, we have the polyhedron generated by the 12 vectors $\pm P_\parallel P_\parallel'(e_i)$, at the origin, and this situation is repeated at each node in L. We can measure in a simple way the proximity of E' and E along the direction u_i by a rational number p_i/q_i. When the w_i's belong to P, say $w_i = n_{1i}e_1 + \ldots + n_{6i}e_6$, we have: $p_i/q_i = 1/2 - 5N_{2i}/4N_{1i}$, with N_{2i} = the sum of the six squares n_{ji}^2 and $N_{1i} = n_{1i}(-n_{2i} - n_{3i} - n_{4i} - n_{5i} - n_{6i}) + n_{2i}(-n_{3i} + n_{4i} + n_{5i} - n_{6i}) + n_{3i}(-n_{4i} + n_{5i} + n_{6i}) + n_{4i}(-n_{5i} + n_{6i}) - n_{5i}n_{6i}$. For each i = 1 to 3, we have the equivalence: $\|w_i - u_i\| \ll 1 \iff 0 < p_i/q_i - \tau \ll 1$, where τ is the golden mean $(1 + \sqrt{5})/2$.

This comparison between sets of atom positions and discrete periodic sets of vertices is made in the following way: fix once for all $0 < \varepsilon \ll 1$, for instance $\varepsilon = 10^{-2}$ times the smallest lattice parameter $\|u_i\|$ (it is a "variable" constant adapted to the precision we have about atom positions in unit cell), we say that a crystallographic structure is "covered by coincidence by L and E' " when, for each atom position x, there exists a unique vertex y belonging to $P_\parallel P_\parallel'(L)$ such that $\|x - y\| \leq \varepsilon$, where $\{u_1, u_2, u_3\}$ represents the unit cell. In this case, we attribute to y the nature of the atom located at the site x. For instance, for $c(Ti_2Ni)$, some vertices will be attributed the "value" Ti, others Ni. We cannot hope that each node y in $P_\parallel P_\parallel'(L)$ is close to an atom position; of course, some y's are deprived of any atom affectation. We give them the value void. In this way, we have built a periodic function on $P_\parallel P_\parallel'(L)$ which takes as many values as there are different types of atoms in the approximant structure, plus one which represents the void affectation. Of course, there is no algebraic structure on the set of values taken by this function.

Because of the algebraic properties of $P_\parallel$ and $P_\parallel'$, this 3D periodic function on $P_\parallel P_\parallel'(L)$ can be uniquely lifted up as a 6D periodic function on L. In other words, we can lift up the atomic decoration of the approximant crystalline structure in $\mathbb{R}^6$, on L, once the covering by coincidence of the approximant structure is chosen; that is, we have to decide which E' subspaces, which w_i's, and which lattices L have to be used, with respect to the crystallographic data of the approximant structure we consider. This is the purpose of the algorithm suggested in § 4.2. The atomic decoration of the lattice L is computed as follows: y in $P_\parallel P_\parallel'(L)$ has the same atomic affectation as z in $P_\parallel'(L)$, where z is the unique point such that $z_\parallel = y$. Any x in L such that $P_\parallel'(x) = z$ receives the same atomic affectation as y.

Let us call $\{t_1, t_2, t_3\}$ a basis for the trace lattice $E_\perp' \cap L$. We call DL (for Decoration Lattice) the sublattice of L bearing all the information about the atomic decoration of the approximant structure: it is defined by its generators $\{t_1, t_2, t_3, w_1, w_2, w_3\}$ and its fundamental domain contains all atom positions lifted up in $\mathbb{R}^6$ on L. Note that the w_i's are taken in $E' \cap L$ but do not constitute necessarily a basis of this trace lattice in E'. This lattice is generally not stable under G and its fundamental domain, together with its collection of atoms, cannot be considered as an atomic surface in $\mathbb{R}^6$ (§ 4.1).

3.3 Pseudo-icosahedral symmetry

As above, let us consider a G-invariant lattice L and E' generated by the w_i's. The lattice $\Sigma \mathbb{Z}u_i$, with i = 1 to 3, admits a reciprocal lattice whose generators u_i^* are defined by $u_i.u_j^* = \delta_{ij}$, in E. The u_i^*'s are therefore $\mathbb{Q}$-linear combinations of the $e_{i\parallel}$'s and are lifted up uniquely in $\mathbb{R}^6$. Let us call w_i^* the vectors such that $P_\parallel(w_i^*) = u_i^*$ for i = 1 to 3. $H = \Sigma \mathbb{Z}w_j^*$, with i = 1 to 3, is now the reciprocal lattice of the tiling, viewed in $\mathbb{R}^6$, given by E' and the w_i's.

In /27/, we have tiled the unit cell $\{u_i\}$ of the approximant tiling by non distorted rhombohedra by setting a strip (TR'+E') around E' and we have analyzed the pseudo-icosahedral character of diffraction peaks in Fourier space with a local potential equal to zero everywhere except at the vertices of rhombohedra where we take it equal to 1. Here, the unit cell is tiled periodically by the lattice $P_{\parallel}P'_{\parallel}(L)$. With the same local potential form, conclusions can be shown to be identical to Prop. 2 in /27/: pseudo-icosahedral symmetry in (3D) reciprocal space can be obtained iff E' lies within a narrow 6D solid angle around E.

When this condition is satisfied for E', intense peaks are aligned on a lattice in reciprocal space and present a pseudo-icosahedral character, that is, positions and intensities of peaks are almost invariant under G (simultaneous double approximation). Conversely, such an almost invariance allows us to locate E' in $\mathbb{R}^6$, close to E.

Let us recall briefly how pseudo-icosahedral symmetry is described: G acts on the structure factor via the module of the Fourier transform of the set of vertices within the unit cell of the periodic tiling and the phase of the FT is taken into account by homogeneous and nonhomogeneous Diophantine approximation /28/ with $120 \cdot s$ linear forms describing the action of G, and s the number of vertices of the lattice $P_{\parallel}P'_{\parallel}(L)$ inside the unit cell. This provides almost invariance in intensities.

Diophantine approximation gives also the existence of a vector g in the G-invariant primitive lattice in $\mathbb{R}^6$ generated by the icosahedral orbit of H and its projection on $E'_{\perp}$, $P'_{\perp}(H)$; the set of exact icosahedral positions of intense peaks (at a given distance of the origin) in reciprocal space is $\rho_{\parallel}(G)P_{\parallel}(g)$. The corresponding set of true Bragg peak positions is $\rho_{\parallel}(G)P_{\parallel}(g^H)$, where g^H means the projection of g on H parallel to $E'_{\perp}$.

In this paper, when we speak of an approximant tiling, E', w_i's, we consider that it arises from a rational cut E' lying in a narrow solid angle around E of predefinite aperture, for instance $\{v \in \mathbb{R}^6 / \|v_{\perp}\|/\|v_{\parallel}\| < 5 \cdot 10^{-2}\}$.

When we speak of an approximant structure, it is to indicate that we consider a crystalline phase (for instance a FK phase) and that the experimentally observed distribution of Bragg peaks presents, at a certain distance from the origin in reciprocal space, a pseudo-icosahedral character. If we want to find an approximant tiling E', w_i's, which covers by coincidence an approximant structure, E' has first to be chosen in a narrow solid angle around E. This is only a consequence of the above proposition. The aperture angle of the solid angle around E can be computed from the observed maximal distortions of diffraction spots.

3.4 Distortions and indexation of distortions

With the same notations, the distortions of Bragg peaks, attached to the exact icosahedral positions given above, are: $dis_t = \rho_{\parallel}(t)P_{\parallel}(g^H - g)$, with t running in G.

Assume that for each Bragg peak, the intensity is slightly spread in a spherical way, as in a Gaussian distribution centered at the peak position. When the approximant tiling is oriented along pseudo -5, -2, -3 symmetry elements under the beam, we then observe misalignments and shifts of diffraction peaks in 2D sections in Fourier space, as in electron microscopy in planes tangent to the Ewald sphere. Actually, we observe the projections of the distortion vectors dis_t on the observation plane.

The experimental situation concerning distortions is far from being simple, with complex shapes and splitting of peaks /29,30/. Knowles /31/ made an analysis of this problem in the AlLiCu system. Some authors interpret peak splitting as a consequence of a macroscopically multitwinned structure built on an approximant tiling which, when endowed with atoms, gives rise to a rhombohedral structure of space group $R\bar{3}m$ /32/. We limit ourselves to approximant structures with well punctuated Bragg peaks.

Distortions can be indexed by 6 integers, which are the coordinates of the vectors dis_t in the basis (= canonical basis of $\mathbb{R}^6$ up to a scaling factor) of the primitive lattice generated by $\rho(G)H$ and $P'_{\perp}(\rho(G)H)$. The symmetry operations of the set of distortions can be computed as well as those which result after projection of the distortions on a 2D "observation plane".

4. Calculation of atomic surfaces and decorations of quasilattices

4.1 Atomic surfaces

An atomic surface is here by definition the irreducible fundamental domain of a G-invariant lattice L in the set (*), punctuated by atoms (the terminology "surface" is of course not correct). This means that it is a finite collection of atoms in $\mathbb{R}^6$, on the lattice I, which is attached to one node of L (it is usual in three dimensions for a motif), for which there exists no other lattice in (*) containing L, and different from L, having a strictly smaller part of this collection as a motif. In other words, we want L to be effectively the icosahedral Bravais lattice of the quasicrystal formed in E from this atomic decoration of $\mathbb{R}^6$.

Assume now L = I, P or 2F and that we have found a "good covering by coincidence" of the investigated approximant structure; a dense periodic atomic decoration is now set up on L, and admits as set of translation vectors the lattice we have called DL (DL depends upon the choice of L). Calculating the atomic surfaces associated with this approximant structure then means two things:

1) Firstly, the determination of a G-invariant lattice in (*) whose fundamental domain is compatible with the atomic decoration of $\mathbb{R}^6$. This signifies that we are looking for the densest and greatest G-invariant lattice in (*) included in DL. The fact that DL contains G-invariant lattices in (*) comes from Lemma 2, p66 in /33/; indeed, if m denotes the covolume of DL in L, we have $L \supset DL \supset mL$, and since the covolume is an integer, mL belongs to (*). Let us call GDL the densest and greatest element of (*) included in DL; we have: $I \supset L \supset DL \supset GDL \supset mL$. From § 2.3, we know that if there are several possibilities for GDL, they are G-equivalent and therefore give rise to the same decorated quasicrystal in E by projection (by $P_{\parallel}$). Of course, GDL can be of P, I or F symmetry type, different from that of L; since this construction makes it the icosahedral Bravais lattice of the quasicrystal we form, its characterization gives directly the symmetry type of the quasicrystal built with L and the approximant crystalline structure. The above sequence of inclusions allows the determination of the generators of GDL by the computer by investigating elements of (*) between L and mL which are included in DL (m is the ratio of the volumes of the unit cells of DL and L).

2) Secondly, the calculation of the atom positions within the unit cell of GDL from those which lie inside a fundamental domain of DL. It amounts to a change of coordinates system for the (6D) atom positions within a unit cell of GDL. Such a decorated unit cell of GDL is an atomic surface in the sense defined above.

The fundamental domain of DL is included in that of the lattice GDL as many times as the value of the covolume (let us call it v; it is an integer) of GDL in DL; this gives a "multiplicity", ≡ v, to each atom and each atomic cluster of the approximant structure lifted up in $\mathbb{R}^6$ in the construction process of the quasilattice once L is given. We limit ourselves to L = P, I or 2F because of the classification up to quasi-homothety of the elements of (*). For each L, we obtain a lattice GDL and therefore three quasicrystals, which can be different.

4.2 Algorithm for the determination of a suitable covering by coincidence of a given approximant structure

On one hand, we have an approximant crystalline structure, such as a Frank-Kasper phase, which exhibits a pseudo-icosahedral character at a certain distance of the origin in Fourier space, and a large fraction of atom sites with coordination 12, centers of generally slightly distorted icosahedra which are connected by interpenetration or by sharing edges or vertices /20,34/. On the other hand, any rational cut E' in a narrow 6D solid angle around E gives rise to pseudo-icosahedral periodic tilings in E in which superimposed imbricated slightly distorted icosahedra of various sizes can be found. If $P'_{\parallel}$ is the projector associated with E', the set of vertices of these icosahedra is the lattice $P_{\parallel}P'_{\parallel}(L)$, with L = P, I or 2F (§ 3.2).

We would like to adapt an approximant tiling to an approximant structure; that is once L is fixed (step 0), we are looking for a rational rank 3 projector $P'_{\parallel}$ of image called E', and three linearly independent vectors, the w_i's, such that the covolume of the lattice $\Sigma \mathbb{Z}u_i$, i = 1 to 3, in $P_{\parallel}P'_{\parallel}(L)$ (= ratio of their respective fundamental domain volumes) is as small as possible and that $P_{\parallel}P'_{\parallel}(L)$ is maximally occupied by the atoms of the approximant structure (by coincidence), with $w_i \in E' \cap L$. This procedure ensures a maximum occupation of L. Such rational cuts E' of course exist and we suggest the following algorithm to find them by computer:

step 0 : choice of L = P, I or 2F.

step 1 : (rough fit of maximal distortions in Fourier space) choice of a narrow solid angle $\delta\Omega$ around E in $\mathbb{R}^6$.

step 2 : list the rational cuts E', w_i's, inside $\delta\Omega$ such that $w_i \in E' \cap L$ and the u_i's with $u_i = P_{\parallel}(w_i)$, represent the unit cell up to a scaling factor, of the approximant structure, whatever its Bravais lattice. Practically, we fit the angle relations between the u_i's as they are in the crystal system and both ratios of lattice parameters $\|u_2\|/\|u_1\|$, $\|u_3\|/\|u_1\|$ to their analogues b/a, c/a. The projection method imposes that both squares $(b/a)^2$ and $(c/a)^2$ should belong to $\mathbb{Z}[(1+\sqrt{5})/2]$ whereas these elements naturally belong to $\mathbb{Q}$. Since $\mathbb{Q}$ and $\mathbb{Z}[(1+\sqrt{5})/2]$ are both dense in $\mathbb{R}$, there is no difficulty to express these ratios as elements of $\mathbb{Z}[(1+\sqrt{5})/2]$, by taking approximants. Rational cuts are classified by increasing density of points of $P_{\parallel}P'_{\parallel}(L)$ inside the unit cell $\{u_i\}$, with their associated projectors $P'_{\parallel}$ and the three rational numbers p_i/q_i reflecting pseudo-icosahedral symmetry.

step 3 : take the first E' in the list and compare $P_{\parallel}P'_{\parallel}(L)$ with the atom positions in the unit cell of the structure. If a covering by coincidence is not performed, we take the following rational cut E', w_i's, in the list and make again the comparison. More algebraically, coincidence means that the fit takes into account atom orbits under the action of the space group of the structure, and that positional parameters are approximated by elements of $\mathbb{Z}[(1+\sqrt{5})/2]$.

When a covering by coincidence is found, we have modelized the approximant structure; experimental distortions are automatically fitted. We now build the atomic surfaces. In short, to summarize, we have:

step 4 : lifting of the atomic decoration on L in $\mathbb{R}^6$.

step 5 : determination of the decoration lattice DL.

step 6 : determination of the lattice GDL from DL.

step 7 : calculation of the atomic surface associated with the crystalline structure.

step 8 : projection of this atomic decoration on E by a strip method to form a decorated quasicrystal. We discuss this point below (§ 4.4).

4.3 Comments

We make lists of w_i's on L expressed by their coordinates on the basis of the e_i's, within $\delta\Omega$. We take 3-tuples of them, construct the list of rational cuts. The determination of projector matrices is performed a posteriori once E' is defined by the w_i's.

The research of rational cuts in the list (step 3) stops by itself. Indeed, looking for rational cuts of higher and higher density of points within the unit cell placed on a lattice of greater and greater "mesh" makes coincidence occur after a finite number of iterations at step 3. If we continue the process, we would obtain the same decorations up to scaling factors.

Once a covering by coincidence is determined, the metrics of $\mathbb{R}^6$ is fixed: the lengths $\|u_i\|$ are rescaled in nm (they are the lattice parameters), which allows the determination of the lattice parameters of L and GDL. The resulting quasicrystal built by this method inherits the local environments of the approximant structure at the same scale, but are undistorted. Indeed, $P'_{\parallel}$ and $P_{\parallel}$ are projectors, i.e. contracting operators of $\mathbb{R}^6$, and therefore, keep memory of local environments; when a local cluster of the approximant structure is lifted up on L as a local cluster, which does not of course happen systematically, it remains a local cluster when selected by the strip and projected by $P_{\parallel}$ on E, that is of perfect icosahedral symmetry when it is "complete", with almost the same distances between atoms as in the approximant structure but undistorted.

It is obvious that a solution E', w_i's, given by the above algorithm depends upon the choice of L, for L = P, I or 2F. We then form three types of atomic decoration and therefore, three quasicrystals of possibly different icosahedral Bravais lattice and complexity (§ 4.1).

Once we have modelized the given approximant structure by a rational cut, and made scaling identifications, we can lift up dislocations in $\mathbb{R}^6$ and study them explicitely within the general framework developed by Kléman /35/, and project them on E together with the atomic decoration.

4.4 Decorated quasilattices and strip geometry

Assume now that an atomic decoration of L (with L =P, I or 2F) is determined which possesses GDL as G-invariant icosahedral Bravais lattice. We try to construct a quasicrystal by the strip + projection method in E (using $P_{\parallel}$). There are alternative ways to tackle this problem:

1) The first approach consists in setting a G-invariant cut volume K in $E_{\perp}$ in order to select decorated nodes of L and project them on E as usual. The choice of K requires a careful analysis of the short bonds problem which is inherent to the use of a strip /10/. The strip is here a cylinder E+K and its surfaces are parallel to E. When the strip is adjusted to respect minimal distances between atoms in the quasicrystal, some (infinitely many) parts of the quasicrystal will be compact and others of smaller density. These density fluctuations reflect the variations of density of the atomic surfaces, and are aperiodic since GDL and E are in incommensurate positions. The whole stability of the quasicrystal can then be investigated by Landau theory /36,37/. Comparison can be made in this way between the three types of quasicrystals, at 0 Kelvin, and afterwards at various temperatures to eventually predict phase transitions between them. In the system AlMnSi /38/, "premonitory rearrangements" upon heating in the i-phase were observed, which can in fact be interpreted as phase transitions between quasicrystalline phases associated with the same approximant crystalline phase (α-AlMnSi).

This method provides G-invariant power spectra in Fourier space, that is distortions in Fourier space are undistorted, and local configurations, in real space, existing in the approximant structure become parts of perfect icosahedral units.

Density fluctuations can be accompanied by stoichiometry fluctuations at the scale of the lattice parameter of GDL, though an average value of the stoichiometry is obviously not far from that of the approximant structure. Such chemical variations can explain the fine speckling (≈5 nm) observed on the i-phase in many systems /39/ by dark field imaging in electron microscopy.

This approach is now followed in the system TiNi, where c(Ti_2Ni) and i(Ti_2Ni) coexist, since we have a good knowledge of the distorted icosahedra within the unit cell of c(Ti_2Ni) /40,41/.

2) The geometry of the strip can be computed in order to obtain everywhere the same compacity in the quasicrystal, respecting minimal distances as well. Such a construction for the strip is not trivial.

5. Conclusion

The present work is an attempt to calculate atomic surfaces in $\mathbb{R}^6$ associated with a known approximant crystalline structure using periodic tilings supplied by the projection method from $\mathbb{R}^6$. Quasicrystals are formed from such atomic decorations of $\mathbb{R}^6$. A numerical application of this method is under way in the system TiNi.

Acknowledgments

A. Loiseau is thanked for having suggested the problem and for providing very useful experimental support. M. Duneau is acknowledged for very stimulating discussions.

References

1. D. Shechtmann, I. Blech, D. Gratias, J.W. Cahn, Phys. Rev. Lett. 53, (1984), 1951.
2. D. Gratias, p 83 in /14/.
3. J.W. Cahn, D. Gratias, B. Mozer, Phys. Rev. B (to appear).
4. D. Gratias, J.W. Cahn, B. Mozer, Phys. Rev. B 38, (1988), 1643.
5. T. Ishimasa, Y.Fukano, M. Tsuchimori, Phil. Mag. Lett. 58, (1988), 157.
6. F. Gähler, E3 in /15/.
7. C. Janot, J.M. Dubois, J. Pannetier, M. de Boissieu, R. Fruchart, p 107 in /14/.
8. B. Bouchet-Fabre, M. Laridjani, A. Chenoufi, J. Dixmier, p 136 in /14/.
9. P. Bak, Scripta Met. 20, (1986), 1199.
10. M. Duneau, C. Oguey, preprint 1988.
11. P.J. Steinhardt, S. Ostlund, The Physics of Quasicrystals (World Scientific), Singapore, (1987).
12. D. Gratias, L. Michel, Int. Workshop on Aperiodic Crystals, J. Physique France, Colloque C3, 47, (1986).
13. K.H. Kuo, Proc. of an Int. Conf. on Quasicrystals, Beijing, Mat. Sci. Forum, 22-24, (1987).
14. C. Janot, J.M. Dubois, Proc. of the ILL/Codest Workshop Quasicrystalline Materials, Grenoble, World Scientific, Singapore, (1988).
15. J.M. Dubois, Troisième Colloque sur les quasicristaux, INPL, Vandoeuvre les Nancy, (1989).
16. D. Gratias, p 83 in /21/.
17. C.L. Henley, V. Elser, Phil. Mag. B 54, (1986), L101.
18. A. Katz, p 195 in /14/.
19. M. Audier, P. Guyot, p 405 in /12/.
20. K.H.Kuo, p 131 in /13/.
21. C. Godrèche, Du cristal à l'amorphe, Les Editions de Physique, Les Ulis, (1988).
22. M. Duneau, A. Katz, Phys. Rev. Lett. 54, (1985), 2688.
23. M. Duneau, p 157 in /21/.
24. D. Martinais, Thèse de Doctorat, Spécialité Mathématiques, Université Paris VII, (June 1988).
25. P. Cartier, C.R. Hebd. Séan. Acad. Sci. B 304, (1987), 789.
26. S. Ebalard, F. Spaepen, J. Mat. Res. (1988), in the press.
27. J.L. Verger-Gaugry, J. Physique France 49, (1988), 1867.
28. R. Descombes, Elements de Théorie des Nombres, PUF, Paris, (1986).
29. P.A. Bancel, P.A. Heiney, p 341 in /12/.
30. K. Urban, J. Mayer, M. Rapp, M. Wilkens, A. Csanady, J. Fidler, p 465 in /12/.
31. K.M. Knowles, p 158 in /14/.
32. M. Audier, P. Guyot, Acta Met. 36, (1988), 1321.
33. H. Cohn, Advanced Number Theory, Dover, New York, (1980).
34. D.P. Shoemaker, C.B. Shoemaker, p 67 in /13/.
35. M. Kléman, p 318 in /14/.
36. P.J. Steinhardt, p 23 in /13/.
37. T. Janssen, p 327 in /14/.
38. J.M. Dubois, J.Pannetier, C. Janot, p 311 in /13/.
39. K.M. Knowles, W.M. Stobbs, J. Microsc. 146, (1987), 267.
40. A. François, DEA ONERA, Université Paris VI, (June 1988).
41. Z. Zhang, H.Q. Ye, K.H. Kuo, Phil. Mag. A 52, (1985), L49.

Part III

Spectral Problems, Automata and Substitutions

Spectral Properties of Schrödinger's Operator with a Thue-Morse Potential

J. Bellissard

Université de Provence & Centre de Physique Théorique,
CNRS-Luminy, Case 907, F-13288 Marseille Cedex 09, France

Abstract. The one dimensional discrete Schrödinger operator with a potential given by a Thue-Morse sequence has a Cantor spectrum, the gap boundaries of which are computed explicitly. The integrated density of states on the gaps are given by 1/2 or by $k/3.2^N$ where k and N are integers and k is not divisible by 3. At small coupling constant V, the gap widths scale according to $V^{Ln4/Ln3}$. Eigenfunctions for energies on the gap boundaries diverge faster than any polynomial at infinity.

1. Introduction

The discovery in 1984 of the icosahedral phase of the AlMn alloy by Schechtman et al. [1] and its interpretation as a quasiperiodic crystal gave rise to many works on tiling the plane or the space. One possible construction is provided by "inflation rules": they may be described through a finite alphabet a, b, c,... together with a substitution rule σ associating a finite word with each letter. Each letter is supposed to represent an elementary piece of the tiling and the substitution rule tells us how to associate them to grow the crystal [2]. One can also interpret the letters as representing different types of atoms in a 1D alloy and the rule tells us how to order them along a chain: namely one creates the infinite word by taking the limit of the words $\sigma^n(a)$ as $n \to \infty$. Among the possible sequences obtained in this way is the famous Thue-Morse one [3,4]. It is made of two letters a, b and the rule is $\sigma(a)=ab$, $\sigma(b)=ba$. In this way we get the sequence $\underline{u}$=abbabaabbaababba.... This sequence in not almost periodic [5,6].

We consider here the 1D discrete Schrödinger operator with a Thue-Morse potential defined on the discrete 1-D lattice **Z** as

$$H\psi(n) = \psi(n+1) + \psi(n-1) + V(n)\,\psi(n) \tag{1}$$

where $V(n) = V(-n-1)$ and $V(n) = V$ whenever the nth letter of $\underline{u}$ is a and $V(n) = -V$ otherwise (by convention $u_0 = a$).

Several works have recently been published dealing with the Thue-Morse potential : optical properties of Morse superlattices [8] and numerical calculation of the vibration spectrum of a Morse alloy [9,10]. The property (iii) below has already been rigorously obtained by Axel and Peyrière [11] . A paper by Luck [12] provides us with several interesting numerical results and conjectures on potentials given by automatic sequences.

In the present paper we will emphasize the following properties:

(i) As soon as $V > 0$, its energy spectrum Σ is a Cantor set.

(ii) The integrated density of states (Fig.1) in gaps takes on the values 1/2 or $k/3.2^N$ where $k, N \in \mathbf{N}$, k is not a multiple of 3, and $k \le 3.2^N$. All such values appear and no others.

Springer Proceedings in Physics, Vol. 47 Number Theory and Physics
Editors: J.M. Luck · P. Moussa · M. Waldschmidt

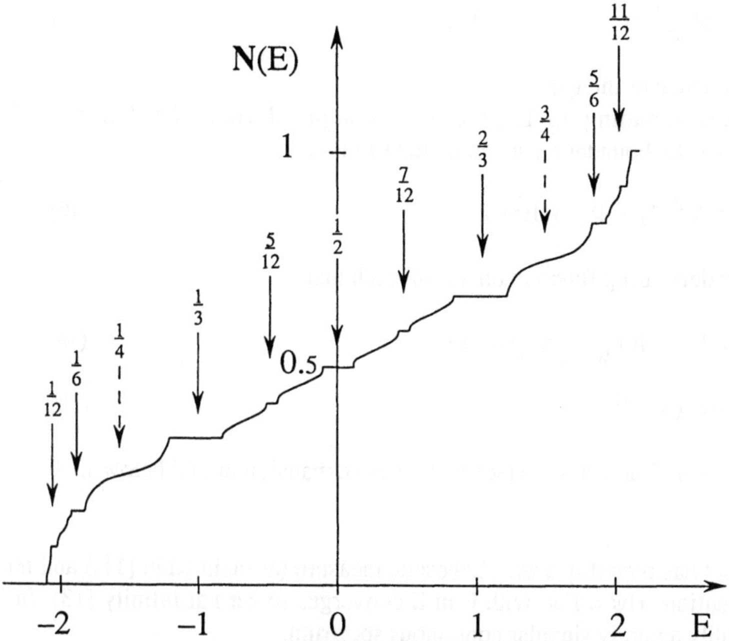

Fig.1- The integrated density of states N(E) for V = .5. Only the main gaps and some of their preimages are visible here. The corresponding indicated values of N(E) are multiples of 1/12. The dashed arrows indicate the position of the corresponding closed gaps.

(iii) There is a countable dense set of energy values in the spectrum $E_0(\underline{\sigma},V)$, such that

$$E_0(\underline{\sigma},V) = \sigma_n\sqrt{(2+\sigma_{n-1}\sqrt{(2+\ldots+\sigma_1\sqrt{(2+2\sigma_0)})\ldots)}} + O(V^2) \qquad (2)$$

as $V \approx 0$, where $\sigma_i = \pm 1$, for which the corresponding eigensolutions are periodic. All but $E_0(\underline{\sigma},V) = \pm 2, 0 + O(V^2)$, are doubly degenerated (see [11]). These energies correspond to "closed gaps".

(iv) For $V \approx 0$ gaps open at the values $E(\underline{\sigma}) = \sigma_n\sqrt{(2+\sigma_{n-1}\sqrt{(2+\ldots+\sigma_1\sqrt{(2+\sigma_0)})\ldots)}}$ of the energy where the σ's take on values ± 1 and n is arbitrary.

(v) For each $\underline{\sigma}$ the width of the gap corresponding to the energy $E(\underline{\sigma})$ satisfies the scaling law [see also [12]]:

$$\delta(\underline{\sigma}) \approx C(\underline{\sigma})\, V^{\mathrm{Ln}4/\mathrm{Ln}3}/b(V) \qquad \text{for some } C(\underline{\sigma}) > 0 \qquad (3)$$

where $b(V)=b(V/3)$ is continuous and bounded away from zero. A numerical study of b gives

$$b(V) \approx 3.8606895\pm 5.10^{-7} + (5.45\pm .05)10^{-5} \cos\pi\{\mathrm{Ln}4/\mathrm{Ln}3\ \ln_2(2/V) - 1.307\pm .001\}. \qquad (4)$$

(vi) If E belongs to the boundary of one of the previous gaps, the corresponding eigensolution of (1) admits the asymptotic behavior

$$\psi(n_k) \approx \text{const.}\ \exp(a|n_k|^{1/2}) \qquad \text{as } k\to\infty \tag{5}$$

where n_k is some subsequence of integer.

(vii) The gap corresponding to $E(\underline{\sigma})$ can be computed from the "main gap" corresponding to $E(0) = -1$. Its boundary is given by the formula

$$4V^2 = (E^2-V^2)(E^2-V^2-4) + \Phi_\infty(4-E^2+V^2) \tag{6}$$

where Φ_∞ is the unique decreasing function on $(2,\infty)$ such that

$$\lim_{w\to\infty} \Phi_\infty(w) = 2, \qquad \lim_{w\to 2} \Phi_\infty(w) = \infty. \tag{7a}$$

$$\Phi_\infty(\Phi_\infty(w)) = \Phi_\infty(w)\,(w-2)^2. \tag{7b}$$

This function is analytic at w=2 and at w=∞ (see its Laurent expansion in §IV below). ◊

Remark. This Cantor set has probably a zero Lebesgue measure (as claimed in [11]) and no eigensolution of the equation $H\psi = E\psi$ with E in Σ converges to zero at infinity [13]. In other words H has probably a purely singular continuous spectrum. ◊

This model provides an interesting example of a Schrödinger operator with a non-almost -periodic potential and nevertheless having a Cantor spectrum. It is believed that replacing the Thue-Morse sequence $\underline{u}$ by any sequence coming from a (primitive) substitution, will give rise to similar properties [12,14].

As we will see, the spectrum of our model is the set of stable points of a 2D mapping given by transfer matrices. This is very much like the case of the Hamiltonians associated with Julia sets [16], with a 1D quasicrystal [17,18], or with a Sierpinsky gasket [19] (see also [20] for other examples). However unlike these models, the gap opening at small coupling is not classical: it involves a scaling law which forbids using perturbation theory. In [12] Luck gives a conjecture about such a behavior, which seems to appear for potentials given by automatic sequences. The wave functions corresponding to energies on the gap boundaries diverge faster than any polynomial at infinity even though it is known that, for almost all energies in the spectrum with respect to the spectral measure, the wave functions are bounded by a polynomial of degree larger than but arbitrary close to one [21].

Our last comment concerns the gap labelling given by (iii). Unlike the result claimed in [11], the integrated density of states is not given by dyadic numbers on the gaps. Should the Thue-Morse sequence be almost periodic, dyadic numbers would emerge, but because of the almost periodicity breaking, an extra 1/3 appears coming from the mean value of the sign change V(n+1)/V(n) which is exactly 1/3. Abstract results coming from K-theory [22] explain this remark.

We will only give hints for the proofs for they are too long to be presented here.

2. Dynamics of transfer matrices

The Schrödinger equation $H\psi = E\psi$ can be written in terms of transfer matrices $T(n)$ as

$$\begin{bmatrix} \psi(n+1) \\ \psi(n) \end{bmatrix} = \begin{bmatrix} E-V(n) & -1 \\ 1 & 0 \end{bmatrix} \begin{bmatrix} \psi(n) \\ \psi(n-1) \end{bmatrix} = T(n) \begin{bmatrix} \psi(n) \\ \psi(n-1) \end{bmatrix}. \tag{8}$$

We set $A=T(0)$ and $B=T(1)$. By construction $T(n)=A$ whenever $u_n=a$ and $T(n)=B$ whenever $u_n=b$. In particular the transfer matrix $M(N)=T(N-1)T(N-2)\ldots T(1)T(0)$ from 0 to $N-1$ is given by the mirror image of $\underline{u}$ if A replaces a and B replaces b. Let us recall that this sequence is not almost periodic in the weakest possible sense [3,4,5,6,7], that no finite subword v of u can appear in the combination vvv inside u and that the letters appearing at ranks $2^{N-1},\ldots, 2^N-1$ are obtained from those at ranks $0,\ldots, 2^{N-1}-1$ by substituting a for b and vice versa. We set

$$A(n) = T(2^n-1)T(2^n-2)\ldots T(1)T(0). \qquad A(0) = A, \tag{9a}$$
$$B(n) = T(2^{n+1}-1)T(2^{n+1}-2)\ldots T(2^n+1)T(2^n). \qquad B(0) = B. \tag{9b}$$

From the definition of the Thue-Morse sequence $B(n)$ is obtained from $A(n)$ by replacing A by B and viceversa leading to the recursion

$$A(n+1) = B(n)\,A(n), \qquad B(n+1) = A(n)\,B(n). \tag{10}$$

Any 2x2 matrix M with determinant one satisfies

$$M^2 - M.\mathrm{Tr}(M) + 1 = 0. \tag{11}$$

Therefore setting $x_n = \mathrm{Tr}(A(n)) = \mathrm{Tr}(B(n))$ (for $n\geq1$) we get the following recursion [9]

$$x_{n+2} = x_{n+1}\,x_n^2 - 2x_n^2 + 2, \qquad (n \geq 1). \tag{12}$$

Clearly $x_{n+1} = x_n^2-2$ implies $x_{n+2} = x_{n+1}^2-2$. This entitles us to define the new variable $v_n = x_n^2-2-x_{n+1}$ which coincides with $(1/2)\,\mathrm{Tr}\{(A(n)-B(n))^2\}$ giving

$$\begin{aligned} v_{n+1} &= v_n\,(v_n + 4 - x_n^2), & x_{n+1} &= x_n^2 - 2 - v_n \quad (n\geq1), \\ v_1 &= 4V^2, & x_1 &= E^2 - V^2 - 2. \end{aligned} \tag{13}$$

It follows that the sequence so defined starts on the axis $v_1= 4V^2$. It will be convenient to introduce the third variable $w = (v + 4 - x^2)$ giving rise to the mappings f from $\mathbf{R}^2$ into itself and $\underline{f}$ from the surface $\Sigma = \{ (x,v,w) \in \mathbf{R}^3;\ w =v+4-x^2 \}$ into itself defined by

$$f(x,v) = (x^2-2-v,\ v(v+4-x^2)), \qquad \underline{f}(x,v,w) = (2-w,\ vw,\ x^2w). \tag{14}$$

◊

3. External boundaries and closed gaps

Definition. A point (x,v) in Σ is called *unstable* whenever there is a neighbourhood $\mathcal{U}$ of (x,v) and an integer n_0 such that for all (x_1,v_1) in $\mathcal{U}$, and all $N \geq n_0$, the iterated $(x_N,v_N)=f^N(x_1,v_1)$ satisfies $|x_N| > 2$. ◊

Proposition 1. Energies E corresponding to unstable points are out of the spectrum of H. ◊

Proof. Thanks to eq.(8)&(9), A(N) is the Floquet matrix of the periodic potential V_N of period 4^N and equal to V on the interval $[0,4^N-1]$. $|x_N| > 2$ means that E belongs to a spectral gap of the periodic approximation H_N of H. If (x,v) is unstable there is a neighbourhood $\mathcal{U}$ of E and an integer n_0 such that for any $N \geq n_0$, $\mathcal{U}$ is contained in a gap of H_N. Since H is the strong limit of the H_N's $\mathcal{U}$ is also contained in a gap of H [15]. ◊

From eq.(14) the axis $L_0=\{(x,0);\, x\in\mathbf{R}\}$ and the parabola $L_1=\{(x,x^2-4);\, x\in\mathbf{R}\}$ (see Fig.2) are two f-invariant curves. $f(L_1)$ reduces to the fixpoint (2, 0). These curves separate the plane

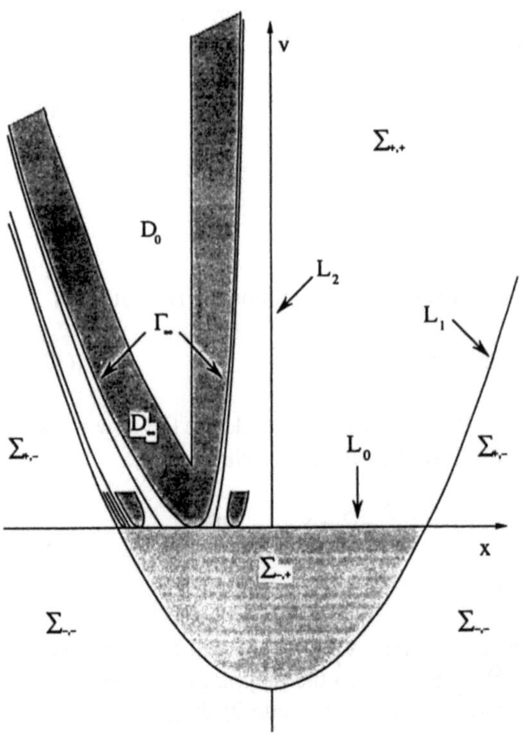

Fig.2- Stability domain in the (x,v)-plane for the dynamical map f coming from transfer matrices. The lines L_0, L_1, L_2, the regions $\Sigma_{\pm,\pm}$ (see §3), the domain D_0, the main gap D_∞ and its boundary Γ_∞ (see §4) are indicated. Some of its first preimages have been sketched. The lines separating them are preimages of the vertical axis L_2.

into four open regions $\Sigma_{\varepsilon,\varepsilon'}=\{(x,v);\ \varepsilon v > 0, \varepsilon' w > 0\}$ where $\varepsilon, \varepsilon'=\pm$. It can be checked directly that the closures of $\Sigma_{\pm,+}$ are f–invariant whereas those of $\Sigma_{\pm,-}$ are exchanged by f. On the other hand the image of the line $L_2=\{(x,v)\ ;\ x=0\}$ is the parabola L_1.

Now $\Sigma'_{\pm,-} = \Sigma_{\pm,-}\setminus L_2$ is made of unstable points. For if $(x, v) \in \Sigma'_{\pm,-}$, its image by f is (x',v') such that $x'=2-w > 2$ and $w'=x^2w < 0$. On the other hand if $x > 2$ and if $w < 0$ we get $x'=2-w > 2$ and $w'=x^2w < 4w < 0$. By continuity of f and by recursion we get the result.

We now remark that the set $L_2\cap\Sigma_{-,-}$ has no real points in its inverse image by f. Therefore there is no way to reach such a set from an initial condition (x_1,v_1) with $v_1 > 0$. In much the same way no point from $\Sigma_{-,+}$ can be reached from such an initial point. Thus starting from (x_1,v_1) with $v_1 > 0$, stability holds only if $w_1 \geq 0$, namely if $x_1^2 \leq 4+v_1$ (Fig.2). Hence from eq.(13) the energy spectrum is contained in the set of energies such that

$$(V^2+1)^{1/2} - 1 \leq |E| \leq (V^2+1)^{1/2} + 1. \tag{15}$$

Moreover the boundaries $E=\pm\{(V^2+1)^{1/2}\pm 1\}$ belong to the spectrum. They correspond to $L_1\cap\Sigma_{+,+}$ namely to $v_1 = x_1^2 - 4$, $w_1 = 0$. Their inverse image is given by $L_2\cap\Sigma_{+,+}$, namely by $x_1=0$. More generally if there is n such that $x_n = \mathrm{Tr}(A(n)) = \mathrm{Tr}(B(n)) = 0$, one gets $A(n)^2 = B(n)^2 = -1$ which implies $A(n+2) = B(n+2) = 1$. It simply means that the corresponding wave functions are periodic of period 2^{n+2}, and that the corresponding energies are "closed gaps" for they are doubly degenerate. In the (x,v) plane these values are represented by curves contained in Σ_{++}, the intercept of which at $v = 0$ being given by the inverse images of the fixpoint $x=2$ by the mapping $x \to x^2-2$. In terms of (E,V) they are given at $V \approx 0$ by

$$E_0(\underline{\sigma},V) = \sigma_{n+1}\sqrt{(2+\sigma_n\sqrt{(2+\ldots + \sigma_1\sqrt{(2+2\sigma_0)})\ldots)}}+O\,(V^2). \qquad \sigma_i = \pm 1. \tag{16}$$

◊

4. The main gap

We now construct a gap inside the region $\Sigma_{+,+}$ which will generate all the other gaps in the spectrum (see Fig.2). The set $D_0 = \{(x,v);\ x<-1,\ w=v+4-x^2>4\}$ is unstable. For if $(x_n,v_n,w_n) = \underline{f}^n(x,v,w)$ is the nth iterated of a point in D_0, we get by recursion $w_n>4^n$, $x_n<-2$ for $n \geq 1$. Let now $\underline{h}^{(-)}$ be the right inverse of $\underline{f}$ defined on $\underline{D} =\{x<0,\ v\geq 0,\ w\geq 0\}$ by $\underline{h}^{(-)}(x,v,w)=(-\{w/(2-x)\}^{1/2}, v/(2-x), 2-x)$. Clearly $\underline{h}^{(-)}(\underline{D})$ is included in $\underline{D}$. If h_- is the corresponding right inverse of f, we define $D_n=h_-(D_{n-1})$ for $n \geq 1$. It is easy to check that D_n is an increasing sequence of open unstable subsets of D and their union D_∞ is an open unstable subset of D, which we will call the *main gap*. The boundary Γ_∞ of D_∞ belongs to the spectrum. If it is parametrized by the mapping $v=x^2-4+\Phi_\infty(2-x)$ a simple calculation shows that Φ_∞ is the unique continuous decreasing function from $(2,\infty)$ into itself such that

$$\Phi_\infty(\Phi_\infty(w)) = \Phi_\infty(w)(w-2)^2. \tag{17}$$

It is tedious but straightforward to prove rigorously that Φ_∞ satisfies the following properties:

(i) Φ_∞ is continuously differentiable, decreasing and convex from $(2,\infty)$ onto itself.

(ii) Φ_∞ admits convergent Laurent expansions around $w = 2$ and $t = \infty$ given by

$$\Phi_\infty(t) = 2 + (t-2)^{-1} - 1/4\,(t-2)^{-2} + 9/32\,(t-2)^{-3} - 1/4\,(t-2)^{-4} + O((t-2)^{-5}), \qquad (18)$$

$$\Phi_\infty(w) = 2(w-2)^{-1} + 1/2 + 5/16(w-2)^2 + 37/256(w-2)^4 + 9/256(w-2)^6 + O((w-2)^8). \qquad (19)$$

(iii) The curve Γ_∞ is f-invariant and the restriction of f on Γ_∞ is conjugate to Φ_∞.

(iv) The only fixpoint of Φ_∞ on $(2,\infty)$ is $w = 3$ and is unstable whereas Φ_∞ admits a unique attracting periodic orbit of period 2 given by $\{2,\infty\}$.

(v) Φ_∞ admits the asymptotic expansion around $w = 3$

$$\Phi_\infty(3-\varepsilon) = 3 + 2\varepsilon + a(\varepsilon)|\varepsilon|^{Ln3/Ln2} + O(|\varepsilon|^2), \qquad a(\varepsilon) = a(-\varepsilon/2) > 0. \qquad (20)$$

where $a(\varepsilon)$ is a continuous bounded function on $[-1,1]\backslash\{0\}$.

(vi) A numerical calculation shows that $a(\varepsilon)$ is approximately

$$a(\varepsilon) \approx 3.8606895 \pm 5.10^{-7} + (5.45 \pm .05)10^{-5} \cos\pi\{\ln_2(\varepsilon) - 1.307 \pm .001\}. \qquad (21)$$

Using eq.(6) we get $b(V) = a((2V)^{Ln4/Ln3})$ in eq.(3) leading to eq.(4). ◊

The asymptotic behavior of wave functions corresponding to energies on the gap boundaries is given by the asymptotic behaviour of the sequence x_n on the curve Γ_∞. By (iii) it is entirely dominated by the iteration of Φ_∞. Using (ii) & (iv) it follows that if $-1<x_1<0$, x_{2n} diverges like $\exp\{2^n\}$ and $x_{2n+1} \approx 2/x_{2n}^2$. Since A(2n) is the transfer matrix for an interval of length $L_{2n} = 2^{2n}$ we will have $x_{2n} \approx \exp\{\sqrt{L_{2n}}\}$. A tedious calculation permits one to show rigorously that the wave function $\psi(L_{2n})$ itself behaves like $\exp\{\sqrt{L_{2n}}\}$.

5. Spectrum and gap labelling

The maps $\underline{h}^{(\pm)}(x,v,w) = (\pm\{w/(2-x)\}^{1/2}, v/(2-x), 2-x)$ are the two right inverses of $\underline{f}$. More generally for each finite sequence $\underline{\sigma} = (\sigma_1,\sigma_2,\dots,\sigma_r)$ of signs, we get the iterated inverse $\underline{h}^{(\underline{\sigma})} = \underline{h}^{(\sigma_1)}o\underline{h}^{(\sigma_2)}o\dots o\underline{h}^{(\sigma_r)}$ (it is not defined everywhere). Taking the preimages $\underline{h}^{(\underline{\sigma})}(D_\infty)$ in $\Sigma_{+,+}$ we get an infinite family of gaps. Since $\underline{f}$ reduces to the mapping $h(x) = x^2-2$ on the axis $\{v = 0\}$ the bottoms of these gaps are precisely given by (see §I(v))

$$E(\underline{\sigma}) = \sigma_n\sqrt{(2+\sigma_{n-1}\sqrt{(2+\dots+\sigma_1\sqrt{(2+\sigma_0)})\dots)}}. \qquad \sigma_i = \pm 1. \qquad (22)$$

These points are dense in the interval [-2,2]. Since gap boundaries are continuous with respect to V, the gaps themselves are dense in the interval $[-(v+2)^{1/2}, (v+2)^{1/2}]$ and the spectrum is contained in a Cantor set. On the other hand, the preimages $\underline{h}^{(\underline{\sigma})}(L_1)$ of the parabola $\{w=0\}$ are all contained in the spectrum. They cut the horizontal axis $\{v = 0\}$ at the preimages of x=2 under h. One checks that the axis x=0 is the first preimage of L_1 and is surrounded by the two first preimages D_∞ and $-D_\infty$ of D_∞. Recursively we see that for a given r the bottoms of the gaps corresponding to the possible values of $\underline{\sigma} = (\sigma_1,\sigma_2,\dots,\sigma_r)$ are separated by a corresponding number of preimages of the fixpoint x=2.

It is known [20] that the integrated density of states N(E) is constant on gaps and continuous with respect to V even at V=0. It is therefore sufficient to compute it at V=0 and for energies at which gaps open. For V = 0, it is given by $N(E) = 1/\pi \cos^{-1}\{-E/2\}$ and the middle gap corresponds to E=0 namely to N(E=0)=1/2. On the other hand if $E = 2\cos\pi(1-N)$ then $\pm\sqrt{(2+E)} = 2\cos\pi(1-(1\pm N)/2)$. Since the other gaps open at energies

$E(\underline{\sigma})=\sigma_n\sqrt{(2+\sigma_{n-1}\sqrt{(2+\ldots\sqrt{(2+\sigma_0)}))}}$ it follows that

$$N(E(\underline{\sigma})) = 1/2 + \sigma_n/4 +\ldots+ \sigma_n\sigma_{n-1}\ldots\sigma_1/2^{n+1} + \sigma_n\sigma_{n-1}\ldots\sigma_1\sigma_0/3.2^{n+1}. \qquad (23)$$

which is equal to $k/3.2^{n+1}$ where k is not a multiple of 3 and $0 \leq k \leq 3.2^{n+1}$ (Fig.1).◊

Remark. With some work one can actually show , using the method provided by J.Moser [23], that gaps will open at the values $k/2^N$ under a perturbation of the Thue-Morse potential by a generic limit periodic potential with frequency module generated by the powers of 1/2. ◊

6. There are no other gaps: a proof using K-theory

The previous arguments do not prove that we have found all gaps. It could happen that other type of gaps exist corresponding to other labels . We intend to sketch a proof here that it is not so. Our argument is based upon the analysis in terms of C*-algebras described in [22].

We recall that there is a natural dynamical system (Ω, ϕ) called the "hull of H" associated with our hamiltonian H where Ω is a compact metrisable space and ϕ a homeomorphism of Ω. The space Ω is abstractly obtained as the strong closure of the family $\{H_a; a\in \mathbf{Z}\}$ of operators obtained by translating H in $\mathcal{B}\{l^2(\mathbf{Z})\}$ and the map ϕ is nothing but the translation by one. Since the kinetic part, namely the free discrete laplacian on **Z**, is translation invariant, the hull is homeomorphic to the closure of the family $\{V(.-a); a\in \mathbf{Z}\}$ in the metrisable compact space $\{-V,V\}^{\mathbf{Z}}$ where the homeomorphism is still the translation by one. This set is itself the hull of the Thue-Morse sequence. It has been proved [24] that this hull is described as follows: Ω is homeomorphic to the cartesian product $\mathbf{Z}[1/2]\times\{-1,1\}$ of the compact completely disconnected ring $\mathbf{Z}[1/2]$ of dyadic integers and of the cyclic group $\{-1,1\}$ of order 2. The map ϕ is then given by

$$\phi(z, \sigma) = (z+1, (-)^{N(z)}\sigma), \qquad (24)$$

where if $z = \varepsilon_0 + \varepsilon_1 2 +\ldots+ \varepsilon_k 2^k +\ldots$ (where $\varepsilon_k=0$ or 1 for all $k \geq 0$) is the dyadic expansion of z, N(z) is its order (mod. 2) namely the smallest value of k such that $\varepsilon_k\neq 0$. We will set N(0)=0(mod.2). By the very definition of the Thue-Morse sequence, we get $V(n)=(-)^{N(n)}V(n-1)$ namely , since $N(-n) = N(n)$,

$$V(n) = f(\phi^{-n}(0,1)), \qquad f(z,\sigma) = V\sigma. \qquad (25)$$

Remark. Had we chosen N(0)=1(mod.2) the potential V(n) would have been odd under the symetry n->(-n-1). On the other hand this dynamical system is minimal (namely every orbit is dense) and uniquely ergodic (namely it admits a unique invariant measure) [14]. ◊

The C*-algebra generated by the family $\{H_a; a\in \mathbf{Z}\}$ is then isomorphic to the cross-product $C(\Omega)\times_\phi\mathbf{Z}$ of the C*-algebra $C(\Omega)$ of continuous functions on the hull, by the action of **Z** through ϕ [25]. An element A of $C(\Omega)\times_\phi\mathbf{Z}$ can be represented by a kernel $A(z,\sigma;n)$ depending continuously upon $(z,\sigma)\in\Omega$ for each $n\in\mathbf{Z}$. The corresponding operator acting on $l^2(\mathbf{Z})$ is given by the matrix

$$< n \mid \pi_{(z,\sigma)}(A) \mid n' > = A(\phi^{-n}(z,\sigma); n'-n). \tag{26}$$

It is then clear that H is an element of this C*-algebra . Moreover this algebra has a trace τ given by the trace per unit length, namely

$$\tau(A) = \int_{\Omega} d\mu\, A(z,\sigma; 0) = \lim_{N\to\infty} 1/(2N+1)\, \mathrm{Tr}(\chi_{[-N,N]}\pi_{(z,\sigma)}(A)). \tag{27}$$

Here the limit is uniform in (z,σ), $\chi_{[-N,N]}$ is the projection onto the subspace $l^2([-N,N])$ of $l^2(\mathbf{Z})$, and μ is the unique ergodic invariant probability measure on Ω.

We have shown in [22] that the integrated density of state (IDS) of a self-adjoint element H of $\mathcal{C}(\Omega)\times_\phi \mathbf{Z}$ is given by Shubin's formula

$$N(E) = \tau(\chi(H \le E)), \tag{28}$$

where $\chi(H\le E)$ represents the eigenprojection of H onto energies smaller than or equal to E. If H is a discrete 1D Schrödinger operator, the IDS is a continuous non-decreasing function of the energy E. Whenever E belongs to a spectral gap of H the projection $P=\chi(H\le E)$ belongs to the C*-algebra. The value of N(E) is then constant whenever E varies in the gap. Moreover it will not change if P is replaced by an equivalent projection Q, namely such that there is U in the C*-algebra satisfying $UU^*= Q$, $U^*U = P$. Thus N(E) depends only upon the equivalent class [P] of P. These equivalent classes, after some standard manipulations [26], are the elements of the countable abelian group $K_0(\mathcal{C}(\Omega)\times_\phi \mathbf{Z})$. Standard methods of K-theory permit one to show the following result [27]:

Theorem. (1°) If P is a projection of $\mathcal{C}(\Omega)\times_\phi \mathbf{Z}$ its trace takes on values in the set $S\cap[0,1]$; S is the set of real numbers given by $\int_\Omega d\mu\, f(z,\sigma)$ where f runs into the set of integer valued continuous functions on Ω .
(2°) The set S coincides with $\mathbf{Z}(1/2)/3 = \{k/3.2^N ; k\in \mathbf{Z}, N\in \mathbf{N}\}$. ◊

Sketch of the proof. (1°) is a standard result valid for the cross-product of a compact completely disconnected space by **Z** [see 28].
(2°) is a property of the Thue-Morse sequence, and similar results can be obtained for a large class of substitution [14,28]. We will refer the reader to [14] for a complete description of the recipe given here.

We have said that the space Ω can be seen as a compact subspace of $\{a,b\}^{\mathbf{N}}$, namely the closure of the family $\underline{u}^{(n)}=u_n u_{n+1}\cdots u_{n+k}\cdots$ arising from the translations of the Thue-Morse sequence $\underline{u}$. Any continuous integer valued function f on Ω is a cylindrical function, namely it depends only upon a finite number of components of $\underline{x}\in\Omega$. Therefore it is a finite linear combination with integer coefficients of characteristic functions of cylinder sets of the form

$$[\underline{b}] = \{ \underline{x}\in\Omega ; x_0=b_0,\ldots,x_j=b_j\} \text{ if } \underline{b} \text{ is the finite word } \underline{b} = b_0\ldots b_j.$$

Using (1°) it is therefore enough to compute the frequency of apparition $\mu([\underline{b}])$ of each finite word $\underline{b}$ in the Thue-Morse sequence $\underline{u}$. Thank to the unique ergodicity of our system this is given by $\mu([\underline{b}]) = \lim_{N\to\infty} 1/(N+1)\, L_{\underline{b}}(u_n u_{n+1}\cdots u_{n+N})$ which is uniform in n. Here $L_{\underline{b}}(\underline{c})$ represents the occurrence number of the word $\underline{b}$ in the word $\underline{c}$. This number can be computed using a specific recipe. For the Thue-Morse sequence it is given as follows:

- $A = \{a,b\}$ is the alphabet of our substitution. We let A_j be the set of words of length j that we consider as a new alphabet.

- given the substitution $\sigma(a) = ab$, $\sigma(b) = ba$, we define on A_j the substitution σ_j by the following rule: if $\underline{a} \in A_j$ and $\sigma(\underline{a})=b_0b_1\ldots b_{2j-1}$ then $\sigma_j(\underline{a})=(b_0\ldots b_{j-1})(b_1\ldots b_j)$.

- let M_j be the occurrence matrix of σ_j defined by $M_j(\underline{a},\underline{b})=L_{\underline{a}}(\sigma_j(\underline{b}))$ if $\underline{a}, \underline{b} \in A_j$. One can show that its highest eigenvalue is equal to 2, namely the length of the substitution σ. By Perron-Frobenius's theorem the corresponding eigenvector V_j has positive coordinates. We normalize it by $\Sigma_{\underline{a}} V_j(\underline{a})=1$.

- it can be shown that V_j $(j \geq 3)$ is given in terms of $V_2=(1/6,1/3,1/3,1/6)$ through an integer valued rectangular matrix $M_{j,2,p}$ (with $j \leq 1+2^p$) by the relation

$$V_j=2^{-p} M_{j,2,p} V_2 \, . \tag{29}$$

- the occurrence number $\mu([\underline{b}])$ of the word $\underline{b}$ in $\underline{u}$ is then given by $V_{|\underline{b}|}(\underline{b})$ where $|\underline{b}|$ is the length of $\underline{b}$ giving $\mu([\underline{b}]) = k/3.2^N$ for some $k,N \in \mathbf{N}$. Thus if f is a continuous integer valued function on Ω one also gets $\int_\Omega d\mu \, f = k/3.2^N$ for some $N \in \mathbf{N}$ and $k \in \mathbf{Z}$.◊

Acknowledgements. I would like to thank R. Dumont for pointing out the Thue-Morse sequence, and G.Janssen for the references [9,11] and F.Delyon for his calculation [13]. This work has been done in 1988 partially at the Centre de Physique Théorique of Marseille, and at the Mathematics Department at Caltech under the USNSF grant DMS-8416049.

REFERENCES

[1] D.SHECHTMAN, I.BLECH, D.GRATIAS, J.V.CAHN, Metallic Phase with Long-Range Orientational Order and No Translational Symmetry, *Phys. Rev. Letter*, 53, (1984), 1951-1953.

[2] International Workshop on Aperiodic Crystals, D. Gratias, L. Michel, Eds *J. de Phys.* C3, 47, (1986), 181-187.

[3] A.THUE, Über die gegenseitige Lage gleicher Teile gewisser Zeichenreichen, *Videnskabsselskabets Skrifter I Mat. nat.*, *Christiania*, (1906).

[4] M.MORSE, Recurrent geodesics on a surface of negative curvature, *Trans. Amer. Math. Soc.*, 22, (1921), 84-100.

[5] K.MAHLER, On the translation properties of a simple class of arithmetical functions, *J. Math. and Phys.*, 6, (1927), 158-163.

[6] S.KAKUTANI, Strictly ergodic symbolic dynamical systems, *Proc. of the 6th Berkeley symposium on Mathematical Statistics and Probability*, vol.2, Univ. of California Press (1972).

[7] G.CHRISTOL, T.KAMAE, M.MENDES-FRANCE, G.RAUZY, Suites algébriques, automates et substitutions, *Bull. Soc. Math. de France*, 108, (1980), 401-419.

[8] R.MERLIN, K.BAJEMA, J.NAGLE, K.PLOOG, Raman scattering by acoustic phonons and structural properties of Fibonacci, Thue-Morse and random superlattices, *J. de Phys.*, C-5, 48, (1987), 503-506.

[9] F.AXEL, J.P.ALLOUCHE, M.KLEMAN, M.MENDES-FRANCE, J.PEYRIERE, Vibrational modes in a one-dimensional "quasi alloy", the Morse case, *J. de Phys.* C3, 47, (1986), 181-187.

[10] R.RIKLUND, M.SEVERIN, YOUYAN LIU, The Thue-Morse aperiodic crystal, a link between the Fibonacci quasicrystal and the periodic crystal, *Int. J. Mod. Phys.*, B1, (1987), 121-132.
[11] F.AXEL , J.PEYRIERE , *C.R.Acad. Sci. Paris*, 306, serie II, (1988), 179-182.
[12] J.M.LUCK, Cantor spectra and scaling of gap widths in deterministic aperiodic systems, *Phys. Rev.*, to appear.
[13] F.DELYON, Private communication.
[14] M.QUEFFELEC, Substitution Dynamical Systems-Spectral Analysis, *Lecture Notes in Mathematics* , 1294, (1987).
[15] M.REED, B.SIMON, Methods of Modern Mathematical Physics, vol.I, Acad. Press, (1972).
[16] J.BELLISSARD, D.BESSIS, P.MOUSSA , Chaotic States of Almost Periodic Schrödinger Operator, *Phys. Rev. Lett.*, 49, (1982), 701-704.
[17] L.P.KADANOFF, M.KOHMOTO, C.TANG, Localization problem in one dimension. Mapping and escape, *Phys. Rev. Lett.*, 50, (1983), 1870-1873.
[18] S.OSTLUNDT, R.PANDIT, D.RAND, H.SCHELLNHUBER, E.D.SIGGIA, One dimensional Schrödinger equation with an almost periodic potential, *Phys. Rev. Lett.*, 50, (1983), 1873-1875.
[19] R.RAMMAL , Spectrum of harmonic excitations on a fractal, *J. de Physique*, 45, (1984), 191-206.
[20] J.BELLISSARD, Almost periodicity in Solid State Physics and C*-Algebras,*The Harald Bohr Centenary* , Berg C. & Flugede F. (Eds), Royal Danish Acad. Science, Copenhagen (1989).
[21] B.SIMON , Fifteen Problems in Mathematical Physics, *Perspective in Mathematics , Anniversary of Oberwolfach*, (1984) , Birkhäuser, Springer-Verlag.
[22] J.BELLISSARD , K-Theory of C*-Algebras in Solid State Physics , *Lecture Notes in Physics* , 257, (1986), 99-156.
[23] J.MOSER, An example of a Schrödinger operator with an almost periodic potential and a nowhere dense spectrum, *Commun. Math. Helv.* , 56, (1981), 198.
[24] G.RAUZY , private communication.
[25] G.PEDERSEN, C*-Algebras and their automorphism group, Academic Press, London, New York, (1979).
[26] B.BLACKADAR, K-Theory for Operator Algebras, Springer-Verlag , (1986).
[27] I thank J.B.Bost for providing a simple proof of it which will be reproduced in [28]
[28] J. BELLISSARD, Théorie des nombres et Physique, les Houches 1989, in preparation.

On the Non-commutative Torus of Real Dimension Two

P. Cohen

Laboratoire de Mathématiques Fondamentales, Equipe de Recherche Associée au CNRS 747, UER 48, Université Pierre et Marie Curie, 4, Place Jussieu, F-75230 Paris Cedex 05, France

Abstract. The aim of the following notes is to recall how, according to the theory of Alain Connes, one may introduce a differentiable structure on a dense sub-algebra of the irrational rotation C*-algebra introduced in the lectures of Jean Bellissard. This sub-algebra may be viewed as a non-commutative generalisation of the classical torus of real dimension two. We compute the variation of the value at the origin of the meromorphic continuation of the zeta-function of the Laplacian on this algebra as the metric on the non-commutative torus varies within a given conformal class. This manifests the involved intervention into pseudo-differential computations of the non-commutativity, even for the simplest of non-commutative differential objects.

An excellent reference for the type of calculation in view is an article of Parker and Rosenberg [1]. Recall that for a classical Riemann surface Σ with metric g, to the Laplacian $\Box_g = d^*d$, where d is the de Rham differential operator acting on the commutative algebra of smooth functions on the Riemann surface, one associates the zeta function

$$\zeta(s) = \sum \lambda_j{}^{-s}, \quad \mathrm{Re}(s) > 1 ,$$

where the summation is over the non-zero eigenvalues λ_j of $\Box_g$. The meromorphic continuation of $\zeta(s)$ to $s = 0$ gives the important information

$$\zeta(0) = (1/24\pi) \int_\Sigma S - \mathrm{Card}\{j|\lambda_j = 0\} = (i/12)\, c_1(\Sigma) ,$$

where S is the scalar curvature and

$$c_1(\Sigma) = (1/2\pi i) \int_\Sigma \partial\bar{\partial} \log(g) \qquad (i = \sqrt{-1})$$

the first Chern number (= Euler-Poincaré characteristic); this vanishes when Σ is the classical 2-torus $\mathbf{R}^2/\mathbf{Z}^2$, for example, and is an invariant within the conformal class of the metric, that is, under the transformation $g \to e^f g$ for f a smooth real valued function on Σ. Moreover recall that one definition of the

Springer Proceedings in Physics, Vol. 47 Number Theory and Physics
Editors: J.M. Luck · P. Moussa · M. Waldschmidt

determinant of the Laplacian $\Box_g$ is given by

$$\log \det \Box_g = -\zeta'(0) .$$

Now fix a real irrational number θ. The C*-algebra A_θ embodies into a single algebra the dynamical system given by an irrational rotation of the circle $\mathbf{S}^1$. That is, one has both an action of $\mathbf{Z}$ and of the algebra $C(\mathbf{S}^1)$ of continuous functions on $\mathbf{S}^1$ given by

$$[1]\, f(s) = f(s-\theta) \quad \text{(irrational rotation)}$$

$$[e^{2\pi i s}]\, f(s) = e^{2\pi i s} f(s) \quad \text{(regular representation)}$$

for all f in $C(\mathbf{S}^1)$. These don't commute; instead, one has

$$[e^{2\pi i s}]\,[1] = e^{2\pi i \theta}[1]\,[e^{2\pi i s}] .$$

We represent these actions on a Hilbert space by passing to $\mathcal{H} = \mathrm{L}^2(\mathbf{R})$, the completion of the algebra of functions on $\mathbf{R}$ of compact support with respect to the inner product

$$< f, g > = \int_{\mathbf{R}} f(s) g(s) ds ,$$

which contains the subspace $\mathcal{S}(\mathbf{R})$ of functions of rapid decay. One defines the following operators on elements $\xi = \xi(s)$ of $\mathcal{S}(\mathbf{R})$:

$$(U\xi)(s) = \xi(s-\theta)$$

$$(V\xi)(s) = e^{2\pi i s}\xi(s)$$

which satisfy

$$VU = e^{2\pi i \theta} UV , \; U^* = U^{-1} , \; V^* = V^{-1} .$$

The norm closure of this algebra is the C*-algebra A_θ. Notice that if we replace θ by zero, U by $e^{2\pi i x}$ and V by $e^{2\pi i y}$ in the above definitions we recover the algebra of continuous functions on $\mathbf{R}^2/\mathbf{Z}^2$.

If one wishes to express the elements of A_θ as certain series

$$\sum a(n,m) U^n V^m , \; a(n,m) \in \mathbf{C} ,$$

where the summation is over the elements (n, m) of $\mathbf{Z}^2$, one finds that the C* or continuity condition is hard to express as a condition on the coefficients. On the other hand, the smoothness condition is easy to control. To see this, we introduce another dynamical system given by the action of

$$\mathbf{T}^2 = \{z \in \mathbf{C} |\ |z| = 1\}^2$$

on A_θ by the 1-parameter groups of automorphisms $\{\alpha_s\}, \{\beta_t\}$ given by

$$\alpha_s(U) = \exp(2\pi i s)U \ , \ \alpha_s(V) = V \ , \ (s \in \mathbf{R}) \ ,$$
$$\beta_t(U) = U \ , \ \beta_t(V) = \exp(2\pi i t)V \ , \ (t \in \mathbf{R}) \ .$$

We define the sub-algebra $A_\theta{}^\infty$ of smooth elements of A_θ to be those x in A_θ such that the mapping

$$\mathbf{R}^2 \rightarrow A_\theta \ , \ (s,t) \rightarrow \alpha_s\beta_t(x)$$

between Banach spaces is smooth. Expressed as a condition on the coefficients, this imposes that they be of rapid decay, that is, that $\{|n|^k|m|^q\ |a(n,m)|\}$ be bounded for any positive k, q.

The derivations associated to the above groups of automorphisms are given by

$$\delta_1(U) = 2\pi i U \ , \ \delta_1(V) = 0 \ ,$$
$$\delta_2(U) = 0 \ , \ \delta_2(V) = 2\pi i V \ ,$$

and are the analogues of the differential operators $\partial/\partial x$ and $\partial/\partial y$ on the smooth functions on $\mathbf{R}^2/\mathbf{Z}^2$. One also has inner derivations arising from commutators which are not seen in the commutative case. These are the derivations associated to the 1-parameter families $\{\sigma_t\}, t \in \mathbf{R}$ of inner automorphisms

$$\sigma_t(x) = e^{ift}xe^{-ift}$$

for $f = f^*$ a non-constant self-adjoint element of $A_\theta{}^\infty$.

The above defines a dynamical system $(A_\theta{}^\infty, \mathbf{R}, \sigma)$, so that, although we shall not pursue this here, one may in fact work with the crossed product algebra $B(\theta, f) = A_\theta{}^\infty \times_\sigma \mathbf{R}$.

The derivation corresponding to the above group is given by $i \log \Delta$, where

$$\Delta(x) = e^{-f}xe^{f}$$

and

$$(\log \Delta)(x) = [f,\ x]\ ,\ x \in A_\theta{}^\infty\ .$$

As θ is supposed irrational, there is a unique trace τ_0 on A_θ determined by the orthogonality properties

$$\tau_0(U^n V^m) = 0 \text{ if } (n,m) \neq (0,0)\ ,\ \text{and } \tau_0(1) = 1\ .$$

We can construct a Hilbert space $\mathcal{H}_0$ from A_θ by completing with respect to the inner product

$$< a, b > = \tau_0(b^* a)\ .$$

Using the derivations δ_1, δ_2 we can construct a complex structure by defining

$$\partial = \delta_1 + i\delta_2\ ,\ \partial^* = \delta_1 - i\delta_2$$

where (extending ∂, ∂^* to unbounded operators on $\mathcal{H}_0$) ∂^* is the adjoint of ∂ with respect to the metric defined by τ_0.

In this non-commutative context, to vary the metric in its conformal class has the following meaning: for the given trace τ_0 above, one looks at the family of states φ_f , $f = f^*$ in $A_\theta{}^\infty$, defined on $A_\theta{}^\infty$ by

$$\varphi(a) = \tau_0(ae^{-f})\ ,\quad a \in A_\theta{}^\infty\ .$$

Note that, whereas for τ_0 we have the relation

$$\tau_0(b^* a) = \tau_0(ab^*)\ ,$$

for φ we have

$$\varphi(ab^*) = \varphi(b^* e^{-f} a e^{f}) = \varphi\left(b^* \sigma_{\sqrt{-1}}(a)\right).$$

If we denote by $\mathcal{H}_f$ the Hilbert space obtained by the same construction as used for $\mathcal{H}_0$ from $A_\theta{}^\infty$ and φ_f, and let the operator

$$\partial\ :\ \mathcal{H}_f \to \mathcal{H}_0$$

be as before, but now with domain $\mathcal{H}_f$, we have the corresponding adjoint:

$$\partial_f = \mathrm{R}(e^f)\partial^*$$

where $\mathrm{R}(\cdot)$ denotes multiplication on the right. Introduce the Laplacian

$$\Box_f = \partial_f^* \partial = \mathrm{R}(k)\ \Box_0\ \mathrm{R}(k)$$

where $k = e^{f/2}$. This is of the form T^*T with $T = \partial \cdot \mathrm{R}(k)$.

One works with the ordinary trace of the operator

$$e^{-tT^*T}\ ,\ t>0\ ,$$

and its Mellin transform. Namely, the relevant zeta function to study is the one defined for $\mathrm{Re}(s) > 1$ by

$$\begin{aligned}\zeta(s) &= \frac{1}{\Gamma(s)}\int_0^\infty \mathrm{Trace}^+\left(e^{-tT^*T}\right)t^s\frac{dt}{t}\\ &= \mathrm{Trace}\left((T^*T)^{-s}\right)\ ,\end{aligned}$$

where

$$\begin{aligned}T^*T &= \mathrm{R}(k)(\delta_1^2+\delta_2^2)\mathrm{R}(k)\\ \mathrm{Trace}^+\left(e^{-tT^*T}\right) &= \mathrm{Trace}\left(e^{-tT^*T}\right) - \mathrm{Dim\ Ker}\ (T)\ ,\end{aligned}$$

and by meromorphic continuation for other values of s. We want to study $\zeta(0)$ and $\zeta'(0)$ as k varies; as in the commutative case, $\zeta(0)$ captures the curvature whereas $\zeta'(0)$ gives $-\log\det\Box_f$.

The calculation is achieved by using Cauchy's formula:

$$e^{-tT^*T} = \frac{-1}{2\pi i}\int_{\mathrm{C}} e^{-t\lambda}(T^*T-\lambda)^{-1}d\lambda\ ,\ t>0\ ,$$

where the path C encircles the positive real axis in the anti-clockwise direction, approximating $(T^*T-\lambda)^{-1}$ sufficiently well, following for example the treatment of Gilkey [2], and making the necessary modifications relevant to the non-commutative case. The techniques involve passing by Fourier transformation to the algebra of symbols which here are functions on $\mathbf{R}^2$ with values in the non-commutative algebra $A_\theta{}^\infty$. For example, the principal symbol of $(\lambda - T^*T)^{-1}$, $\lambda \in \mathbf{C}-\mathbf{R}^+$, exists and equals $(\lambda - k^2|\xi|^2)^{-1}$, $\xi \in \mathbf{R}^2$. Two features in particular emerge during the calculation. One is the complication introduced by the non-commutativity; for example, multiplication on the right by $(k^2|\xi|^2+1)^{-1}$ is the same as multiplication on the left by $(k^2|\xi|^2\Delta+1)^{-1}$, where $\Delta(x) = k^{-2}xk^2$. The other is that there is no angular dependence: the formulae depend only on $|\xi|$ (this explains the remark made earlier about the appearance of the dynamical system $(A_\theta{}^\infty, \mathbf{R}, \sigma)$, where $\sigma_t(x) = k^{2it}xk^{-2it}$).

These calculations are the subject of joint work with A. Connes [3]. We quote here the result for $\zeta(0)$, which is zero in the commutative case, but here gives a result of some complexity:

$$\zeta(0) = \tau_0\left(h(\theta,k)\right)$$

where

$$\begin{aligned} h(\theta,k) = & -(2\pi/3)(k^{-1}\delta_j(k))(\delta_j(k)k^{-1}) + (2\pi/6)(k^{-1}\delta_j^2(k)) \\ & -(2\pi)\mathcal{D}_1(k^{-1}\delta_j(k))(\delta_j(k)k^{-1}) \\ & -(4\pi)\mathcal{D}_2(1+\Delta^{1/2})(k^{-1}\delta_j(k))(\delta_j(k)k^{-1}) \\ & -(2\pi)\mathcal{D}_3(1+\Delta^{1/2})^2(k^{-1}\delta_j(k))(\delta_j(k)k^{-1}) \, . \end{aligned}$$

Here summation over $j = 1,2$ for repeated indices is understood, and $\mathcal{D}_m$ $(m \geq 1)$ stands for the modified logarithm

$$\mathcal{D}_m = (\Delta - 1)^{-(m+1)} \left(\log \Delta - \sum_{j=1}^{m} (-1)^{(j+1)} \frac{(\Delta - 1)^j}{j} \right) .$$

The result for $\zeta'(0)$ is much more complicated and will not be stated here.

References

[1] T. Parker and S. Rosenberg, *Invariants of Conformal Laplacians*, Journal of Differential Geometry, vol. 25, 199–222, 1987.

[2] P. B. Gilkey, *The Index Theorem and the Heat Equation*, Publish or Perish, Boston, 1974.

[3] P. Cohen and A. Connes, in preparation.

Topological Invariants Associated with Quasi-Periodic Quantum Hamiltonians

H. Kunz
Institut de Physique Théorique, Ecole Polytechnique Fédérale de Lausanne, P.H.B., Ecublens, CH-1015 Lausanne, Switzerland

Abstract. It is shown that the integers which label the gaps of a quasi-periodic Schrödinger operator are Chern numbers of certain unitary vector bundles, and describe physically adiabatic charge transport in these systems.

Apparently, the first examples of a topological invariant associated with a Schrödinger equation were discovered independently by a mathematician, S. Novikov [1], and a physicist, D. Thouless [2]. The Schrödinger equation considered was describing the motion of a particle, confined to a plane, under the influence of a magnetic field and a periodic potential. In this case, the topological invariant corresponds to a physical quantity, measurable with a very high degree of accuracy, the Hall resistance.

We would like here to discuss analogous phenomena for the Schrödinger equation describing the motion of a particle in a quasi-periodic potential.

Some time ago, Johnson and Moser [3] proved the following theorem : Consider the hamiltonian

$$H = -\left(\frac{1}{2}\right) \frac{d^2}{dx^2} + V(x) , \tag{1}$$

where $V(x)$ is a quasi-periodic potential

$$V(x) = \sum_{n \in \mathbf{Z}^m} a(n) \exp(2\pi i x \, (n, \omega)) , \tag{2}$$

the frequencies ω being rationally independent. Then if the energy μ is in a gap of H, i.e. $\mu \in \rho(H)$, the so-called integrated density of states $D(\mu)$ is in the frequency module of V, i.e.,

$$D(\mu) = \sum_{j=1}^{m} N_j \omega_j , \tag{3}$$

with $N_j \in \mathbf{Z}$.

Such a result was also proved and generalised to the multidimensional case by Bellissard et al. [4], using C*-algebra techniques.

It seems natural to ask the following question : can we identify individually each of the integers N_j and what is their physical meaning? The answer, as we shall see, is the following : these integers are topological invariants (Chern numbers of a certain unitary vector bundle). Physically they can be interpreted as the number of electric charges transported, in response to an adiabatic variation of some parameters in the hamiltonian. Such a point of view leads in fact to a new proof of the Johnson-Moser theorem and its extensions to more than one dimension.

We will need an extended version of the adiabatic theorem in quantum mechanics, which is the following : Let $H(t/T)$ be a time dependent hamiltonian with $0 \leq t \leq T$. The adiabatic limit will correspond to T going to infinity. Assume that the energy μ is in a gap of $H(s)$ for all $s \in [0,1]$ and consider the following spectral projector:

$$P(s) = E_{H(s)}(-\infty, \mu) . \tag{4}$$

We want to look at the behaviour of the density matrix $r(t)$, which is the solution of the equation

$$i\,\dot{r} = \left[H(\frac{t}{T}), r \right] , \tag{5}$$

with $r(0) = P(0)$. Then if $\dot{H}(0) = 0$, the adiabatic theorem tells us that

$$r(t) = P(\frac{t}{T}) + \frac{1}{T} \oint_\Gamma \frac{dz}{2\pi} \dot{G}_z(\frac{t}{T}) G_z(\frac{t}{T}) + O(\frac{1}{T^2}) \tag{6}$$

where

$$G_Z(s) = (z - H(s))^{-1} , \tag{7}$$

Γ being a contour in the z plane, encircling the origin and going through the point μ on the real axis.

As Berry [5] taught us, interesting topological effects can be expected if, in the adiabatic process, the parameters on which the hamiltonian depends make a loop, so that $H(0) = H(1)$.

We consider now the following class of hamiltonians, on $L^2(\mathbf{R}^d, d^d x)$

$$H = -\frac{1}{2}\Delta + V_\varphi(x) , \tag{8}$$

with the quasi-periodic potential

$$V_\varphi = \sum_{n \in \mathbf{Z}^m} a(n) \exp\left[2\pi i(n, \Omega x) + 2\pi i(n, \varphi)\right] , \tag{9}$$

satisfying the conditions :

1) The $m \times d$ matrix Ω is such that if $n \in \mathbf{Z}^m$, $\Omega^{\mathrm{T}} n = 0$ implies $n = 0$

2) $a^*(n) = a(-n)$

3) $\sum_{n\in\mathbf{Z}^m} |n|^3|a(n)| \leq \infty$, where $|n| = \sup_i |n_i|$.

When the energy μ is in a gap, the adiabatic process we will consider is the following : increase the phase φ_j by 1. In this way we make a loop in the parameter space. More precisely, let

$$\varphi_j(\frac{t}{T}) = \varphi_j(0) + \psi(\frac{t}{T}) , \tag{10}$$

with $\dot{\psi}(0) = 0$, $\psi(0) = 0$ and $\psi(1) = 1$.

This will generate a current in the α direction and a corresponding charge transport, defined by

$$Q(j\alpha) = \lim_{T\to\infty} \int_0^T M[p_\alpha(r(t) - r(0))]\, dt , \tag{11}$$

where

$$p_\alpha = \frac{1}{i}\frac{\partial}{\partial x_\alpha} , \tag{12}$$

and for suitable operators A of kernel $A(x, y)$ the average $M(A)$ is defined by

$$M(A) = \int_{\mathbf{T}^m} d\varphi\, A_\varphi(0,0) . \tag{13}$$

The adiabatic theorem (6), then gives the following expression for $Q(j\alpha)$, with $Q(j\alpha) = -Q(\alpha j)$:

$$Q(\alpha j) = iM(P[P_\alpha, P_j]) , \tag{14}$$

where

$$P = E_H(-\infty, \mu) , \tag{15}$$

$$P_\alpha = \oint_\Gamma \frac{dz}{2\pi i} G_z p_\alpha G_z , \tag{16}$$

$$P_j = \frac{\partial P}{\partial \varphi_j} , j = 1, \ldots, m . \tag{17}$$

At this stage it appears useful, when $d \geq 2$, to introduce the following quantities:

$$Q(\alpha j\beta k) = \frac{i^2}{2!}\sum_\pi (-1)^\pi M(PP_{\pi(\alpha)}P_{\pi(j)}P_{\pi(\beta)}P_{\pi(k)}) , \tag{18}$$

$$Q(\alpha j\beta k\gamma l) = \frac{i^3}{3!}\sum_\pi (-1)^\pi M(PP_{\pi(\alpha)}P_{\pi(j)}P_{\pi(\beta)}P_{\pi(k)}P_{\pi(\gamma)}P_{\pi(l)}) , \tag{19}$$

π denoting a permutation of the indices appearing in the corresponding Q.

In this context, the integrated density of states $D(\mu)$ is simply given by

$$D(\mu) = M(P) . \tag{20}$$

We can prove the following theorem [6] :

Theorem If $\mu \in \rho(H)$, then :

1) In one dimension, charge transport is quantised, i.e. $Q(1j) \in \mathbf{Z}$, and $Q(1j)$ is the 1st Chern number of a unitary vector bundle on a 2-torus. Moreover, the integrated density of states is given by

$$D(\mu) = \sum_j \Omega(j1)\, Q(1j)$$

2) In two dimensions, charge transport is weakly quantised, i. e. it is given by

$$Q(\alpha j) = \sum_k \Omega(k\beta) Q(\beta k\alpha j)$$

where $Q(\beta k\alpha j) \in \mathbf{Z}$ and is the 2d Chern number of a unitary vector bundle on a 4-torus.

3) In three dimensions, the adiabatic charge transport is again weakly quantised, i.e. it is given by

$$Q(\alpha j) = \sum_{kl} \Omega(k\beta)\, \Omega(l\gamma) Q(\gamma lk\beta\alpha j)$$

where $Q(\gamma lk\beta\alpha j) \in \mathbf{Z}$, and is the 3rd Chern number of a unitary vector bundle on a 6-torus. The integrated density of states is given by

$$D(\mu) = \sum_{jkl} \Omega(l\gamma)\, \Omega(j\alpha)\Omega(k\beta) Q(\gamma lk\beta\alpha j) .$$

We may note that since the integrated density of states in a gap will not vanish, at least one of the topological numbers is not zero, so that the corresponding bundle will not be trivial. Previously, Thouless [8] had considered a one-dimensional electronic system acted on by a periodic potential which varies adiabatically and periodically in time and discovered that the charge transport induced by such a varying potential is quantised, when the energy of the electrons lies in a gap.

Finally, this approach has been used [6,7] to prove the quantization of the Hall conductivity and to give an associated charge transport, when the magnetic flux through a unit cell is irrational. The original work of Thouless et al.

[2] considered the case of a rational flux. In fact, the problem of a particle in a magnetic field and a periodic potential in two dimensions is somehow similar to a one-dimensional problem in a quasi-periodic potential with two basic frequencies.

References

[1] Novikov S. P. : Sov. Math. Dokl. **23**, 298 (1981)

[2] Thouless D. J. , Kohmoto M. , Nightingale P. P. , Den-Nijs M. : Phys. Rev. Lett. **49**, 1405 (1982)

[3] Johnson R. , Moser J. : Comm. Math. Phys. : **84**, 403 (1982)

[4] Bellissard J. , Lima R. , Testard D. : Lectures Notes on Recent Results in Mathematics and Physics, edited by L. Streit (World Scientific, Singapore, 1985), Vol.1

[5] Berry M. : Proc. Roy. Soc. London **A 392**, 45 (1984)

[6] Kunz H. : Phys. Rev. Lett. **57**, 1095 (1986)

[7] Kunz H.: Adiabatic charge transport and topological invariants for electrons in a quasi-periodic potential and a magnetic field, preprint (1989)

[8] Thouless D. J. : Phys. Rev. **B27**, 6083 (1983)

Spectra of Some Almost Periodic Operators

A. Sütő

Université de Lausanne, Institut de Physique Théorique,
CH-1015 Lausanne, Switzerland

Abstract. Two examples of almost periodic Jacobi matrices, representing respectively the so-called Fibonacci and a Hierarchical Hamiltonian, are discussed. Each of them has a purely singular continuous spectrum supported by a Cantor set.

1 Introduction

The one dimensional one-electron Schrödinger equation, $(-d^2/dx^2 + V(x))\,\psi(x) = E\,\psi(x)$ is certainly one of the best known differential equations that has ever been studied in the physics literature. When the potential $V(x)$ is a piecewise continuous periodic function, the equation has a countable set of stability intervals separated by gaps [1]. The stability intervals define the one-electron energy spectrum of a one dimensional crystalline solid. The spectrum is known to be purely absolutely continuous [1]; each stability interval is a "band" in the spectrum, and the solutions with E in a band are periodic apart from a phase factor: they are called the Bloch waves.

The case of nonperiodic potentials is substantially more difficult. It is known from the theory of inverse scattering [2] that all kinds of spectra may occur. Recently, physicists have been interested mainly in two types of potentials: the one with some randomness, the other one with almost periodic features. Random potentials are crucial in electron localization [3] (for a review see [4]), while almost periodic potentials appear in rather different fields like quasicrystals, electrons subjected to competing incommensurate forces, stability problems in classical mechanics, etc., reviewed in [5, 6, 7].

The potential acting on the electron is supposed to be created by a lattice of positively charged ions. The wave functions are therefore sharply peaked on the ionic sites. This suggests replacing the Schrödinger equation by its discrete, "tight binding" version,

$$\psi(n+1) + \psi(n-1) + V(n)\psi(n) = E\,\psi(n), \tag{1}$$

which is more accessible for numerical work and is simpler to tackle rigorously. As compared with the original equation, notice a shift by 2 in the energy and an immaterial change of sign: If the left-hand-side of Eq. (1) is written as $((H_0 + V)\psi)(n)$, then the operator $H_0 + V$ is unitarily equivalent with $-H_0 + V$. The operator $H_0 + V$ can be represented by an infinite tridiagonal matrix H, called a Jacobi matrix, with elements $H_{ij} = \delta_{i,j+1} + \delta_{i+1,j} + V(i)\,\delta_{ij}$. If V is periodic then the spectrum of $H_0 + V$ is still purely absolutely continuous and consists of a finite number of bands within which the solutions are Bloch waves. Bounded random potentials give rise to dense pure point spectra with, at most, a finite number of gaps and exponentially localized eigenstates

Springer Proceedings in Physics, Vol. 47 Number Theory and Physics
Editors: J.M. Luck · P. Moussa · M. Waldschmidt

[8, 9, 10]. Almost periodic potentials are midway between the periodic and the random potentials, and they may lead practically to any kind of spectra. Apart from general results, there is a small number of examples which have been studied in detail. The best known is the Almost Mathieu operator $H_0 + V_\omega : V_\omega(n) = \mu \cos 2\pi(\alpha n + \omega)$ where α is an irrational number. (For two recent developments see [11, 12]).

Probably the most interesting feature of Eq. (1) with an almost periodic potential is that the bands may completely disappear from the spectrum. Since the spectrum contains no isolated point either ([13], see also [6]), this means that in such a case it is a Cantor set: a closed set without intervals and isolated points. If this Cantor set is of zero Lebesgue measure then all the spectral measures of $H_0 + V$ are singular; if, in addition, the spectrum is purely continuous then we conclude that it is purely singular continuous.

Below, we discuss two examples in which these exotic properties appear transparently and can be proved with mathematical rigor. The work on the Fibonacci Hamiltonian is published in two papers [14, 15]. The results on the Hierarchical Hamiltonian are due to a collaboration with Hervé Kunz and Roberto Livi [16].

2 Fibonacci Hamiltonian

The potential is given by

$$V_\omega(n) = \mu\, \chi_{[1-\alpha,1[}(< n\alpha + \omega >) \tag{2}$$

where χ_I is the characteristic function of the interval I, α is the inverse golden mean $(\sqrt{5}-1)/2$, $\omega \in [0,1[$ and $< \cdot >$ means fractional part. The potential amplitude $\mu \neq 0$ and can be taken positive: $H_0 - V = -(-H_0 + V)$ so that for the spectra as sets (denoted by σ) we get $\sigma(H_0 - V) = -\sigma(H_0 + V)$.

Physicists usually consider $V_0 \equiv V$ only; the embedding of V into a family (V_ω), called the hull of V, is both natural and technically motivated: Random and almost periodic potentials are also ergodic in the sense that the sequence $\{V_\omega(n)\}_{n=-\infty}^{\infty}$ corresponds to consecutive values of a function along a trajectory of an ergodic dynamical system (the rotation of the circle through $2\pi\alpha$, in (2) and in the Almost Mathieu case). Different trajectories correspond to different $\omega's$. The embedding is sometimes necessary because certain properties are valid only for almost every ω (with respect to the invariant measure of the dynamics) and are difficult to verify for a given ω.

The origins of the Fibonacci Hamiltonian go back to papers by Kohmoto et al. [17, 18] and Ostlund et al. [19], in which all the properties we are going to discuss were predicted by numerical and heuristic arguments. The form (2) of the potential is not very revealing. We gain a better insight by rewriting (2) as

$$V_\omega(n) = \mu([(n+1)\alpha + \omega] - [n\alpha + \omega]) \tag{3}$$

where $[x]$ is the integer part of x. From (3), one can easily see the remarkable symmetries of $V_0 \equiv V$:

$$\begin{aligned} V(-n) &= V(n-1), \quad n \geq 2 \\ V(F_k + n) &= V(n), \quad k \geq 3, \quad 1 \leq n \leq F_k. \end{aligned} \tag{4}$$

The second line describes the almost periodicity of $V(n)$ and shows the role of shifts by Fibonacci numbers F_k ($F_0 = F_1 = 1$, $F_{k+1} = F_k + F_{k-1}$). Another way to display the almost periodicity of $V(n)$ is to write it in the form

$$\mu^{-1}V(n) = \alpha + \sum_{k=1}^{\infty} \frac{1}{\pi k}(\sin 2\pi k(n+1)\alpha - \sin 2\pi k n\alpha), \tag{5}$$

valid for any irrational α and integer n, unless $n = 0, -1$.

Equation (1) is usually solved by the use of transfer matrices. If $\psi(n)$ is a solution of Eq. (1), we can write

$$\begin{pmatrix} \psi(n+1) \\ \psi(n) \end{pmatrix} = T(n) \begin{pmatrix} \psi(1) \\ \psi(0) \end{pmatrix} \tag{6}$$

with

$$T(n) = \begin{pmatrix} E - V(n) & -1 \\ 1 & 0 \end{pmatrix} \cdots \begin{pmatrix} E - V(1) & -1 \\ 1 & 0 \end{pmatrix}. \tag{7}$$

In the case when V is periodic with period length L, the spectrum of $H_0 + V$ is $\{E \in \mathbb{R} |\, |\mathrm{tr}\, T(L)| \leq 2\} = \{E | \mathrm{tr}\, T(n) \text{ is a bounded sequence}\}$. One may then guess (as in [17–19]) that the spectrum of the Fibonacci Hamiltonian $H = H_0 + V$ is

$$\sigma(H) = B \equiv \{E | \mathrm{tr}\, T(n) \text{ is bounded as } n \to \infty\}. \tag{8}$$

One may also wonder whether the constructive form of the spectral condition, $\sigma(H) = \{E |\, |\mathrm{tr}\, T(L)| \leq 2\}$, for periodic potentials has its analogue in the Fibonacci case. In the first rigorous work on the model, Casdagli [20] showed that for $\mu \geq 16$, B is a Cantor set of zero Lebesgue measure; however, the proof of (8) and a constructive spectral condition were obtained only later [14]. Denote $M_n = T(F_n)$ and $\tau_n = \mathrm{tr}\, M_n$. From (4), one obtains

$$M_{n+1} = M_{n-1} M_n, \tag{9}$$

and this equation together with $\det M_n = 1$ yields

$$\tau_{n+1} = \tau_n \tau_{n-1} - \tau_{n-2}, \qquad n \geq 1, \tag{10}$$

with the initial conditions $\tau_{-1} = 2$, $\tau_0 = E$, $\tau_1 = E - \mu$. Thus, τ_n is an $F_n{}^{\text{th}}$ degree polynomial of E. The iteration (10) has an invariant

$$\tau_{n+1}^2 + \tau_n^2 + \tau_{n-1}^2 - \tau_{n+1}\tau_n\tau_{n-1} = \mu^2 + 4. \tag{11}$$

The formulas (9)–(11) were obtained in [17, 19]. They are used in [14] to prove

Theorem 1 . *Let $\sigma(H)$ denote the spectrum of $H_0 + V_0$, $\rho(H) = \mathbb{R} \setminus \sigma(H)$. Then*

$$\sigma(H) = \{E \in \mathbb{R} | \{\tau_n\} \text{ is bounded}\}$$

$$\rho(H) = \bigcup_{n=1}^{\infty} \{E |\, |\tau_{n-1}| \leq 2,\ |\tau_n| > 2,\ |\tau_{n+1}| > 2\} \cup \{|E| > 2,\ |E - \mu| > 2\}. \tag{12}$$

In [14] there is also a simple proof that for $|\mu| \geq 4$ the spectrum is a Cantor set.

The asymptotic behaviour of the solutions of Eq. (1) is characterized by the Lyapunov exponent. It is defined as

$$\gamma(E,\omega) = \lim_{|n|\to\infty} \frac{1}{|n|} \ell n ||T_\omega(n)|| \tag{13}$$

where $T_\omega(n)$ denotes the transfer matrix (7) for the equation $(H_0 + V_\omega)\psi = E\psi$. In fact, the limit in (13) does not always exist; however, if the potential is ergodic then for any fixed E it exists for almost every ω, and gives some number $\gamma(E) \geq 0$ independent of ω [21].

For the Fibonacci potential V_0 one can show [14, 15] that if $\{\tau_n\}$ is bounded then Eq. (1) has no decaying solution at the infinity, and $\gamma(E,0) = 0$. This implies the following

Proposition 1 . *The spectrum of H_0+V_0 is purely continuous. For $E \in \sigma(H)$, $\gamma(E,0) = 0$.*

In case of ergodic potentials, knowledge concerning the Lyapunov exponent permits conclusions about spectral measures.

Theorem (Kotani [22], Simon [23]). *If $\gamma(E) = 0$ on a subset A of $\mathbb{R}$ with positive Lebesgue measure, then for a.e. ω the absolutely continuous spectrum of $H_0 + V_\omega$ is non-empty in A.*

On the other hand, we have

Theorem (Kotani [24]). *Let V_ω be nonperiodic ergodic and taking values from a finite set. Then for a.e. ω, $H_0 + V_\omega$ has no absolutely continuous spectrum.*

Since the potential (2) is two-valued and ergodic, the above two theorems can be confronted by observing [15]:

Lemma . *Assume that for a given ω, $\gamma(E,\omega)$ exists for $E \in A' \subset \mathbb{R}$. Then $\gamma(E,\omega) = \gamma(E)$ for $E \in A = A' \setminus S$, where S is a set of zero Lebesgue measure: $|S| = 0$.*

Thus, we find that $\gamma(E) = 0$ if $E \in \sigma(H) \setminus S$, where $|S| = 0$. Were $\sigma(H)$ of positive Lebesgue measure, the above two theorems would contradict each other. In the closed set $\sigma(H)$ there is no isolated point (since the spectrum is purely continuous) and no interval (since $|\sigma(H)| = 0$). Hence we have

Theorem 2. *For any $\mu \neq 0$, $\sigma(H)$ is a Cantor set of zero Lebesgue measure. The spectrum of $H_0 + V_0$ is purely singular continuous.*

Recently, the above results could be extended to potentials (2) with any irrational α [25].

3 Hierarchical Hamiltonian

Hierarchical operators have a long history. A certainly incomplete list of references can be found in [16]. Our immediate predecessors are Schneider et al. [26] and Livi et al. [27] who wrote down the trace recurrence (Eq. (19) below) and performed the first nonrigorous analysis. For a fixed positive number, R, the potential is given by

$$V(2^m(2\ell+1)) = \mu \sum_{k=0}^{m-1} R^k, \quad m \geq 0, \; \ell \in \mathbb{Z}, \tag{14}$$

$$V(0) = \begin{cases} \mu/(1-R), & R < 1 \\ \infty, & R \geq 1. \end{cases}$$

If $R < 1$, the sequence $V(n)$ is limit periodic and has a natural ergodic embedding (V_ω):

$$V_\omega(n) = \mu(1 - |n+\omega|_{1/R})/(1-R). \tag{15}$$

Here ω is a 2-adic integer and $|\cdot|_s$ for $s > 1$ is a valuation, replacing the modulus: For $\omega = \sum_0^\infty \omega_k 2^k$, $\omega_k \in \{0,1\}$, let $\mathrm{ord}(\omega) = \min\{k|\omega_k = 1\}$; then $|\omega|_s = s^{-\mathrm{ord}(\omega)}$. The potential (14) corresponds to $\omega = 0$. Using the valuation, it has the form of a Coulomb potential in 1, 2 and 3 dimensions for $R < 1$, $= 1$ and > 1, respectively:

$$V(n) = \begin{cases} \mu(1-|n|_{1/R})/(1-R), & R < 1, \\ -\mu \log_2 |n|_2, & R = 1, \\ \mu(1-1/|n|_R)/(1-R), & R > 1, \; n \neq 0. \end{cases}$$

V has a scaling and a translational symmetry,

$$V(2n) = RV(n) + \mu \tag{16}$$

$$V(\ell \cdot 2^n + k) = V(k), \quad k = 1, 2, \ldots, 2^{n-1}, \tag{17}$$

which completely determine the sequence: given $V(1)$, Eqs. (16) and (17) admit a unique solution for $V(n)$, $n \neq 0$. Equation (17) implies a recurrence relation for the transfer matrices: Let $M_n = T(2^n)$ (cf. Eq. (7)), then

$$M_{n+1} = \begin{pmatrix} 1 & -\mu R^n \\ 0 & 1 \end{pmatrix} M_n^2 \tag{18}$$

from which we find that the traces $\tau_n = \mathrm{tr}\, M_n$ satisfy the equation

$$\tau_{n+1} = \tau_n^2 - 2 + R\,\tau_n(\tau_n - \tau_{n-1}^2 + 2), \quad n \geq 1. \tag{19}$$

With the initial condition $\tau_0 = E$, $\tau_1 = E^2 - \mu E - 2$, τ_n becomes a $2^{n\,\mathrm{th}}$ degree polynomial of E. As in the case of the Fibonacci Hamiltonian, we are looking for a spectral condition in terms of the traces τ_n only. Curiously, this is simpler to obtain for $R \geq 1$, when the potential is unbounded:

Theorem 3 . *For $R \geq 1$,*

$$\sigma(H) = \{E | \{\tau_n\} \textit{ is bounded}\},$$

$$\rho(H) = \mathbb{R} \setminus \sigma(H) = \bigcup_{n=1}^{\infty} \{E | \tau_n > \max\{2, \; \tau_{n-1}^2 - 2\}\}.$$

Observe the analogy with Theorem 1.

Theorem 4 .
For $R < 1$,

$$\sigma(H) = \{E | \{\tau_n\} \text{ is bounded or } \tau_n \to \infty \text{ and } \tau_n/\tau_{n-1}^2 \to R\},$$

$$\rho(H) = \bigcup_{n=1}^{\infty} \{E | \tau_n > \max\{1/R, \; R\tau_{n-1}^2\}\}.$$

As in the case of the Fibonacci potential, the boundedness of $\{\tau_n\}$ implies that Eq. (1) has no decaying solution and that $\gamma_V(E)$, the Lyapunov exponent calculated with V given by (14), vanishes. Furthermore, since $H_0 + V$ with unbounded V cannot have any absolutely continuous spectrum [28], we obtain

Proposition 2 . *For $R \geq 1$, the spectrum of $H_0 + V$ is purely singular continuous, and $\gamma_V(E) = 0$ for $E \in \sigma(H)$.*

If $R < 1$ then still most of the spectrum is continuous and $\gamma_V(E) = 0$ in it. There is, probably, a unique eigenvalue at the upper ($\mu > 0$) or lower ($\mu < 0$) border of the spectrum.

Theorem 5 . *For any $R > 0$, the spectrum is a Cantor set: The zeros of the polynomials $\tau_n(E)$ are at the border of spectral gaps, and they form a dense set in the spectrum.*

Due to the explicit knowledge about gap edges, the hierarchy of spectral gaps can be easily studied. The interested reader may find details in [16].

Acknowledgements. The author would like to thank the organizers for making possible his participation at the School. Financial support from the School and from the University of Lausanne is gratefully acknowledged.

References

1. Reed, M., Simon, B.: Methods in modern mathematical physics, Vol. IV. New York: Academic Press 1978, Theorem XIII.90.

2. Gel'fand, M., Levitan, B.M.: On the determination of a differential equation from its spectral function. Amer. Math. Soc. Transl. **1**, 253–304 (1955).

3. Anderson, P.W.: Absence of a diffusion in certain random lattices. Phys. Rev. **109**, 1492 (1958).

4. Erdös, P., Herndon, R.C.: Theories of electrons in one-dimensional disordered systems. Adv. Phys. **31**, 65–163 (1982).

5. Simon, B.: Almost periodic Schrödinger operators: A review. Adv. Appl. Math. **3**, 463–490 (1982).

6. Cycon, H.L., Froese, R.G., Kirsch, W., Simon, B.: Schrödinger operators. Berlin, Heidelberg, New York: Springer 1987, Chaps. 9 and 10.

7. Bellissard, J.: Almost periodicity in solid state physics and C* algebras. Talk given at the Harald Bohr Centenary Conference on Almost Periodic Functions, April 25, 1987.

8. Kunz, H., Souillard, B.: Sur le spectre des opérateurs aux différences finies aléatoires. Commun. Math. Phys. **78**, 201–246 (1980).

9. Fröhlich, J., Martinelli, F., Scoppola, E., Spencer, T.: Constructive proof of localization in the Anderson tight binding model. Commun. Math. Phys. **101**, 21–46 (1985).

10. Carmona, R., Klein, A., Martinelli, F.: Anderson localization for Bernoulli and other singular potentials. Commun. Math. Phys. **108**, 41–66 (1987).

11. Fröhlich, J., Spencer, T., Wittwer, P.: Localization for a class of one dimensional quasi-periodic Schrödinger operators. Preprint 1988.

12. Chulaevsky, V., Delyon, F.: Purely absolutely continuous spectrum for almost Mathieu operators. Preprint 1989.

13. Pastur, L.: Spectral properties of disordered systems in the one-body approximation. Commun. Math. Phys. **75**, 179–196 (1980).

14. Sütő, A.: The spectrum of a quasiperiodic Schrödinger operator. Commun. Math. Phys. **111**, 409–415 (1987).

15. Sütő, A., Singular continuous spectrum on a Cantor set of zero Lebesgue measure for the Fibonacci Hamiltonian. J. Stat. Phys. **56**, 525–531 (1989).

16. Kunz,H., Livi, R., Sütő, A.: Cantor spectrum and singular continuity for a hierarchical Hamiltonian. Commun. Math. Phys. **122** (1989).

17. Kohmoto, M., Kadanoff, L.P., Tang, C.: Localization problem in one dimension: mapping and escape. Phys. Rev. Lett. **50**, 1870–1872 (1983).

18. Kohmoto, M., Oono, Y.: Cantor spectrum for an almost periodic Schrödinger equation and a dynamical map. Phys. Lett. **102A**, 145–148 (1984).

19. Ostlund, S., Pandit, R., Rand, D., Schnellnhuber, H.J., Siggia, E.D.: One-dimensional Schrödinger equation with an almost periodic potential. Phys. Rev. Lett. **50**, 1873–1876 (1983).

20. Casdagli, M.: Symbolic dynamics for the renormalization map of a quasiperiodic Schrödinger equation. Commun. Math. Phys. **107**, 295–318 (1986).

21. Furstenberg, H., Kesten H.: Products of random matrices. Ann. Math. Stat. **31**, 457–469 (1960).

22. Kotani, Sh.: Lyapunov indices determine absolutely continuous spectra of stationary random one-dimensional Schrödinger operators. In Stochastic Analysis, ed. by K. Ito (North-Holland, Amsterdam 1984) pp. 225–247.

23. Simon, B.: Kotani theory of one-dimensional stochastic Jacobi matrices. Commun. Math. Phys. **89**, 227–234 (1983).

24. Kotani, Sh.: Jacobi matrices with random potentials taking finitely many values. Preprint (1988).

25. Bellissard, J., Jochum, B., Scoppola, E., Testard, D.: Spectral properties of one dimensional quasi-crystals. Preprint 1989.

26. Schneider, T., Wurtz, D., Politi, A., Zannetti, M.: Schrödinger problem for hierarchical heterostructures. Phys. Rev. **B36**, 1789–1792 (1987).

27. Livi, R., Maritan, A., Ruffo, S.: The spectrum of a one dimensional hierarchical model. J. Stat. Phys. **52**, 595–608 (1988).

28. Simon, B., Spencer, T., to be published.

The Quantum Hall Effect and the Schrödinger Equation with Competing Periods

D.J. Thouless

Cavendish Laboratory, Cambridge, UK*
*Permanent address: Dept. of Physics FM-15, University of Washington, Seattle, WA98195, USA

Abstract It is shown how the Schrödinger equation for an electron in a two-dimensional periodic potential and a strong magnetic field can be reduced to a one-dimensional probe, in a potential with two different periods. The quantum Hall conductance is a topological invariant for this problem. This idea has been generalized to cover more realistic examples of the quantum Hall effect. This survey ends with a discussion of some striking but unexplained numerical results for the spectral measure for such a problem, and with some remarks on the fractional quantum Hall effect.

1. Electrons in a magnetic field and the one-dimensional quasiperiodic equation

When I was a graduate student at Cornell, one of the few faculty members who moved easily between the worlds of mathematics and physics was Mark Kac. I remember him defining the difference between a mathematician and a physicist by saying that a mathematician is interested in the complex properties of simple systems, but a physicist is interested in the simple properties of complex systems. While I was listening to talks given by physicists here I was not sure that their interests all fitted this description, but my own do. The problem of electrons moving simultaneously in a magnetic field and a periodic potential is a very complicated one that has intrigued physicists over the past forty years, but the results that I am going to talk about involve very simple aspects of mathematics, including number theory.

For the idealised case of free electrons in a uniform magnetic field B the energy levels are highly degenerate, forming Landau levels separated by $\hbar$ times the cyclotron frequency. The form of the wave functions depends on the vector potential that is used to represent the magnetic field, but with Landau's choice of gauge, a vector potential in the y-direction, the form of the wave function is

$$\exp(i\kappa_2 y) f_n(x/l - \kappa_2 l), \tag{1}$$

where f_n is an Hermite function and l is the magnetic length $(\hbar/eB)^{\frac{1}{2}}$. The energy depends on the integer n, but is independent of the value of the continuous parameter κ_2. A weak periodic potential of the form

$$2V_1' \cos(2\pi x/a) + 2V_2' \cos(2\pi y/b) \tag{2}$$

†Permanent address

Springer Proceedings in Physics, Vol. 47 Number Theory and Physics
Editors: J.M. Luck · P. Moussa · M. Waldschmidt

breaks the degeneracy of the Landau levels. To lowest order in degenerate perturbation theory the x-dependent term gives the energy a sinusoidal dependence on κ_2 with period a/l^2, while the y-dependent term mixes states whose value of κ_2 differs by $2\pi/b$. The resulting equation can be represented in the form

$$V_2 c_{n-1} + 2V_1 \cos(2\pi n/\phi + k_2)c_n + V_2 c_{n+1} = Ec_n, \tag{3}$$

where E is the energy eigenvalue, c_n is the amplitude of the component with κ_2 equal to $2\pi n/b + \phi k_2/b$, and ϕ is the number of flux quanta per unit cell, given by

$$\phi = abBe/h = ab/2\pi l^2. \tag{4}$$

This equation was derived by Harper[1], now at Edinburgh University, who was a student of Peierls at Birmingham at the time. He derived it for a tight binding model perturbed by a weak magnetic field, which, oddly enough, gives exactly the same form of equation with ϕ replaced by its reciprocal.

The spectrum of this equation (the union over all values of k_2) behaves very strangely as ϕ is varied. If $\phi = p/q$ is rational it can readily be seen that eq. 3 is periodic with period p; it can be reduced to a finite matrix problem by the use of Bloch waves (the Floquet theorem), and the spectrum has p distinct bands. This situation was explored in 1964 by Azbel[2], who showed the importance of the continued fraction expansion of $1/\phi$, and observed that for ϕ irrational the spectrum is a Cantor set, with gaps interrupting all intervals of the spectrum. Beautiful pictures of this were obtained by Hofstadter[3], then a graduate student of Wannier at the University of Oregon; such pictures have already been shown by Belissard at this meeting.

Some important results were obtained for this problem by Aubry and André [4]. They observed that the Fourier transform of this equation is essentially the same equation with V_1 and V_2 interchanged; in terms of the original problem of an electron in a magnetic field this can be achieved simply by taking the vector potential to be in the x direction rather than in the y direction. They could then show that for ϕ a good irrational the Lyapunov exponent of the equation is $\max[0, \ln(V_1/V_2)]$. A positive Lyapunov exponent implies that the eigenstates are exponentially localized and the spectrum is discrete (but dense), while it is thought that the spectrum is continuous and the eigenstates extended in most cases that the exponent is zero. It is also known[4,5] that the measure of the spectrum for this model is $4|V_1 - V_2|$. There is a critical point at $V_1 = V_2$, where the Lyapunov exponent changes from zero to positive, and where the measure of the spectrum is zero; the spectrum at this critical point must be singular continuous. In the magnetic model this critical point corresponds to a square symmetric periodic potential; in the asymmetric cases the eigenstates are either extended in the x direction and localized in the y direction or vice versa.

It has been shown [6] that if ϕ is a Liouville number rather than a generic irrational there are no eigenstates however large V_1/V_2 is. This is because such a number is close enough to a rational at each stage of approximation that the deviations from rationality are not enough to localize the wave function.

2. Hall conductance as a topological invariant

The discovery of the quantum Hall effect, and in particular Laughlin's [7] explanation of it, made us look again at this old problem. This work was done in collaboration with all the postdocs in our group at the University of Washington, Kohmoto, Nightingale and den Nijs. Laughlin's argument persuaded us that the Hall conductance must be integer whenever the Fermi energy was in an energy gap, so that each subband, or each energy gap, could be characterized by an integer Hall conductance. How then is the quantum unit of Hall conductance for a single Landau level split up between its p subbands in such a way that each subband itself carries an integer? One guess, supported by a few simple examples, was that the Hall conductance was all associated with the central subband, or, in the case of even p, the touching pair of central subbands, but a few more examples soon showed that this was a bad guess. We tried laboriously to calculate a two-dimensional integral to get the Hall conductance for each band, knowing that if we did the numerical integration right we should get an integer. In order to avoid doing this kind of boring work we did some thinking, and came up with a theory[8] that was sufficiently elegant that it had to have a wider field of applicability than the artificial problem of electrons in an ideal periodic potential.

We found that the Hall conductance could be written as se^2/h, where

$$s = \frac{1}{2\pi}\int_{T^2} d^2k \sum_\alpha \mathrm{Im}\left(\frac{\partial\phi_\alpha}{\partial k_1}, \frac{\partial\phi_\alpha}{\partial k_2}\right). \tag{5}$$

Here the integral is over the Brillouin zone, the two-torus of possible Bloch wavevectors, the sum is over the subbands α occupied by the electrons, and the integrand involves the scalar product of the two partial derivatives of the wave function in the Brillouin zone. We showed that this is an integer by means that every physicist knows. An application of Stokes' theorem turns the integral into the line integral of the gradient of a phase round the perimeter of the Brillouin zone; this phase is the Berry phase[9]. The equivalence of the eigenfunctions on opposite points of the edge of the Brillouin zone allows one to deduce that the phase must change by an integer multiple of 2π round this perimeter. We also deduced, by an unnecessarily clumsy argument that has since been made much neater[5,10], that if the filled subbands occupy a proportion f of the Landau level, then we have

$$f = s + t/\phi, \quad \text{where} \quad s, t \in Z. \tag{6}$$

This relation, which was also derived by Středa[11], is essentially the gap-labelling theorem of Johnson and Moser[12]. From the point of view of differential geometry this is the topological invariant that defines the first Chern class of the mapping of the torus (the Brillouin zone) on the complex projective space of eigenfunctions of the Hamiltonian.

Figure 1 shows how the Hall conductance is distributed between the subbands in a sequence of rational approximations to ϕ equal to the golden mean. The largest gaps,

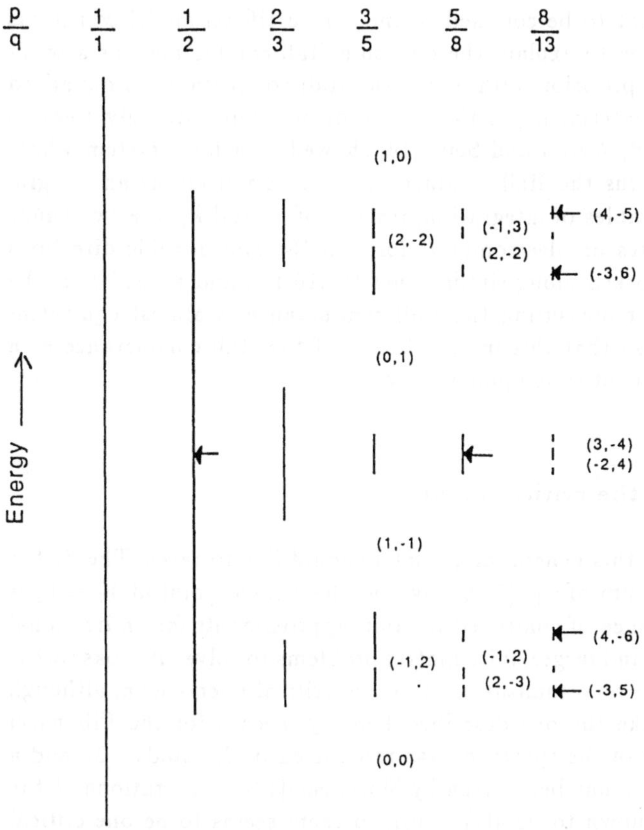

Figure 1. Energy bands for a sequence of approximants p/q to the golden mean. The gaps are labelled with (s,t) from eq. 6, where s is the Hall conductance in units of e^2/h. The arrows show gaps that are too small to be seen.

which appear in very low order approximations, have small Hall conductance, but the small gaps that appear at a late stage in the approximation may have rather large Hall conductances associated with them; in the case of $\phi = 8/13$ the subbands have a Hall conductance of either +5 or -3.

I have said that s is the Hall conductance. but I have not given a direct interpretation of t. If one takes the other limit I mentioned earlier, of a tight-binding model perturbed by a weak magnetic field, then the roles of s and t are interchanged, just as ϕ is replaced by its reciprocal.

In general the bands are distinct for the model I have been describing, because the bands are determined by the eigenvalues of a complex matrix that depends on the two parameters k_1, k_2; the Von Neumann and Wigner theorem[13] tells us that generically the eigenvalues will not cross as these two parameters are varied. I have heard that Sinai says that a third incommensurate period leads to a spectrum with only a finite number of gaps, which is what one would expect, since the third incommensurate period allows a third parameter to vary, so that level crossing will be expected to occur.

Equation (5) is too elegant to be confined to the very artificial model which has been discussed so far. It seems to explain the quantum Hall conductance as a topological quantum number; the precision with which the Hall conductance is quantized and its insensitivity to the material properties of the device point strongly towards such an interpretation. Indeed, Avron and Seiler[14] showed that for a system whose topology in real space is a torus the Hall conductance can be written as an integral of just this form, but the variables of integration, instead of k_1 and k_2, are two gauge fields whose variation generates an electromotive force in the two possible directions round the torus. We have worked along similar lines [15,16,17], and argued that the normal experimental setup for measuring the Hall conductance is almost equivalent to a torus with a hole in it, so that this interpretation of the Hall conductance as a topological quantum number is of wide applicability.

3. Spectral measure near the critical point

There are two other topics in this general area that I should like to raise. The first is the question of how the spectrum of eq. (3) behaves as the critical point at $V_1 = V_2$ is approached. If we use a sequence of continued fraction approximants for an irrational ϕ, we get a sequence of larger and larger finite matrix problems to solve. It looks rather like a finite size problem in the renormalization theory of critical phenomena, although it is not a clean procedure like the one described here by Yoccoz for the Fibonacci chain. The self-similar nature of the spectrum was recognized by Hofstadter[3], and a detailed renormalization theory has been given by Wilkinson[18]. For irrational ϕ the measure of the spectrum is known to be $4|V_1 - V_2|$, so there seems to be one critical exponent whose value is unity. For rational $\phi = p/q$ numerical work[5] showed that for $V_1 = V_2$ the measure of the spectrum approaches zero like $9.33V_1/p$, more or less whatever the value of q. In fact, in all the examples I worked out the measure for odd values of p was a little below the limiting value, while the measure for even values of p was a little higher. I have by now been able to determine the numerical value of the coefficient of p^{-1} to an accuracy of a few parts in 10^8, but I have little idea of why this should be so, except that it is a sort of finite size scaling amplitude. One expects one-dimensional quantum problems to be analogous to two-dimensional classical problems, and we know that two-dimensional critical phenomena can be related to conformally invariant theories and Virasoro algebras, but I still do not know where my 9.3299489 comes from.

Since I first wrote this paragraph I have succeeded in evaluating the coefficient analytically, and found that it is equal to $32G/\pi$, where G is Catalan's constant.

4. Fractional quantum Hall effect

Finally, I should say a few things about the fractional quantum Hall effect. The topological arguments for an integer value of the Hall conductance seem so general that it is an embarrassment that experimentalists see fractional values such as 1/3 and 2/3 whose precision is also very high (a few parts in 10^5[19]). It is now almost universally accepted that Laughlin's form of the ground state wave function[20], which leads to a cusp in the energy at $1/(2m+1)$ filling of the Landau level, and to a gap separating the ground state from the quasiparticle excitations, is correct. It has been argued in various places [15,21,22] that a fractional value of the Hall conductance must imply a $2m+1$ fold degeneracy of the ground state. If there are, for example, three degenerate levels, then the states in eq. 5 need not return to equivalent states on opposite sides of the region of integration, but may involve a different linear combination of the three degenerate states. One has therefore to extend the range of integration to a surface of three times the area before one can conclude that the integral is an integer multiple of 2π. In this way one can extend the topological argument to get exact fractions as well as integers for the Hall conductance. The difficulty of this approach is that neither the Laughlin wave function for a disk[20], nor Haldane's equivalent expression for the wave function on a spherical surface[23] appears to have any degeneracy. After a lot of discussion with Avron about this problem I think I know how to overcome this objection to the idea of a discrete degeneracy.

Acknowledgements

This work was supported in part by the US National Science Foundation under Grant No. 86-13598.

References

1. P. G. Harper, Proc. Phys. Soc. (London) **A65**, 317 (1955).
2. M. Ya. Azbel, Zh. Eksp. Teor. Fiz. **46**, 929 (1964) [Soviet Phys. JETP **19**, 634 (1964)].
3. D. Hofstadter, Phys. Rev. B **14**, 2239 (1976).
4. S. Aubry and G. André, Ann. Israel Phys. Soc. **3**, 133 (1980).
5. D. J. Thouless, Phys. Rev. **28**, 4272 (1983).
6. A. Ya. Gordon, Uspekhi Mat. Nauk **31**, 257 (1976); J. Avron and B. Simon, Bull. Am. Math. Soc. **6**, 81 (1982).
7. R. B. Laughlin, Phys. Rev. B **23**, 5632 (1981).
8. D. J. Thouless, M. Kohmoto, M. P. Nightingale and M. den Nijs, Phys. Rev. Lett. **49**, 405 (1982).
9. M. V. Berry, Proc. Roy. Soc. A **392**, 45 (1984).
10. J. Avron, R. Seiler and B. Simon, Phys. Rev. Lett. **51**, 51 (1983).
11. P. Středa, J. Phys. C **15**, L1299 (1982).
12. R. Johnson and J. Moser, Commun. Math. Phys. **84**, 403 (1982).
13. J. von Neumann and E. P. Wigner, Z. Phys. **30**, 467 (1929).
14. J. Avron and R. Seiler, Phys. Rev. Lett. **54**, 259 (1985).
15. Q. Niu, D. J. Thouless and Y.-S. Wu, Phys. Rev. **B31**, 3372 (1985).
16. Q. Niu and D. J. Thouless, Phys. Rev. B **35**, 2188 (1987).

17. D. J. Thouless, in *The Quantum Hall Effect*, ed. R. E. Prange and S. M. Girvin (Springer-Verlag 1987), pp. 101-16.
18. M. Wilkinson, Proc. Roy. Soc. A **391**, 305 (1984); J. Phys. A **20**, 1761 (1987).
19. A. M. Chang, P. Berglund, D. C. Tsui, H. L. Störmer, J. C. M. Hwang, Phys. Rev. Lett. **53**, 997 (1984).
20. R. B. Laughlin, Phys. Rev. Lett. **50**, 1395 (1983).
21. R. Tao and Y.-S. Wu, Phys. Rev. B **30**, 1097 (1984).
22. R. Tao and F. D. M. Haldane, Phys. Rev. B **33**, 3844 (1986).
23. F. D. M. Haldane, Phys. Rev. Lett. **51**, 605 (1983).

Finite Automata in 1-D and 2-D Physics

J.-P. Allouche

CNRS, URA 0226, Mathématiques et Informatique,
351, cours de la Libération, F-33405 Talence Cedex, France

Abstract. We first recall the definition, the main arithmetic property and examples of applications to physics of one-dimensional finite automata. Then, explaining how to generalize this notion to the multidimensional case we give two applications of 2-D finite automata to physics.

1. Introduction

In an attempt to describe structures intermediate between order and chaos, one can use finite automata : these automata, which are different from the cellular automata have many nice arithmetic properties. As they are actually equivalent to the substitutions of constant length, we are going to use the formalism of substitutions. In what follows we shall first describe the one-dimensional substitutions of constant length and give examples of applications to 1-D physics. Then we shall show how to generalize substitutions to the 2-D (or n-D) case, and give two applications : one of them concerns the construction of fractals (and is due to J. Shallit), the other one gives a construction of the regular Robinson (aperiodic) tiling of the plane by means of 2-D substitutions (and is due to O. Salon and the author).

2. 1-D substitutions of constant length

Let us recall that a substitution of length k (or a k-substitution, or a uniform morphism of length k) on an alphabet (i.e. a finite set) A is a map from A to the set of words on A with k letters. For instance let A be the set {a,b,c,d} and define the 2-substitution σ by :

$$\sigma(a) = ab$$
$$\sigma(b) = cb$$
$$\sigma(c) = ad$$
$$\sigma(d) = cd.$$

If the image of a certain letter by the substitution is a word beginning by this letter, one can construct by iteration an infinite sequence which is a fixed point of the substitution. In the above example, starting with the letter a, one obtains, replacing at each step each letter by its image by σ :

$$a \longrightarrow ab \longrightarrow abcb \longrightarrow abcbadcb \longrightarrow abcbadcbabcdadcb...$$

An infinite sequence is said to be generated by a k-substitution (one also says that the sequence is k-automatic) if it can be obtained by renaming the letters of a fixed point of a k-substitution. In the previous example, if one replaces a and b by 1 and c and d by 0, one obtains the infinite sequence

$$1101100111001001...,$$

which is generated by 2-substitution (or 2-automatic). The reader might have recognized the regular paperfolding sequence discussed by Mendès France in this volume [1].

An ultimately periodic sequence is k-automatic (for every $k \geq 2$). One of the simplest examples of non-periodic automatic sequences is the Prouhet-Thue-Morse sequence which is the fixed point (beginning by 0) of the 2-substitution $0 \rightarrow 01$, $1 \rightarrow 10$; this sequence begins as follows :

$$0110100110010110100 1... .$$

3. Arithmetic properties of 1-D substitutions of constant length

Automatic sequences have many arithmetic properties, an important one is the theorem of Christol, Kamae, Mendès France and Rauzy [2] :

THEOREM : *Let* (a_n) *be a sequence with value in* $[0,p-1]$, *where* p *is a prime number. This sequence is p-automatic (i.e. generated by a p-substitution) if and only if the formal power series* $\sum_0^{+\infty} a_n X^n$ *is algebraic over the field of rational fractions modulo* p, $\mathbb{F}_p(X)$.

For other arithmetic properties of automatic sequences, see the survey of the author [3], see also the paper of Dumont in this volume [4].

4. Examples of applications of 1-D substitutions of constant length to physics

The automatic sequences being to some extent intermediate between order and chaos, one might want to use them in physics ; actually these sequences are deterministic, but they are complex enough to provide us with interesting properties of physical models. Let us give some examples of applications to physics :

- Mendès France and the author showed the relationship between the Rudin-Shapiro sequence and the 1-D Ising model (at imaginary temperature) in [5] ; they also studied the 1-D Ising model with impurities in the case where the sequence of impurities is generated by a substitution of constant length [6].

- Axel, Kleman, Mendès France, Peyrière and the author studied the 1-D mass and spring chain, where all the springs are identical and where there are two kinds of masses distributed according to the Thue-Morse sequence [7]. This first work used a "trace formula"given by Peyrière and the author in [8] and has been completed by Axel and Peyrière in [9] and [10].

- Other applications of substitutions to physics have been given by Aubry, Godrèche, Luck, and Vallet in [11], [12], [13], [14], [15], [16], [17], [18].

For related problems one can read the papers of Katz and Duneau [19], Godrèche [20], Mosseri [21], Moussa [22] in this volume ; one can also read the more complete review of Axel [23].

5. 2-D uniform substitutions

It is possible to generalize the notion of 1-D substitution of constant length to 2 (or more) dimensions. For instance let A be the alphabet {a,b,c} and define the "2×2-substitution" σ by :

$$\sigma(a) = \begin{matrix} a & b \\ c & a \end{matrix}$$

$$\sigma(b) = \begin{matrix} a & c \\ b & c \end{matrix}$$

$$\sigma(c) = \begin{matrix} b & c \\ b & a \end{matrix}$$

Iterating σ, one obtains, replacing each letter by its image by σ :

$$a \longrightarrow \begin{matrix} a & b \\ c & a \end{matrix} \longrightarrow \begin{matrix} abac \\ cabc \\ bcab \\ baca \end{matrix} \ \ldots$$

Replacing now, say a and b by 0 and c by 1, one obtains the (double) sequence :

$$\begin{array}{l} 0001\ \ldots \\ 1001 \\ 0100 \\ 0010 \\ \cdot \\ \cdot \\ \cdot \end{array}$$

which is said to be generated by a 2×2-substitution (one also says that this sequence is 2-automatic).

6. Arithmetic properties of n-D uniform substitutions

The theorem of Christol, Kamae, Mendès France and Rauzy has been generalized by Salon in [24] :

THEOREM : *Let* $(a(n_1,\ldots,n_r))$ *be a sequence with values in* $[0, p-1]^r$, *where* p *is a prime number. The sequence* $(a(n_1,\ldots,n_r))$ *is generated by a* $p\times p\times\ldots\times p$*-substitution if and only if the formal series* $\sum_{n_1\ldots n_r} a(n_1,\ldots,n_r)X_1^{n_1}\ldots X_r^{n_r}$ *is algebraic over the field of rational functions modulo* p, $\mathbb{F}_p(X_1,\ldots,X_r)$.

7. Applications of 2-D substitutions to physics

7.1 Construction of fractals

Shallit gave in [24] (see also [25]) a way of constructing classical fractals by means of k×k-substitutions. Let us give some pictures of his [these pictures are taken from the appendix of [24] where all explanations can be found ; for another way of constructing Sierpinski sets see also Geronimo [26]].

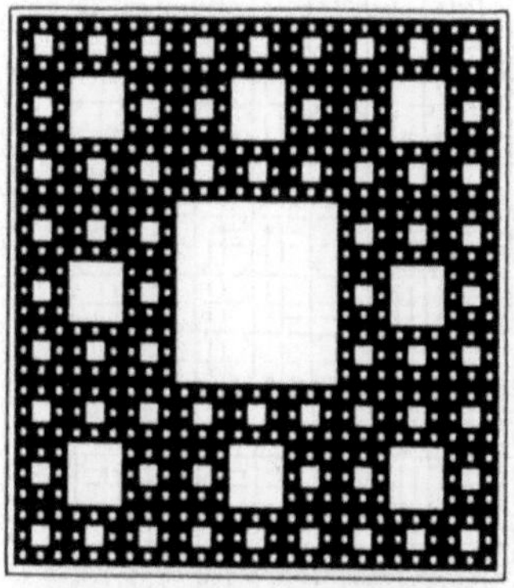

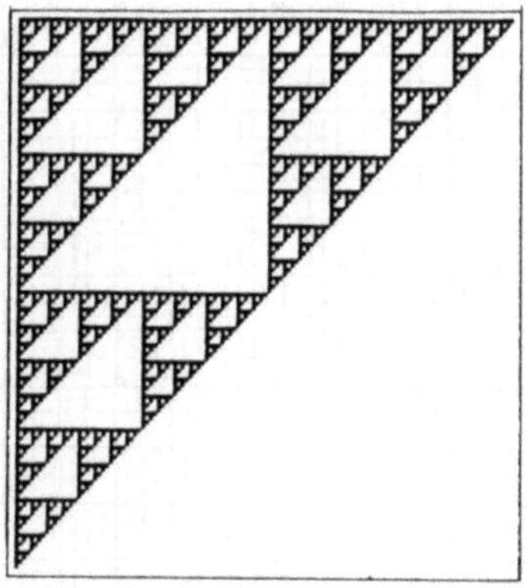

7.2 Robinson tilings and substitutions

It is well known that the Penrose tiling can be described by means of the (non-constant length) substitution $0 \to 01$, $1 \to 0$, (see also the paper of Dekking [27]). This tiling cannot be described by a k×k-substitution as there is no underlying square lattice. Actually other periodic tilings of the plane are of interest, and among them the Robinson tiling. The Robinson tiles are of six types and it is possible to tile the plane in a regular aperiodic way using these six tiles and the tiles deduced from them by rotation or symmetry. The pictures below are taken from the book of Grünbaum and Shephard [28] and from [30] :

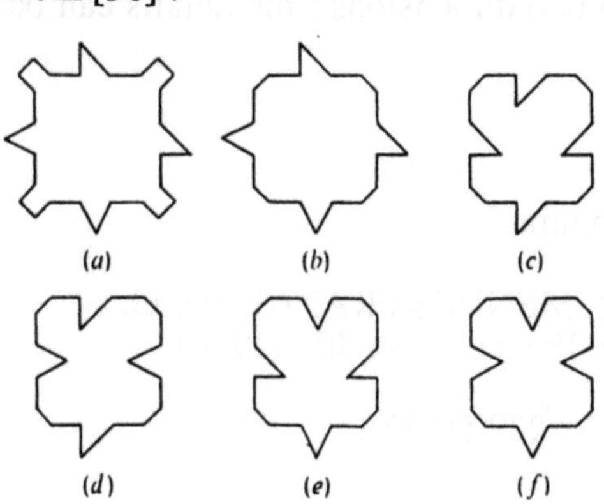

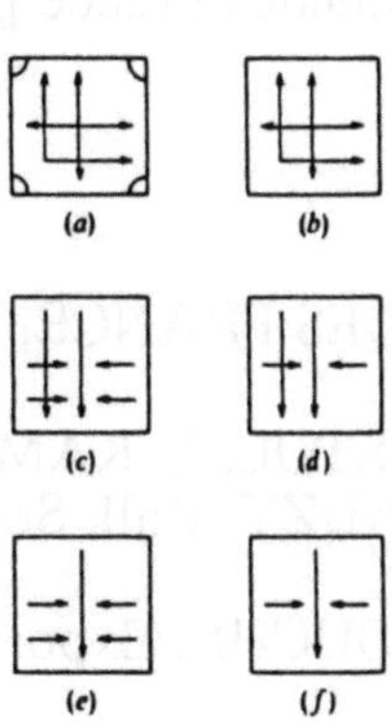

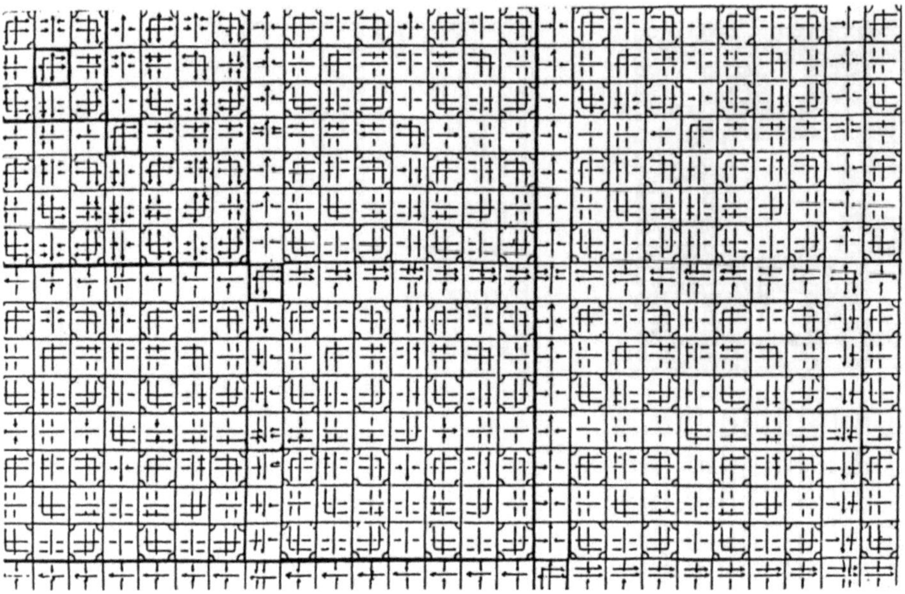

As the Robinson tiling contains an underlying square lattice, it seems appealing to try and describe it in terms of 2-D substitutions. Actually this is possible and given by the following theorem of Salon and the author (see [29], see also [30]) :

THEOREM : *The (regular) Robinson tiling of the plane can be generated by a* 2×2*-substitution.*

The alphabet for the substitution is huge , but what is interesting is first of all the theoretical possibility of generating the Robinson tiling by a 2x2-substitution, i.e. an inflation rule, second that this tiling can be generated by a "perturbed symmetry operation", generalizing the paperfolding (see Blanchard and Mendès France [31]) to two dimensions : the details can be found in [30].

[1] M. MENDES FRANCE, This volume.

[2] G. CHRISTOL, T. KAMAE, M. MENDES FRANCE and G. RAUZY, Bull. Soc.Math. France, 108 : 401 (1980).

[3] J.P. ALLOUCHE, Expo. Math. 5 : 239 (1987).

[4] J.M. DUMONT, This volume.

[5] J.P. ALLOUCHE and M. MENDES FRANCE, Bull. Soc. Math. France 113 : 273 (1985).

[6] J.P. ALLOUCHE and M. MENDES FRANCE, J. Stat. Phys., 42 : 809 (1986) ; and J. de Physique C3 : 63 (1986).

[7] F. AXEL, J.P. ALLOUCHE, M. KLEMAN, M. MENDES FRANCE and J.PEYRIERE, J. de Physique C3 : 181 (1986).

[8] J.P. ALLOUCHE and J. PEYRIERE, C.R. Acad. Sci. Paris, t. 302, Sér. II N° 18, 1986.

[9] F. AXEL and J. PEYRIERE, C.R. Acad. Sci. Paris, t. 306, Sér. II, 1986, p. 179.

[10] F. AXEL and J. PEYRIERE, Spectrum and extended states in a harmonic chain with controlled desorder : effects of the Thue-Morse symmetry.Preprint submitted to J. Stat. Phys., April 1989.

[11] S. AUBRY and C. GODRECHE, J. Phys. (Paris) Coll C3 (1986).

[12] S. AUBRY, C. GODRECHE and F. VALLET, J. Phys. (Paris) 48 : 327 (1987).

[13] J.M. LUCK, J. Phys. A 20 : 1259 (1987).

[14] C. GODRECHE, J.M. LUCK and F. VALLET, J. Phys. A 20 : 4483 (1987).

[15] S. AUBRY, C. GODRECHE and J.M. LUCK, Europhys. Lett. 4 : 639 (1987).

[16] S. AUBRY, C. GODRECHE and J.M. LUCK, J. Stat. Phys. 51 : 1033 (1988).

[17] J.M. LUCK, Phys. Rev. B, to appear, March 15, 1989.

[18] C. GODRECHE and J.M. LUCK, J. Stat. Phys., to appear, April 1989.

[19] A. KATZ and M. DUNEAU, This volume.

[20] C. GODRECHE, This volume.

[21] R. MOSSERI, This volume.

[22] P. MOUSSA, This volume.

[23] F. AXEL, Proc. of the 3rd international meeting on quasicrystals ; incommensurate structures in condensed matter, Mexico, 1989.

[24] O. SALON, Sém. de Théorie des Nombres de Bordeaux, Exposé 4, 1986-1987, with an appendix by J. SHALLIT.

[25] J. SHALLIT, University of Chicago, Technical Report 87-010, june 1987.

[26] J.S. GERONIMO, This volume.

[27] F.M. DEKKING, Adv. in Math., 44,1 : 78 (1982).

[28] B. GRUNBAUM and G.C. SHEPHARD, Tiling and patterns, W.H. Freeman & Company, New York, 1987.

[29] J.P. ALLOUCHE and O. SALON, Preprint.

[30] O. SALON, to appear in Sém. de Théorie des Nombres de Bordeaux, 1988-1989.

[31] A. BLANCHARD and M. MENDES FRANCE, Bull. Soc. Math. France, 106 : 325 (1982).

Summation Formulae for Substitutions on a Finite Alphabet

J.M. Dumont

Département de Mathématiques, Faculté des Sciences de Luminy,
Unité de Recherche Associée au CNRS 225, 70, route Léon Lachamp,
F-13288 Marseille Cedex 9, France

Let A be a finite alphabet , σ a substitution over A , $(u_n)_{n \in N}$ a fixed point for σ , and for each $a \in A$, f(a) a real number; $s^f(n) = \sum_{i \le n} f(u_i)$.
We establish, specially in the "marginal case" (second eigenvalue of the matrix of the substitution is one) and under some additional assumptions, asymptotic formulae for

$$\sum_{n<N} s^f(n) \ , \quad \sum_{n<N} (s^f(n) - \gamma \log_\theta n)^2 \quad \text{and} \quad \sum_{n<N} s^f(n)^k \ .$$

The results give answers to problems that appear in some works of C. Godrèche, J.M. Luck and F. Vallet. The calculi rely upon systems of representation of integers.

1.1 Introduction.

Let $A = \{1,..., d\}$ be a finite "alphabet" , A^* the set of words with letters in A , and σ a "morphism" or "substitution" from A^* into itself, i.e. $\sigma(m\,m') = \sigma(m)\,\sigma(m')$ for all words m,m' . σ is entirely defined by its values on the letters of A , and we will suppose

$$\forall a \in A : \sigma(a) \ne \omega \ (\omega = \text{empty word}) .$$

Assume that 1 is a proper prefix of $\sigma(1)$; then for all integers $n \ge 1$, $\sigma^n(1)$ is proper prefix of $\sigma^{n+1}(1)$ and we can define the infinite word u as the limit of the sequence $\sigma^n(1)$ (in the same way as [1]) , which is a fixed point for the extension of σ to $A^{\mathbb{N}}$.
For instance if $A = \{1, 2\}$, $\sigma(1) = 12$, $\sigma(2) = 21$, u is the "Thue-Morse sequence";
if $\sigma(1) = 12$, $\sigma(2) = 1$, u is the "Fibonacci infinite word".
When α is a quadratic algebraic number, $0 < \Delta < 1$, the sequence

$$\chi(n,\alpha,\Delta) = \begin{cases} 1 & \text{if } 0 \le \text{Frac}\,(n\alpha) < \Delta \\ 0 & \text{if } \Delta \le \text{Frac}\,(n\alpha) < 1 \end{cases} \qquad (n \in N)$$

can also be described by a sequence $f(u_n)$ where $u = (u_n)_{n \in N}$ is a fixed point for an appropriate substitution on a finite alphabet A , and $f : A \to \{0,1\}$ (see [7] and [8]).
The substitutions have been studied from different points of view by many authors. See for instance [2-6] .
Let f be an application from A to $\mathbb{R}$, $u = (u_n)_{n \ge 1}$ a fixed point of σ and, for $n \ge 1$,

$$s^f(n) = \sum_{1 \le i \le n} f(u_i) \qquad (1.1)$$

The matter of this paper is to give results and methods for asymptotic evaluations of the quantities

$$s^f(n)\ ,\ \sum_{n<N} s^f(n)\ ,\ \sum_{n<N} (s^f(n) - \gamma \log_\theta n)^2 \quad \text{where}$$

$$\gamma = \lim_{N \to \infty} (N \operatorname{Log}_\theta N)^{-1} \sum_{n<N} s^f(n) \quad \text{and} \quad \sum_{n<N} (s^f(n))^k .$$

For $m \in A^*$, $i \in A$, let $L_i(m)$ count the number of occurrences of the letter i in the word m. The matrix $M = (L_i(\sigma(j)))_{(i,j) \in A^2}$ holds for a fundamental part in such calculations. Let $\{\theta_i / 1 \le i \le d'\}$ be the set of (distinct) eigenvalues of M such that $i \le j \Rightarrow |\theta_i| \ge |\theta_j|$. If M is primitive ($\exists n, M^n > 0$) we have, by the Perron - Frobenius theorem

(H) $\theta_1 \in \mathbb{R}_+$, $i \ge 2 \Rightarrow |\theta_i| < \theta_1$, and there is a unique vector $\Lambda = (\lambda_i)_{i \in A}$.

s.t. $\quad M\Lambda = \theta_1 \Lambda\ ,\ \sum_{i=1}^{d} \lambda_i = 1$.

Hereafter we assume that (H) is true, and that, moreover, $\theta_1 = \theta > 1$, $\theta_2 \in \mathbb{R}$, $i \ge 3 \Rightarrow |\theta_i| < |\theta_2|$.

For $i \in A$ $\quad \lim_{n \to \infty} |\sigma^n(i)| \theta^{-n}$ ($|m|$ = length of the word m) will be written $\varepsilon(i)$ (existence of these limits are insured for M primitive).

1.2 Results for $\theta_2 > 1$ (see [9]) .

Let α be the integer such that $\alpha + 1$ is the order of θ_2 in the minimal polynomial of M and $\beta = \log_\theta(\theta_2)$ $(0 < \beta < 1)$.

Then, there exists a continuous function F , defined for $x > 0$, such that

(i) $s^f(N) = (\Lambda.f)\, N + (\log_\theta N)^\alpha N^\beta F(N) + o((\log N)^\alpha N^\beta)$

(ii) $x > 0 \Rightarrow F(\theta x) = F(x)$

Moreover: $\exists K > 0$ s.t. $x \ge 1 \Rightarrow |F(x) - F(y)| \le K\, |x - y|^\beta$ and, apart from the case $F \equiv 0$, F is nowhere differentiable.

The function $\Phi(x) = x^\beta F(x)$ is proved to be "self-affine" in the base θ .

For instance the sums $S_1(N) = \sum_{n<N} (-1)^{s(3n)}$ and $S_2(N) = \sum_{n<N} (-1)^{r(n)}$ (where $s(n)$ = sum of binary digits of n , $r(n)$ = number of blocks 11 in binary expansion of n)

studied in [10] and [11] respectively are of the form (1.1) (see [12] and [1]).
In both these cases $\theta = 4$, $\alpha = 0$, $\Lambda.f = 0$, $\theta_2 > 1$ ($\theta_2 = 3$ in the first case, $\theta_2 = 2$ in the second).

1.3 Results for $\theta_2 = 1, \alpha = 0, \Lambda.f = 0$.

It is easy to prove that, if $\Lambda.f = 0$, $|\theta_2| = 1$, $\alpha = 0$, we have

$$s^f(n) = O(\log n) .$$

But, generally, $s^f(n) (\log n)^{-1}$ does not have a limit for $n \to \infty$. However, with some additional conditions, we get the following results :

1.3.1 There exists a real number γ such that

$$\sum_{n<N} s^f(n) = \gamma N \log_\theta(N) + O(N) .$$

1.3.2 There exists a real number δ such that

$$\sum_{n<N} (s^f(n) - \gamma \log_\theta(n))^2 = \delta N \log_\theta(N) + O(N) .$$

1.3.3 If γ in (1.3.1) is zero, and $k \in N$:

$$\sum_{n<N} (s^f(n))^{2k} = (2k-1)!!\, \delta^k N (\log_\theta N)^k + O(N(\log N)^{k-1}) .$$

where $(2k - 1)!!$ is the product of odd integers $\leq 2k - 1$.
The result (1.3.1) was already obtained in [13] by using properties of the "columns" of the system of numeration associated with the substitution.
In particular in the case where $\forall i \in A$ $f(\sigma(i)) = i$ (with $f(a_1 a_2 \ldots a_k) = \sum_{i=1}^{k} f(a_i)$) we have

$$\gamma = \theta^{-1} \sum_{mj \leq \sigma(i)} \lambda_i f(m)\, \varepsilon(j) ,$$

the summation holding for all $(m,j,i) \in A^* \times A \times A$ s.t. mj is a prefix of $\sigma(i)$.

For instance : $A = \{1,2\}$ $\sigma(1) = 112$, $\sigma(2) = 221$, $f(1) = -f(2) = 1$, we have $\theta = 3$, $\varepsilon_j = 1$, $\gamma = 0$; if $\sigma(1) = 112$, $\sigma(2) = 212$, f as above, we get : $\gamma = 1/3$.
Here we will develop an alternative proof of (1.3.1) which allows us to calculate easily δ and to prove (1.3.3).

We will give for γ a distinct formula as above; the consistency between the two results is easily obtained, proving first that $\sum_{i \in A} \lambda_i \ \varepsilon(i) = 1$.

The result 1.3.2 was conjectured by C. Godrèche et al. ([7]) . The result 1.3.3 was conjectured by J.M. Luck (private communication) in a more general form : for $\gamma \neq 0$ the l.h.s. of result 1.3.3 has to be replaced by $\sum_{n<N} (s^f(n) - \gamma \log_\theta (n))^{2k}$ and the r.h.s. is the same as above. In other words, the centered moments of the distribution of $s^f(n)$ have the same property as the moments of a centered gaussian distribution.

2. Systems of numeration associated with a substitution.

Here we will remember some useful results proved in [9] . If m, m' are words, the relation m is a prefix of m' is written $m \leq m'$; and $m < m' \Leftrightarrow m \leq m'$ and $m \neq m'$.
Remember that $1 < \sigma(1)$; $u = (u_i)_{i \geq 1}$ fixed point for σ s.t. $u_1 = 1$.

Definition : A sequence $(m_i, a_i)_{i = 0,1,\dots,n}$ in $A^* \times A$ is a admissible $(a \in A)$ iff

(i) $m_n a_n \leq \sigma(a)$

(ii) $1 \leq i \leq n \Rightarrow m_{i-1} a_{i-1} \leq \sigma(a_i)$.

Theorem. Let N be an integer ≥ 1 . Then there exists a unique integer $\nu = \nu(N)$ and a unique 1-admissible sequence $(m_i, a_i)_{i = 0,\dots, \nu}$ such that $m_\nu \neq \omega$ and

$$u_1 u_2 \dots u_N = \sigma^\nu (m_\nu) \ \dots \ \sigma^o (m_o) \ . \tag{2.1}$$

Note that the "digits" m_i belong to a finite set, the set of the proper prefixes of the words $\sigma(i)$, $i \in A$, and are not independent of each other.
For instance, if $A = \{1,2\}$, $\sigma(1) = 12$, $\sigma(2) = 1$: $m_i \in \{\omega,1\}$ and in the "admissible writing" (2.1) , $(m_{i+1},m_i) \neq (1,1)$ for al i , $0 \leq i \leq \nu -1$.

In the case $\theta > 1$, it can be proved

$$\nu (N) - \log_\theta(N)= O(1) \tag{2.2}$$

and, if $\varepsilon(\omega) = 0$, $\varepsilon(a_1 \dots a_k) = \sum_{i=1}^{k} \varepsilon(a_i)$

$$N - \sum_{i=o}^{\nu} \varepsilon(m_i)\theta^i = O(\log N) \qquad \text{if } \theta_2 = 1, \alpha = 0 \ . \tag{2.3}$$

We mention, although it will be not used here, that there exists a system of representations of some real numbers in the form :

If $a \in A$, $x \in [0,\varepsilon(a)[$, there exists $(m_i, a_i)_{i \geq 1}$, $(m_i, a_i) \in A^* \times A$ such that

$$\text{(i)} \quad x = \sum_{i=1}^{\infty} \varepsilon(m_i)\theta^{-i}$$

$$\text{(ii)} \quad m_1 a_1 \leq \sigma(a) \; ; \; i \geq 2 \Rightarrow m_i a_i \leq \sigma(a_{i-1})$$

This representation is unique, with the additional condition

$$\text{(iii)} \quad \forall I, I \in N, \exists i > I, (m_i, a_i) \neq \sigma(a_{i-1}) .$$

Such a representation is used in [9] to demonstrate the properties of the function F in results (1.2).

3. Proof of the results in the case $\theta_2 = 1$, $\alpha = 0$, $\Lambda.f = 0$.

Hereafter we assume that

$$\forall a \in A , \; f(\sigma(a)) = f(a) . \qquad (3.1)$$

This condition is not really restrictive for the results of this section. Indeed one can proves (see [9], lemma 2.2) that, if $\theta_2 = 1$, $\alpha = 0$, $\Lambda.f = 0$, there exist numbers $f'(a)$ $(a \in A)$ such that

$$s^f(n) = s^{f'}(n) + O(1) \text{ and (3.1) true for } f'.$$

3.1 A fundamental lemma

For $x \in \mathbb{R}_+$ let $M(x)$ be the $d \times d$ matrix whose element of the arrow a and column b is

$$M(x) = \sum_{m \in A^* / mb \leq \sigma(a)} x^{f(m)} .$$

(Note that $M(1) = {}^tM$, M as in § 1.1.)

Assume that there exists an open interval $I \subset \mathbb{R}_+$, $1 \in I$, such that for $x \in I$ $M(x)$ has distinct eigenvalues : $\theta(x) = \theta_1(x)$ $(j = 2,..., d')$ with $\forall j \geq 1 : \theta_j(1) = \theta_j$, and that there exist $v_j(x) \in \mathbb{R}^d$ such that

$$M(x)\, v_j(x) = \theta_j\, v_j(x) \quad \text{and } \mathbb{1} = \sum_{j=1}^{d'} v_j(x) \qquad \mathbb{1} = {}^t(1,1,...1) \qquad (3.2)$$

<u>Example</u> : $A = \{1,2,3\}$ $\sigma(1) = 13$ $\sigma(2) = 122$ $\sigma(3) = 12$

$$f(1) = -f(2) = 1 , \; f(3) = 0 . \; M(x) = \begin{pmatrix} 1 & 0 & x \\ 1 & x+1 & 0 \\ 1 & x & 0 \end{pmatrix}$$

$\theta = 1+\sqrt{2}$, $\theta_2 = 1$, $\theta_3 = 1-\sqrt{2}$

$\theta(x) = \frac{1}{2}\left[x+1+(x^2+6x+1)^{1/2}\right]$, $\theta_2(x) = 1$, $\theta_3(x) = \frac{1}{2}\left[x+1-(x^2+6x+1)^{1/2}\right]$.

$M(x)$ is diagonalisable in an open interval containing 1 , so that (3.2) is true. Moreover, let us define, for $m \in A^*$, $1 \le j \le d'$

$$k_{m,j}(x) = \sum_{m'b \le m} x^{f(m')} v_{b,j}(x) \tag{3.3}$$

where $v_j(x) = (v_{b,j}(x))_{1 \le b \le d}$.

Lemma

For $v \in A^*$, $x \in I$, we define $S_v(x) = \sum_{w \in A^* / w < v} x^{f(w)}$.

If $v = \sigma^v(m_v)\,\sigma^{v-1}(m_{v-1}) \dots \sigma^0(m_0)$ $(m_i \in A^*)$ we have

$$S_v(x) = \sum_{j=1}^{d'} \sum_{i=0}^{v} k_{m_i,j}(x)\; x^{f(M_i)} \left[\theta_j(x)\right]^i \tag{3.4}$$

where $M_v = \omega$ and $i < v \Rightarrow M_i = m_v \dots m_{i+1}$.

Proof The definition of $M(x)$, $S_v(x)$, and (3.1) imply that, for all $n \in N$, and $x > 0$
$\left(S_{\sigma^n(a)}(x)\right)_{a \in A} = M(x)\left(S_{\sigma^{n-1}(b)}(x)\right)_{b \in A}$; $\left(S_a(x)\right)_{a \in A} = 1$
[Note that $w < \sigma^n(a) \Leftrightarrow \exists (m,w',b) \in A^* \times A^* \times A$ s.t.
$w = \sigma^{n-1}(m)\,w'$, $mb \le \sigma(a)$, and $w' < \sigma^{n-1}(b)$.]
Thus, by (3.22) , for all $a \in A$:

$$S_{\sigma^n(a)}(x) = \sum_{j=1}^{d'} \theta_j^n(x)\, v_{a,j}(x) . \tag{3.5}$$

Moreover, if $v = \sigma^v(m_v)\,v'$, $v' \in A^*$ we have

$$S_v(x) = \sum_{mb \le m_v} x^{f(m)}\, S_{\sigma^v(b)}(x) + x^{f(m_v)}\, S_{v'}(x) . \tag{3.6}$$

[For (3.6) let us remark that $w < v \Leftrightarrow w = \sigma^v(m_v)\,w'$ with $mb \le m_v$ and $w' < \sigma^v(b)$ or $w = \sigma^v(m_v)w''$ with $w'' < v'$] .

(3.4) is then an easy consequence of (3.5) (3.6) and (3.3) . Let us remark that for all $a \in A$ $v_{a,1}(1) = \varepsilon(a)$, by $x = 1$ in (3.4) and $n \to \infty$; thus $k_{m,1}(1) = \sum_{m'b \le m} \varepsilon(b) = \varepsilon(m)$. (3.7)

3.2 Proof of result 1.3.1

We can now state the result 1.3.1 in a more precise form : Let us suppose that $\theta_j(x)$ and $v_j(x)$ are differentiable, for $x = 1$, and define $\gamma = \theta^{-1}\theta'(1)$.

Then for $\Sigma_1(N) = \sum_{n<N} s^f(n)$, we have

$$\Sigma_1(N) = \gamma N \log_\theta N + O(N) \quad . \tag{3.8}$$

Proof

If $v = u_1 \ldots u_N$ $\Sigma_1(N) = (S_v(x))'_{x=1}$,
then we can use the representation (2.1) of v and the formula (3.4) .
It is easy to prove, using $j > 1 \Rightarrow |\theta_j(1)| \leq 1$, and (2.2) that the terms of $(S_v(x))'_{x=1}$ in (3.4) corresponding to $j \neq 1$ are in $O((\log N)^2)$.

Then, using (3.7) we get

$$\Sigma_1(N) = \theta'(1) \sum_{i=0}^{v} i\,\varepsilon(m_i)\,\theta^{i-1} + \sum_{i=0}^{v} \left[k_{m_i,1}(x)\; x^{f(M_i)}\right]'_{x=1} \theta^{i} + O\left((\log N)^2\right) . \tag{3.9}$$

Thence (3.8) is proved using (3.9) , (2.2) and (2.3)

3.3 Proof of the result 1.3.2

First we calculate $\sum_{n<N} (s^f(n))^2 = (S_v(x))''_{x=1} + (S_v(x))'_{x=1}$ (3.10)

for $v = u_1 \ldots u_N$.

We suppose $\theta_j(x)$ and $v_j(x)$ to be differentiable to order 2 in I .
By (2.2) and (2.3) we have

$$\sum_{i=0}^{v} i^2 \varepsilon(m_i)\,\theta^{i} = -v^2 \sum_{i=0}^{v} \varepsilon(m_i)\,\theta^{i} + 2v \sum_{i=0}^{v} i\,\varepsilon(m_i)\,\theta^{i} + O(N) \; .$$

Using (3.10) , (3.4) and (3.9) we find

$$\sum_{n<N} (s^f(n))^2 = \delta \sum_{i=0}^{v} i\,\varepsilon(m_i)\,\theta^{i} - \gamma^2 v^2 N + 2\gamma v \Sigma_1(N) + O(N) \, , \tag{3.11}$$

where $\delta = \theta^{-1}\theta''(1) - \gamma^2 + \gamma$ (3.12)

Then, for $\Sigma_2(N) = \sum_{n<N} (s^f(n) - \gamma \log_\theta n)^2$ we use a partial summation for

$\sum_{n<N} s^f(n) \log_\theta n$ and $\sum_{n<N} (\log_\theta n)^2$ and we find

$$\Sigma_2(N) = \sum_{n<N} (s^f(n))^2 - 2\gamma \log_\theta N \, \Sigma_1(N) + \gamma^2 N (\log_\theta N)^2 + O(N) . \qquad (3.13)$$

Then (3.11), (3.13), (3.9), (2.2), (2.3) give

$$\Sigma_2(N) = \delta N \log_\theta N + O(N) \quad \text{with } \delta \text{ as in (3.12)} . \qquad (3.14)$$

For instance, in the example of § (3.1) $\gamma = 1/2$ and $\delta = \sqrt{2}/8$.

3.4 Proof of result 1.3.3

Remember that here we assume $\gamma = 0$; thus $\theta'(1) = 0$.

Let k be an integer; we will suppose $\theta_j(x)$, $v_j(x)$ $2k$. differentiable for $x = 1$.

We have (Leibniz formula) for any m. differentiable function φ:

$$(\varphi^i(x))^{(m)} = \sum_{j_1+j_2+\dots+j_i=m} \frac{m!}{j_1!j_2!\dots j_i!} \varphi^{(j_1)}_{(x)} \varphi^{(j_2)}_{(x)} \cdots \varphi^{(j_i)}_{(x)} . \qquad (3.15)$$

We have to calculate $(S_v(x))^{(2k)}_{x=1}$ in (3.4) where $v = u_1 \dots u_N$.

First, using $|\theta_j(1)| \le 1$ for $j > 1$ and (3.14) for $\varphi(x) = \theta_j(x)$, it is easy to prove that the terms of $(S_v(x))^{(2k)}_{x=1}$ corresponding to $j \ne 1$ are in $O((\log N)^{(2k+1)})$.

Now, using (3.15) for $\varphi = \theta$, we need not consider the case where some of the $j_\ell = 1$ (because $\theta'(1) = 0$). And if in the sequence $j_1 \dots j_i$ h integers j_ℓ are distinct from 0 and 1 we have $h \le \left[\frac{m}{2}\right]$.

Moreover there are $\begin{pmatrix} i \\ i-h \end{pmatrix} K \le i^h K$ such sequences with $(i-h)$ zero elements ($K = (m-1)^h$ uniformly bounded in i).

Thence, if $A_{i,m} = (k_{m_i}(x)\, x^{f(Mi)})^{(2k-m)}_{x=1} (\theta^i(x))^{(m)}_{x=1}$ we have for $m < 2k$

$(\theta^i(x))^{(m)}_{x=1} = O(i^{k-1}\theta^i)$ (because $h \le k-1$) and $\sum_{i=0}^{\nu} A_{i,m} = O(\nu^{k-1}\theta^\nu)$ by (2.2).

This proves also that

$$(S_v(x))^{m}_{x=1} = O(N(\log N)^{(k-1)}) \quad \text{for } m < 2k . \qquad (3.16)$$

In the same way, we have

$$(\theta^{i}(x))^{(2k)}_{x=1} = \frac{(2k)!!}{2^k}\binom{i}{k}\theta^{i-k}(\theta''(1))^k + O(i^{k-1}\theta^{i}) , \qquad (3.17)$$

the first term corresponding to the cases where i-k of the j_ℓ are 0 , and the k others are 2, the second term relating to the other cases.

But $\frac{(2k)!}{2^k k!} = (2k-1)!!$. Then, using (3.7) and $\delta = \theta''(1)\theta^{-1}$ (here $\gamma = 0$) we have

$$S_v(x)^{(2k)}_{x=1} = (2k-1)!!\,\delta^k \sum_{i=0}^{v} i_k\, \varepsilon(m_i)\,\theta^{i} + O(N(\mathrm{Log}\,N)^{k-1})$$

The result follows at once using (3.16) , (2.2) and (2.3) .

4. Final Remarks

4.1 It is easy to prove that, generally ($\gamma \neq 0$),

$$\sum_{n<N} (s^f(n))^k = \gamma^k N(\log_\theta N)^k + O(N(\log_\theta N)^{k-1}) \quad .$$

4.2 Concerning the result 1.3.1 , A. Thomas and the author of this paper will demonstrate in a forthcoming paper that the term O(N) is, more precisely,

$$N F(N) + o(N) ,$$

where F(x) is a continuous nowhere differentiable (under some general assumptions) function from $\mathbb{R}_+$ to $\mathbb{R}$, and

$$x > 0 \Rightarrow F(\theta x) = F(x) \ .$$

It would be probably proved that the error term O(N) in the result 1.3.2 is of the same type.

I thank Prof. A. Thomas for his helpful discussions concerning this work.

References

[1] G. CHRISTOL, T. KAMAE, M. MENDES-FRANCE et G. RAUZY : Suites algébriques, automates et substitutions, Bull. Soc. Math. France 108, (1980), 401-419.

[2] F.M. DEKKING : Recurrent sets, Advances in Math, 44 (1982), 78-104.

[3] E. BOMBIERI and J.E. TAYLOR : Which distributions of matter diffract ? An initial investigation , J. Phys. (Paris) Coll. C 3 (1986).

[4] M. QUEFFELEC : Substitutions, Dynamical Systems, Spectral Analysis (Springer, 1987).

[5] C. MAUDUIT : Sur l'ensemble normal des substitutions de longueur quelconque, J. Number Theory 29 (1988), 235-250 .

[6] G. RAUZY : Sequences defined by iterated morphisms, Advances International Workshop on Sequences, Amalfi Coast (1988) , to appear.

[7] C. GODRECHE, J.M. LUCK, and F. VALLET : Quasiperiodicity and types of order : a study in one dimension , J. Phys. A 20 (1987) 4483-4499.

[8] S. AUBRY, C. GODRECHE and J.M. LUCK: Scaling Properties of a structure Intermediate between Quasiperiodic and Random , J. Statistical Phys. 51 (1988) , 1033-1075.

[9] J.M. DUMONT et A. THOMAS : Systèmes de Numération et Fonctions Fractales relatifs aux substitutions, Theor. Comp. Sci. Proceedings of the international conference on Arithmetics and Coding Systems (1987) .

[10] J. COQUET : A summation formula related to the binary digits, Inv. Math 73 (1983) 107-115.

[11] J. BRILLHART, P. ERDOS and P. MORTON : On sums of Rudin-Shapiro coefficients II, Pacific J. Maths 107 (1983)271-323.

[12] A. COBHAM : Uniform tag sequences, Math. Syst. Theory 6 (1972) 164-192.

[13] J.M. DUMONT : Formules sommatoires et systèmes de Numérations liés aux substitutions, Sém. Th. Nombres de Bordeaux (1987-1988) Exp. n° 39.

The Inhomogeneous Ising Chain and Paperfolding

M. Mendès France

Département de Mathématiques, Université Bordeaux I,
F-33405 Talence Cedex, France

Abstract. It is a well known fact that the Ising chain with nearest neighbor interaction cannot have phase transitions. By modifying the concept of phase transition and by allowing complex temperatures we shall discover discontinuities in the behavior of the partition function by giving a geometrical interpretation.

I. The Ising Chain

We are given an infinite sequence $\epsilon = (\epsilon_n), n \geq 0$ of ± 1 and we consider the open Ising chain governed by the Hamiltonian

$$\mathcal{H}(\sigma) = -J \sum_{q=0}^{N-1} \epsilon_q \sigma_q \sigma_{q+1}$$

where $\sigma \in \{-1,+1\}^N, \sigma_N = +1$, and where $J > 0$ is the coupling constant. Notice we have no external field. The partition function

$$Z(\beta, N) = \sum_{\sigma \in \{-1,+1\}^N} exp(-\beta \mathcal{H}(\sigma)) \qquad (\beta = \frac{1}{T} \geq 0)$$

is easily computed (see for example [1] p. 144) :

$$Z(\beta, N) = (2cosh \beta J)^N. \qquad (1)$$

We propose to discuss a geometrical interpretation of $Z(\beta, N)$ when β is a complex number.

2. Paperfolding

Consider a sheet of paper which we fold according to the instructions

$$\epsilon_{N-1}, \epsilon_{N-2}, \cdots, \epsilon_0.$$

This means that we first fold the sheet in two, right half over left if $\epsilon_{N-1} = +1$ and right half under left if $\epsilon_{N-1} = -1$. See Fig. 1.

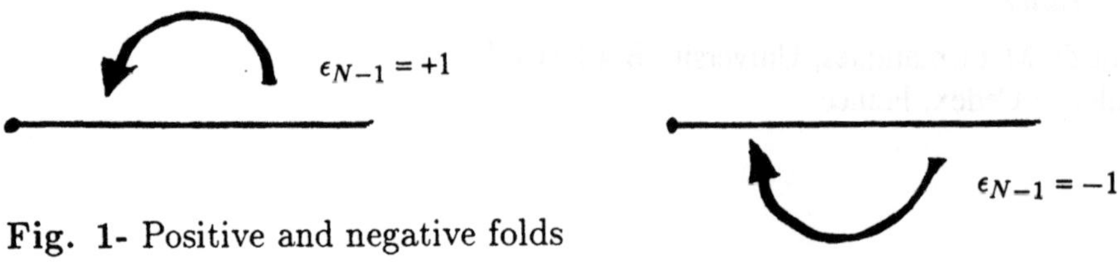

Fig. 1- Positive and negative folds

This fold having been performed we follow instruction ϵ_{N-2} to fold the folded sheet once more, etc. (Fig. 2).

Fig. 2- Folding twice

We assume that at each folding operation, the sheet of paper is stretched to twice its length so that the diameter of the folded edge is unchanged.

After N folding operations (and stretching), we unfold the sheet of paper to the angle $\alpha \in [0, \pi]$, and we observe the edge. For $\alpha = 0$ the sheet stays folded. For $\alpha = \pi$ the sheet is completely flat. For $0 < \alpha < \pi$ the edge is a broken line $\Gamma(\alpha, N)$ containing 2^N sides.

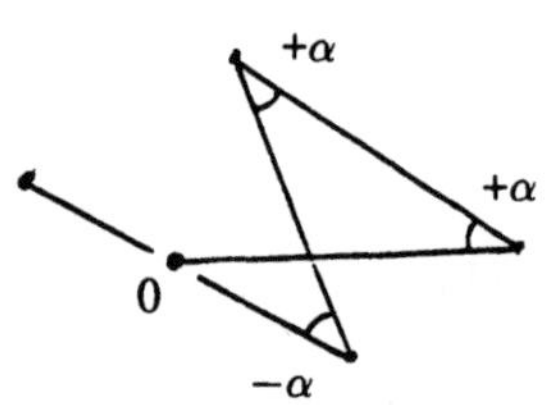

Fig.3- Unfolding to the angle α

The sheet of paper is allowed to go through itself hence provoking multiple points in the curve $\Gamma(\alpha, N)$, (see Fig. 3).

The sequence of signs of the angles $\pm\alpha$ that one encounters as one travels along $\Gamma(\alpha, N)$ is called a paperfolding sequence (PFS for short) of length $2^N - 1$:

$$f(1), f(2), \cdots, f(2^N - 1).$$

The space

$$[\cup_{k=1}^{\infty}\{-1,+1\}^k] \cup \{-1,+1\}^{\mathbb{N}}$$

is compact with respect to the topology according to which two sequences are close together if their terms coincide up to some order. The set

$$\{(f(1), f(2), \cdots, f(2^N - 1))/N \in \mathbb{N}\}$$

has limit points known as infinite PFS. These sequences are easily characterised (see for example [2] or [3]) :

1. $(f(n))$ is $PFS \Rightarrow (f(2n))$ is PFS
2. $(f(n))$ is $PFS \Rightarrow (f(2n+1)) = (-1)^n f(1)$

THEOREM. *Let $(f(n))$ be an infinite PFS generated by the folding instructions $\cdots, \epsilon_3, \epsilon_2, \epsilon_1, \epsilon_0$. Let*

$$n = \sum_{q=0}^{\infty} e_q(n) 2^q \qquad e_q(n) = 0 \text{ or } 1,$$

be any nonnegative integer. Then

$$\sum_{k=1}^{n} f(k) = \frac{1}{2} \sum_{q=0}^{\infty} (1 - \sigma_q(n)\sigma_{q+1}(n))\epsilon_q \tag{2}$$

where

$$\sigma_q(n) = 1 - 2e_q(n) \in \{-1,+1\}.$$

It should be noted that for large $q(q > logn/log2), e_q(n) = 0$ hence

$$\sigma_q(n)\sigma_{q+1}(n) = 1.$$

The sum which appears in the right hand side of formula (2) is thus finite.

Proof of the theorem. Let $\epsilon = (\epsilon_n)$ be the sequence of folding instructions and let $f = f_\epsilon$ be an associated PFS. Define

$$S_\epsilon(x) = \sum_{k=1}^{[x]} f_\epsilon(k), \qquad x > 0.$$

Let T be the erasing operator :

$$T(\ldots, \epsilon_2, \epsilon_1, \epsilon_0) = \ldots \epsilon_2, \epsilon_1 \ ;$$

T can be thought of as the shift. Then

$$\begin{cases} f_\epsilon(2n) = f_{T\epsilon}(n), & n \geq 1 \\ f_\epsilon(2n+1) = (-1)^n f_\epsilon(1) = (-1)^n \epsilon_0, & n \geq 0 \end{cases}$$

and

$$f_{T^q\epsilon}(1) = \epsilon_q, \qquad q = 0, 1, 2, \ldots$$

Now

$$\begin{aligned} S_\epsilon(n) &= \sum_{1 \leq k \leq \frac{n}{2}} f_\epsilon(2k) + \sum_{0 \leq k < \frac{n}{2}} f_\epsilon(2k+1) \\ &= \sum_{1 \leq k \leq \frac{n}{2}} f_{T\epsilon}(k) + f_\epsilon(1) \sum_{1 \leq k < \frac{n}{2}} (-1)^k \\ &= S_{T\epsilon}(\frac{n}{2}) + f_\epsilon(1) \frac{1 - \sigma_0(n)\sigma_1(n)}{2}. \end{aligned}$$

Iterating the formula, we obtain

$$S_\epsilon(n) = \frac{1}{2}(1 - \sigma_0(n)\sigma_1(n)) f_\epsilon(1) + \frac{1}{2}(1 - \sigma_0(\frac{n}{2})\sigma_1(\frac{n}{2})) f_{T\epsilon}(1) + S_{T^2\epsilon}(\frac{n}{2^2})$$

where $\sigma_q(x)$ stands for $\sigma_q([x])$. Noticing that

$$\sigma_q(\frac{n}{2}) = \sigma_{q+1}(n)$$

and iterating Q times we get

$$S_\epsilon(n) = \frac{1}{2} \sum_{q=0}^{Q-1} (1 - \sigma_q(n)\sigma_{q+1}(n)) f_{T^q\epsilon}(1) + S_{T^Q\epsilon}\left(\frac{n}{2^Q}\right).$$

For $n2^{-Q} < 1$,

$$S_{T^Q\epsilon}\left(\frac{n}{2^Q}\right) = 0.$$

Thus $N > logn/log2$ implies

$$\begin{aligned} S_\epsilon(n) &= \frac{1}{2} \sum_{q=0}^{N-1} (1 - \sigma_q(n)\sigma_{q+1}(n)) \epsilon_q \\ &= \frac{1}{2} \sum_{q=0}^{\infty} (1 - \sigma_q(n)\sigma_{q+1}(n)) \epsilon_q. \qquad \text{QED} \end{aligned}$$

3. Geometrical interpretation

We observe that for $n < 2^N$

$$S(n) = S_\epsilon(n) = \frac{1}{2}\sum_{q=0}^{N-1} \epsilon_q - \frac{1}{2}\sum_{q=0}^{N-1} \epsilon_q \sigma_q(n)\sigma_{q+1}(n)$$

and that when n runs through the range $[0, 2^N[$, the vector

$$\sigma = (\sigma_0(n), \sigma_1(n), \ldots, \sigma_{N-1}(n))$$

describes the set $\{-1,+1\}^N$. The bijection $\{0,1,2,\ldots,2^N-1\} \longleftrightarrow \{-1,+1\}^N$ allows us to identify integers $n < 2^N$ with spin configurations $\sigma \in \{-1,+1\}^N$. We then write $\mathcal{H}(\sigma) = \mathcal{H}(n)$. Choosing $J = 1/2$, we have

$$\mathcal{H}(n) = S(n) - \frac{1}{2}\sum_{q=0}^{N-1} \epsilon_q$$

and

$$Z(\beta, N) = \left(exp\frac{1}{2}\beta \sum_{q=0}^{N-1} \epsilon_q \right) \sum_{n=0}^{2^N-1} exp\left(-\frac{1}{2}\beta \sum_{k=1}^{n} f(k) \right).$$

Choose

$$\beta = -2i(\pi - \alpha), \quad 0 \leq \alpha \leq \pi.$$

$Z(\beta, N)$ is a product of the unimodular factor

$$exp(-i(\pi-\alpha)\sum_{q=0}^{N-1} \epsilon_q)$$

with

$$\zeta(\alpha, N) = \sum_{n=0}^{2^N-1} expi(\pi-\alpha)\sum_{k=1}^{n} f(k).$$

The sum $\zeta(\alpha, N)$ represents a broken line $\Gamma(\alpha, N)$ in the complex plane with vertices at the points

$$\sum_{n=0}^{M-1} expi(\pi-\alpha)\sum_{k=1}^{n} f(k), \quad M = 0,1,2,\ldots,2^N-1.$$

$\Gamma(\alpha, N)$ coincides with the curves discussed in 2.

The infinite sum

$$\zeta(\alpha) = \sum_{n=0}^{\infty} expi(\pi - \alpha) \sum_{k=1}^{n} f(k)$$

represents the family of "paperfolding curves" which are limit points of $\Gamma(\alpha, N)$ as N goes to infinity.

We propose to study these curves Γ_α. Let $\delta(\alpha)$ be the diameter of Γ_α :

$$\delta(\alpha) = \sup_{0 \le M \le N < \infty} \left| \sum_{n=M}^{N} expi(\pi - \alpha) \sum_{k=1}^{n} f(k) \right| .$$

Formula (1) implies

$$|\zeta(\alpha, N)| = \left| 2sin\frac{\alpha}{2} \right|^N$$

so that $|\zeta(\alpha, N)|$ tends to 0 or infinity according to whether $\alpha < \pi/3$ or $\alpha > \pi/3$.

As a consequence,

$$\delta(\alpha) = \begin{cases} \text{finite if} \quad 0 \le \alpha < \pi/3 \\ \text{infinite if} \quad \pi/3 \le \alpha \le \pi \end{cases}$$

(if $\alpha < \pi/3$, it is easy to see that

$$\delta(\alpha) \le (1 - 2sin\frac{\alpha}{2})^{-1}).$$

The graph of δ as a function of α is represented in Fig. 4.

Question 1- Is $\alpha \mapsto \delta(\alpha)$ continuous in the range $[0, \pi/3[$?
Question 2- Is it increasing ?
Question 3- It is known that Γ_α is self-avoiding for $\alpha > \frac{\pi}{2}$. For $\alpha < \frac{\pi}{2}$,

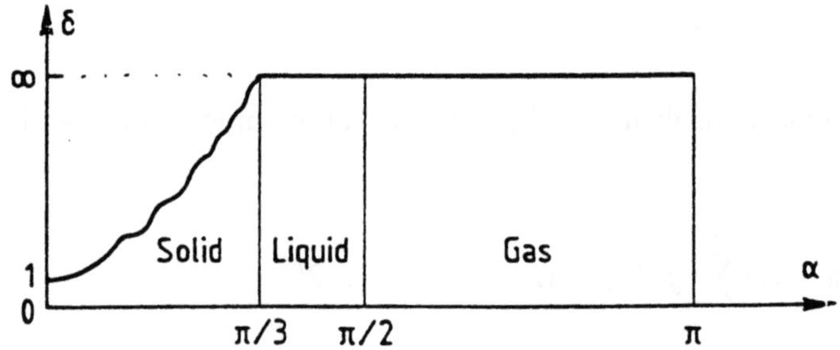

Fig. 4- The curve $\delta(\alpha)$ where it is shown that the liquid phase boils at 90°

Γ_α has infinitely many double points and edges overlap. I would then like to argue as follows :

For $0 \leq \alpha < \pi/3$ the curve Γ_α has a high density of double points and overlapping. For $\alpha = \pi/3$, the density of double points and overlapping decreases abruptly. For $\alpha > \pi/2$, there are no more double points, nor overlapping. (See [4] where the dragon curve is discussed. See also Fig. 5). Hence two phase transitions, one at $\pi/3$ and one at $\pi/2$.

How can I justify my intuitions ?

Question 4- Let $(f(n))$ be an arbitrary sequence of ± 1. Study of the broken line

$$\sum_{n=0}^{\infty} expi(\pi - \alpha) \sum_{k=1}^{n} f(k)$$

as a function of the angle α is required.

Fig. 5- The dragon curve

Question 5- Can one discuss the Ising chain

$$\mathcal{H}(\sigma) = -J \sum_{q=0}^{N-1} \epsilon_q \sigma_q \sigma_{q+1} - H \sum_{q=0}^{N-1} \sigma_q$$

in a similar way ?

4. A final remark

The paperfolding sequence is related to the Rudin-Shapiro sequences (see [2]). On the other hand ALLOUCHE and MENDES FRANCE [5] observed that the Rudin Shapiro sequences can be thought of as an Ising chain with complex temperature. The present article illustrates the then necessary link between the Ising chain and paperfolding :

Paperfolding —— Rudin Shapiro sequences

Ising chain

5. Appendix

In December 1988, T. KAMAE sent me a letter in which he proves that the function $\delta(\alpha)$ is continuous at all points with the possible exception of the α's such that α/π is a rational number p/q where $(p, q) = 1$, p and q odd.

This partially answers question 1. T. KAMAE also establishes that when α/π is irrational, the closure of the curve Γ_α is a disk centered at the origin. G. RAUZY made some computer graphics which illustrate the behavior of $\Gamma(\alpha, N)$ as a function of α and N.

May KAMAE and RAUZY accept my warmest thanks.

REFERENCES

1. C.J. THOMPSON, "Mathematical Statistical Mechanics," Princeton University Press, 1979.
2. M. MENDES FRANCE, G. TENENBAUM, *Dimension des courbes planes, papiers pliés et suites de Rudin-Shapiro*, Bull. Soc. Math. France **190** (1981), 207–215.
3. M. DEKKING, M. MENDES FRANCE, A VAN DER POORTEN, *Folds !*, Math. Intell. **4** (1982), 130–138, 173–181, 190–195.
4. C. DAVIS, D. KNUTH, *Number representation and dragon curves I,II*, Jour. Recreational Math **3** (1970), 61–81, 133–149.
5. J.P. ALLOUCHE, M. MENDES FRANCE, *Suite de Rudin-Shapiro et modèle d'Ising*, Bull. Soc. Math. France **113** (1985), 273–283.

Part IV

Dynamical and Stochastic Systems

A Nonlinear Evolution with Travelling Waves

F.M. Dekking

Delft University of Technology, Department of Mathematics and Informatics,
Julianalaan 132, NL-2628 BL Delft, The Netherlands

Abstract. We consider a discrete time, discrete space nonlinear evolution which describes irregular filament growth, polymers on disordered trees and branching random walk. A limit theorem is proven which indicates travelling wave behaviour of the evolution.

Let $n = 0, 1, 2, \ldots$ be a discrete time parameter. The *state at time* n will be a probability distribution f_n on the integers (i.e. $f_n : \mathbb{Z} \to [0,1]$ with $\sum_k f_n(k) = 1$), or rather the *tails* F_n of this distribution given by $F_n(k) = f_n(k) + f_n(k+1) + \ldots$ for each integer k. The evolution is given by

$$F_{n+1} = \Phi(F_n) \quad n = 0, 1, 2, \ldots$$

for a Φ which maps tails to tails.

In our main example Φ is "quadratic averaging": for a fixed $p \in [0,1]$ the evolution is given by

$$F_{n+1}(k) = pF_n^2(k-1) + (1-p)F_n^2(k) \quad k \in \mathbb{Z}. \tag{1}$$

As initial state we usually take point mass at the origin:

$$F_0(k) = 1 \quad \text{for } k \leq 0, \quad F_0(k) = 0 \quad \text{for } k > 0.$$

Before we discuss the behaviour of F_n as $n \to \infty$, we give three closely related interpretations of equation (1).

I. Irregular filament growth.
At time 0 we are given a single cell of thickness 1. At each time instant cells give birth to exactly two new cells, which may have thickness 1 both or 2 both, with probability $1-p$ and p (for some fixed $p \in [0,1]$), independent of all other cells in the same or previous generations.
The parent cell keeps the same thickness, and is elongated. See Figure 1 for a simulation of this model over 8 time instants (with $p = \frac{1}{2}$). We are interested in the *thinnest spot* of the filament. Let $F_n(k)$ be the probability that the filament has minimal thickness $n + k + 1$ at time n. By conditioning on the value of

Springer Proceedings in Physics, Vol. 47 Number Theory and Physics
Editors: J.M. Luck · P. Moussa · M. Waldschmidt

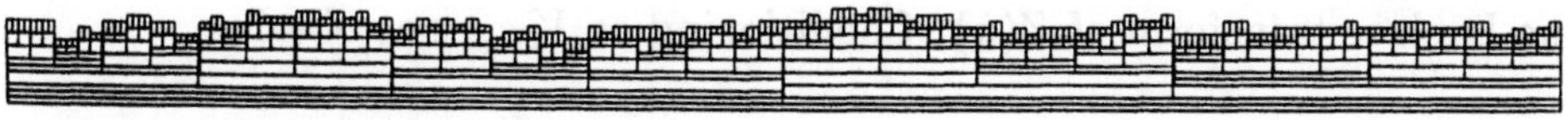

Figure 1. Irregular filament growth.

the thickness of the two cells in the second layer, one finds that the F_n satisfy equation (1).

Getting ahead of things to come we mention here that for $p < \frac{1}{2}$ the minimal thickness behaves in a well-controlled way; for example, if $p = \frac{1}{4}$, then for *all* n the minimal thickness will be either $n+1$ or $n+2$ with probability at least 0.97.

II. Polymers on disordered trees.

This interpretation is from Derrida and Spohn [4]. (I am grateful to H. Kunz for pointing out this paper to me at Les Houches). These authors consider one branch of a Cayley tree (with root 0), where each node has 3 neighbours (this structure is also known as the binary tree), and study self-avoiding walks on this tree. Each bond (i,j) of the tree has been assigned independently a random potential V_{ij}, where $V_{ij} = 0$ with probability $1-p$ and $V_{ij} = 1$ with probability p. A self-avoiding walk w on the tree is simply a sequence of sites $w = (i_1, \ldots, i_n)$, such that $(0, i_1)$ and (i_k, i_{k+1}) are bonds for $k = 1, \ldots, n-1$ (cf. Figure 2). The *energy* $E(w)$ of the walk is defined by

$$E(w) = V_{0,i_1} + V_{i_1 i_2} + \ldots V_{i_{n-1} i_n}$$

if $w = (i_1, \ldots, i_n)$. We denote $|w| = n$ for the length of a walk.

Let

$$Z(n) = \sum_{w:|w|=n} \exp(-\beta E(w))$$

be the partition function of this model (as usual $\beta \in \mathbb{R}$ is the inverse temperature).

We denote by $\mathbb{E}$ the expectation operator (w.r.t. the random potentials).

Let

$$G_n(\lambda) = \mathbb{E} \exp(-\lambda Z(n))$$

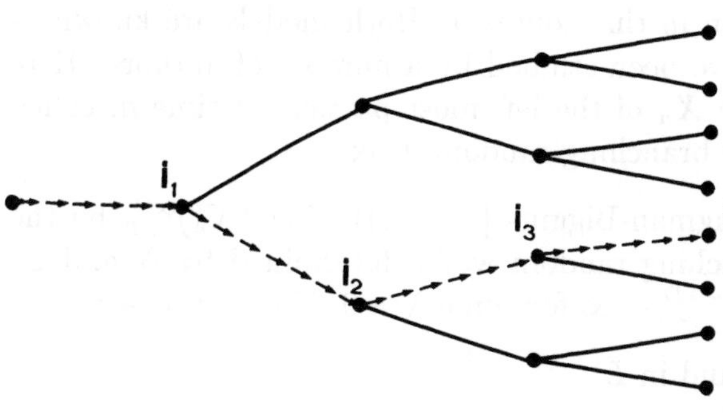

Figure 2. Self-avoiding walk on a binary tree.

be the Laplace transform of $Z(n)$. Conditioning on V_{0,i_1}, one obtains

$$G_{n+1}(\lambda) = (1-p)\mathbb{E}e^{-\lambda[Z'(n)+Z''(n)]} + p\mathbb{E}e^{-\lambda[e^{-\beta}Z'(n)+e^{-\beta}Z''(n)]}$$

where $Z'(n)$ and $Z''(n)$ are independent with distribution $Z(n)$, hence

$$G_{n+1}(\lambda) = (1-p)G_n^2(\lambda) + pG_n^2(e^{-\beta}\lambda).$$

Putting $F_n(k) = G_n(e^{k\beta})$, we see that the F_n satisfy (1).
Karp and Pearl [6] treat the same model, but the energies are instead called *costs*. They consider the problem of finding a cheapest walk of length n, for fixed n. In view of things to come, it is interesting to mention here that they show that for $p < \frac{1}{2}$ there exists an algorithm to find a cheapest walk in linear expected time, but for $p > \frac{1}{2}$, *every* algorithm which guarantees finding an exact cheapest walk, or even a walk within a fixed cost ratio of the cheapest, must run in exponential time.

III. Branching random walk.
Particles are sitting at integer sites. At time n a particle at x moves to $x+1$ with probability p, or stays at x (with probability $1-p$), and then splits into two particles. All movements are independent of each other. We start with one particle at the origin at time 0. Let X_n be the position of the left-most particle at time n (so $X_0 = 0$). Define

$$F_n(k) = \mathbb{P}[X_n \geq k]$$

to be the probability that all particles are beyond $k-1$. Conditioning on the movement of the first particle we see once more that the F_n satisfy (1).

We now generalize III. Random variables $D \geq 0$ and $N \geq 0$ are given. Particles are sitting at integer sites. At time n a particle at x moves to $x + D_{n,x}$, and then splits into $N_{n,x}$ particles, where $D_{n,x}$ is distributed as D, $N_{n,x}$ as N, and both are independent of all the other $D_{n',x'}$ and $N_{n',x'} (n' \leq n, x' \in \mathbb{Z})$.
In a slightly more general model, particles first split into N particles, and then each particle moves according to distributions $D_1, D_2, \ldots, D_N$ (see e.g. [1]; and [3] for the present discussion in this context). Both models are known as branching random walk, and have been studied by a number of authors. Here we are interested in the position X_n of the left-most particle at time n, called the minimal displacement of the branching random walk.

Theorem 1. (Hammersley-Kingman-Biggins [1, 5, 7]). Let $(X_n)_{n=0}^{\infty}$ be the minimal displacement of a branching random walk, determined by N and D. Suppose that $\mathbb{E}N < \infty$, and $\mathbb{E}e^{-\lambda D} < \infty$ for some $\lambda > 0$. Then as $n \to \infty$

$$\tfrac{1}{n}X_n \to \gamma \qquad \text{almost surely and in } L^1$$

where γ is given by

$$\gamma = \inf\{a : \mathbb{E}N\mu(a) > 1\}, \qquad \mu(a) := \inf\{\mathbb{E}e^{(a-\lambda)D} : \lambda \geq 0\}.$$

In our main example ($N \equiv 2, \quad \mathbb{P}[D=1] = 1 - \mathbb{P}[D=0] = p$) we have

$$\mu(a) = (p/a)^a((1-p)/(1-a))^{1-a},$$

hence here $\gamma = 0$ if and only if $p \leq \frac{1}{2}$.
Theorem 1 raises the question whether $X_n - n\gamma$ converges in some sense as $n \to \infty$. The following result (obtained with Bernard Host) sheds some light on this problem.

Theorem 2. If $\mathbb{E}D < \infty$, then $(X_n - \mathbb{E}X_n)_{n=0}^{\infty}$ is tight, i.e.,

$$\forall \epsilon > 0 \quad \exists K \in \mathbb{R} : \mathbb{P}[|X_n - \mathbb{E}X_n| \leq K] > 1 - \epsilon \qquad \text{for } n = 0, 1, 2, \cdots$$

Proof: Note that

$$X_{n+1} = D + \min\nolimits_{1 \leq j \leq N} X_n^{(j)} \tag{2}$$

where D is the displacement of the first particle, N the number of its off-spring, and where the $X_n^{(j)}$ are independent random variables distributed as X_n.
To simplify the exposition we assume that $N \geq 2$ almost surely. Taking expectations in (2), we obtain

$$\begin{aligned}\mathbb{E}X_{n+1} &\leq \mathbb{E}D + \mathbb{E}\min(X_n^{(1)}, X_n^{(2)}) \\ &= \mathbb{E}D + \tfrac{1}{2}\mathbb{E}\{X_n^{(1)} + X_n^{(2)} - |X_n^{(1)} - X_n^{(2)}|\} \\ &= \mathbb{E}D + \mathbb{E}X_n - \tfrac{1}{2}\mathbb{E}|X_n^{(1)} - X_n^{(2)}|.\end{aligned}$$

Since (X_n) is nondecreasing, this implies

$$\mathbb{E}|X_n^{(1)} - X_n^{(2)}| \leq 2\mathbb{E}D.$$

Using the Markov inequality ($\mathbb{P}[|X| \geq a] \leq \mathbb{E}X/a$), and the well-known fact that $\mathbb{E}|X| \leq \mathbb{E}|X' - X''|$ if X is a random variable with mean zero, and X' and X'' are independent random variables distributed as X, we obtain for all $K > 0$

$$\mathbb{P}[|X_n - \mathbb{E}X_n| \geq K] \leq \tfrac{1}{K}\mathbb{E}|X_n - \mathbb{E}X_n| \leq \tfrac{1}{K}\mathbb{E}|X_n^{(1)} - X_n^{(2)}| \leq \tfrac{2}{K}\mathbb{E}D.$$

This obviously implies tightness.

We return to our main example ($N = 2, \mathbb{P}[D=1] = 1 - \mathbb{P}[D=0] = p$).
It is not hard to prove [3] that for $p < \frac{1}{2}$, the X_n remain bounded almost surely, and hence almost surely converge to a random variable X. The $F(k) := \mathbb{P}[X \geq$

$k]$ are obtained by letting $n \to \infty$ in equation (1), and solving for $F(k)$. If

$$\varphi_p(x) = \frac{1-\sqrt{1-4p(1-p)x^2}}{2(1-p)}$$

then $\mathbb{P}[X \geq k] = \varphi_p^k(1)$ for $k = 1, 2, \ldots$, where φ_p^k denotes the kth iterate of the function φ_p.
If $p = \frac{1}{2}$, then X_n does not remain bounded (although $\gamma = 0$), but moves to infinity as $\log\log n / \log 2$. This remarkable result is obtained by Bramson in [2]. Theorem 2 implies that every subsequence of $(X_n - [\mathbb{E}X_n])$ has a subsequence which converges in distribution (here $[\mathbb{E}X_n]$ denotes the largest integer smaller than $\mathbb{E}X_n$). What are the possible limits? It is shown in [3] that those are exactly given by an uncountable family X_α, $\alpha \in (0,1)$, of random variables, which we call *circle laws*. If

$$\varphi(x) = 1 - \sqrt{1 - x^2}$$

then the distribution of the X_α is given by

$$\mathbb{P}[X_\alpha \geq k] = \varphi^k(\alpha) \quad k \in \mathbb{Z}.$$

Finally, if $p > \frac{1}{2}$, then one still has tightness, but we do not know what the possible limit distributions are.

References.

[1] J.D. Biggins, Chernoff's theorem in the branching random walk, J. Appl. Prob. 14(1977), 630-636.

[2] M.D. Bramson, Minimal displacement of branching random walk, Z. Wahrscheinlichkeitstheorie verw. Gebiete 45(1978), 89-108.

[3] F.M. Dekking and B. Host, Limit distributions of minimal displacement of branching random walk (in preparation).

[4] B. Derrida and H. Spohn, Polymers on disordered trees, spinglasses, and traveling waves, J. of Stat. Physics 51 (1988), 817-840.

[5] J.M. Hammersley, Postulates of subadditive processes, Ann. Prob. 2 (1974), 652-680.

[6] R.M. Karp and J. Pearl, Searching for an optimal path in a tree with random costs, Art. Intelligence 21(1983), 99-116.

[7] J.F.C. Kingman, The first birth problem for an age-dependent branching process, Ann.Prob. 3(1975), 790-801.

Iterating Random Maps and Applications

J.S. Geronimo

School of Mathematics, Georgia Institute of Technology, Atlanta, GA 30332, USA

Abstract. A discrete-time Markov process on a locally compact metric space obtained by randomly iterating a finite number of Lipschitz maps is considered. The probability of choosing a particular map at each step is allowed to depend on the current position. Conditions are given that guarantee a unique invariant, attractive measure for the process. An application to real Julia sets associated with polynomials is presented.

1. Introduction

In recent years a particular class of discrete time Markov processes called iterated function systems Barnsley and Demko [1] have been used to construct and discuss the properties of fractals Hutchinson [2], Mandelbrot [3]. These processes arise naturally in other areas such as population dynamics Onicescu and Mihoc [4], learning models Karlin [5], Dubins and Freedman [6], Julia sets Bellissard, et al. [7], Bessis, et al. [8], and computer graphics Diaconis and Shahahahani [9], Demko, et al. [10], Berger and Soner [4]. The particular processes we study here are considered in Barnsley, et al. [12].

Let X be a locally compact metric space with metric d having the additional property that sets of finite diameter in X are relatively compact. Let ω_i, i = 1,2,...,N be Lipschitz functions from X into X with $d(\omega_i(x),\omega_i(y)) \leq s_i d(x,y)$ for all x,y in X. Let p_i: X $\rightarrow$ [0,1] be continuous, i = 1,2,...,N, with $p_i(x) \geq 0$ and $\sum_{i=1}^{N} p_i(x) = 1$ for all x in X. We now construct a random walk on X with the above maps ω_i and probabilities p_i. To this end we define the probability of transfer for a given point x $\in$ X to a Borel subset B $\subseteq$ X as

$$P(x,B) = \sum_{i=1}^{N} p_i(x)\delta_{\omega_i(x)}(B) \tag{1}$$

where δ_z denotes the Dirac measure concentrated at z. Let $\nu \in$ P(X) (the space of Borel probability measures on X) and let $\{z_n^{\nu}$: n = 0,1,...,$\}$ be the Markov stochastic process having initial distribution ν and transistion probability given by (1). Set $\Omega = N^{\mathbb{N}} = \{\underline{i} = \{i_1,i_2,...,\}: 1 \leq i_j \leq N\}$ and let P_x be the probability measure on Ω given by

$P_x\{\underline{i}: j^{th}$ coordinate of $\underline{i}$ is i_j, $j = 1,2,...,n\}$

$$= p_{i_2}(x)p_{i_2}(\omega_{i_1}(x)) \cdots p_{i_n}(\omega_{i_{n-1}} \circ \cdots \circ \omega_{i_1}(x)).$$

Thus the process $\{z_n^x = z_n^{\delta_x}\}$ represented as

$$z_n^x(\underline{i}) = \omega_{i_n} \circ \cdots \circ \omega_{i_1}(x) \tag{2}$$

can be thought of as a random walk on X, using an N-sided die which may change from place to place. The main problem is to show that there is an attractive invariant probability measure for the above Markov process. A measure μ is said to be attractive if for all $f \in C(x)$ (the bounded real-valued continuous function on X),

$$E(f(z_n^\nu)) \rightarrow \int f d\mu \ \forall \ \nu \in P(X)$$

where

$$E(f(z_1^\nu)) = \sum_{i=1}^{N} \int p_i(x) f(\omega_i(x)) d\nu.$$

and stationary if

$$E(f(z_1^\mu)) = \int f d\mu \ \forall \ f \in C(X) \tag{3}$$

Theorem 1 [12]. *Suppose that there exists* $r < 1$ *such that*

$$E_x(\log d(z_1^x, z_1^y)) = \sum_{i=1}^{N} p_i(x) \log(d(\omega_i(x), \omega_i(y)))$$
$$\leq \log(rd(x,y)) \ \forall \ x,y \in X. \tag{4}$$

Furthermore, suppose that $p_i(x) > \delta > 0$ *and* $p_i(x)$ *is Hölder continuous with some positive exponent for all* i *and for all* $x \in X$. *Then there exists an attractive invariant probability measure for the Markov process described above.*

Theorem 1 remains valid under weaker conditions on the $p_i(x)$'s [12]. The case where the $p_i(x)$'s are independent of x was considered by Barnsley and Elton [13].

Example 1. Let $X = R$, $p_1(x) = 2/3$, $\omega_1(x) = \frac{1}{2}x$, $p_2(x) = 1/3$, $\omega_2(x) = 2x + 1$. Then

$$\sum_{i=1}^{2} p_i(x) \log\left(\frac{d(\omega_i(x), \omega_i(y))}{d(x,y)}\right) < -\frac{1}{3} \log 2$$

which shows that the hypotheses of Theorem 1 are satisfied.

Example 2. Let X be the interior and boundary of the triangle whose vertices are at (0,0), (1,0), $(\frac{1}{2},\frac{\sqrt{3}}{2})$. Let $p_1(x,y) = \frac{1}{3} - \frac{x}{6}$, $p_2(x,y) = \frac{1}{3}$, and $p_3(x,y) = \frac{1}{3} + \frac{x}{6}$. Let

$$\omega_1(\bar{x}) = \frac{1}{2}\begin{bmatrix} 1 & 0 \\ 0 & 1 \end{bmatrix}\bar{x}$$

$$\omega_2(\bar{x}) = \frac{1}{2}\begin{bmatrix} 1 & 0 \\ 0 & 1 \end{bmatrix}\bar{x} + \begin{bmatrix} 1 \\ 0 \end{bmatrix}$$

and

$$\omega_3(\bar{x}) = \frac{1}{2}\begin{bmatrix} 1 & 0 \\ 0 & 1 \end{bmatrix}\bar{x} + \begin{bmatrix} 1/2 \\ \sqrt{3}/2 \end{bmatrix}$$

Again by Theorem 1 there is a unique invariant, attractive probability measure associated with the Markov process described earlier.

As evidenced in Example 1 it is not necessary for all the ω_i's to be strict contractions. Condition 4 says that an average contractivity between points is sufficient to give an attractive invariant probability measure μ.

Given the invariant, attractive measure μ of Theorem 1 we would like to know how the support of μ is related to the maps ω_i. Remembering that A = support $\mu = \cap(B \text{ closed: } \mu(B) = 1)$ it follows from the invariance property (3) i.e.,

$$\int_A f d\mu = \sum_{i=1}^{N} \int_A p_i(x) f(\omega_i(x)) d\mu(x), \quad f \in C(A)$$

[11] that $A = \bigcup_i \omega_i(A)$.

Consequently we see in Example 2 A is the Sierpinski gasket. In order to deal with Example 1 we need the following theorem [13, Lemmas 3 and 4].

Theorem 2. *Let* p_1 *and* p_2 *be constant nonzero probabilities summing to one. Let* $\omega_1(x) = a_1 x$ *and* $\omega_2(x) = a_2 x + b$ *satisfy condition* (4). *Suppose* $0 < a_1 < 1 < a_2$ *and* $b > 0$ *then the support of* μ *is equal to the half-infinite interval* $[0,\infty)$.

2. Application to Real Julia Sets

Let T(z) be a monic polynomial of degree N and let A be the set of repulsive fixed points of T and all its iterates. The Julia set J associated with T is the closure of A. Let ω_i, i = 1,2,...,N, be a complete assignment of the inverse branches of T. Let $g \in C(J)$ be Hölder continuous on J such that $g(z) > 0$ and $\sum_{i=1}^{N} g(\omega_i(z)) = 1$ for all $z \in J$. We now prove the following

Theorem 3. *Suppose* J *is a subset of the real line and suppose that* $p_i(x) = g(\omega_i(x))$, $x \in J$, $i = 1,2,\ldots,N$, *where* g *is as above. Then there is a unique invariant measure* μ *supported on* J *having the property that*

$$\int_J f(x)d\mu(x) = \sum_{i=1}^{N} \int_J p_i(x)f(\omega_i(x))d\mu(x) \;\forall\, f \in C(J).$$

Before giving the proof of Theorem 3 we need to introduce a metric. It follows from the properties of J Brolin [14] that J is real if and only if there exist an interval I such that $T^{-1}(I) \subset I$. Suppose I_0 is the smallest interval such that $T^{-1}(I_0) \subset I_0$. Because the properties of J are invariant under translations [14], it is without loss of generality that we can take $I_0 = [-a,a]$. In this case it is not difficult to verify that $T(a) = a$ and $T(-a) = -a$ or $T(-a) = a$ for all N odd or even, respectively. Since the capacity of J is 1 [14], the smallest a may be is 2. We now define a metric on I_0 as follows [15]

$$d(x,y) = |F(x) - F(y)| \;\forall\, x,y \in I_0 \tag{5}$$

where

$$F(x) = \int_{-a}^{x} \frac{d\omega}{\sqrt{a^2 - \omega^2}}\,. \tag{6}$$

Note that the topology induced on I_0 by the above metric is equivalent to the topology induced on I_0 by the Euclidean metric.

Theorem 4. *Let* ω_i, $i = 1,2,\ldots,N$, *be the inverse branches of* $T(z)$ *given above. Let* I_0 *be as above then*

$$d(\omega_i(x),\omega_i(y)) \le \frac{1}{N}\, d(x,y), \quad x,y \in I_0 \tag{7}$$

The proof of Theorem 4 is based on the following.

Lemma 1. For $x \in T^{-1}(I_0)$

$$\frac{1}{N}\frac{|T'(x)|}{\sqrt{a^2 - (T(x))^2}} \ge \frac{1}{\sqrt{a^2 - x^2}} \tag{8}$$

Proof. Let $E = T^{-1}(I_0)$ and let $G_N(z)$ and $G(z)$ be the Green's functions (in the potential theoretic sense) associated with E and I_0 respectively. It is not difficult to verify that

$$G_N(z) = \frac{1}{N}\,\ell n\left|\frac{T(z) + \sqrt{(T(z))^2 - a^2}}{2}\right| - \frac{1}{N}\,\ell n\,\frac{a}{2}$$

and

$$G(z) = \ell n\left|\frac{z + \sqrt{z^2 - a^2}}{2}\right| - \ell n\,\frac{a}{2}.$$

Since $a \geq 2$,

$$G_N(z) = \ell n|z| - \frac{1}{N}\,\ell n(\tfrac{a}{2}) + O(\tfrac{1}{|z|})$$

and

$$G(z) = \ell n|z| - \ell n\,\frac{a}{2} + O(\tfrac{1}{|z|})$$

near z equal to infinity, there exists a neighborhood of infinity where

$$G_N(z) > G(z).$$

Furthermore since $T^{-1}(I_0) \subset I_0$ $G_N(z) \geq G(z)$, $z \in I_0$. Now $G_N(z)$ and $G(z)$ are harmonic on the extended complex plane minus I_0 and $\{\infty\}$. Therefore the above inequalities imply that

$$G_N(z) \geq G(z)\ \forall\ z.$$

Let $z = x + iy$ and divide the above inequality by $y > 0$. Thus

$$\frac{G_N(z)}{y} \geq \frac{G(z)}{y}.$$

Now letting $y \to 0$, $y > 0$ (which is equivalent to taking the normal derivative) with $x \in T^{-1}(I_0) = E$ and using the fact that $G(E) = G_N(E) = 0$ yields the inequality

$$\frac{1}{N}\,\frac{|T'(x)|}{\sqrt{a^2 - (T(x))^2}} \geq \frac{1}{\sqrt{a^2 - x^2}}, \quad x \in T^{-1}(I_0)$$

which is the result to be proved.

Proof of Theorem 4. From (5) we find that

$$\frac{d(\omega_i(x),\omega_i(y))}{d(x,y)} = \frac{|F(\omega_i(x) - F(\omega_i(y))|}{|F(x) - F(y)|}, \quad x,y \in I_0.$$

Consequently, by Cauchy's mean value theorem

$$\frac{d(\omega_i(x),\omega_i(y))}{d(x,y)} = \frac{\sqrt{a^2 - c^2}}{\sqrt{a^2 - (\omega_i(c))^2}} |T^{-1}(\omega_i(c))|$$

for $c \in (x,y)$. Choosing x in Lemma 1 to be $\omega_i(c)$ gives the desired result.

Proof of Theorem 3. By the hypotheses on g and the fact that under the hypotheses on T(x), the ω_i's are contracting Lipschitz maps (Theorem 4). We see that conditions of Theorem 1 are satisfied. Thus the theorem is proved.

Example 3. The particular case $g(z) = \frac{W(z)}{T'(z)}$ where W(z) is a monic polynomial of the same degree as $T'(z)$ chosen so that $\frac{W(z)}{T'(z)} > 0$, $z \in J$, plays an important role in the study of orthogonal polynomials and discrete limit periodic Schrödinger operators on Julia sets Moussa [16], [8], [7], [17].

References

[1] M. F. Barnsley and S. Demko. Iterated Function Systems and the Global Construction of Fractals. Proc. R. Soc. London. Vol. A 399 (1985), 243-275.

[2] J. Hutchinson. Fractals and Self Similarity. Indiana Univ. J. of Math. 30 (1981), 713-747.

[3] B. Mandelbrot. The Fractal Geometry of Nature. 1982. W. H. Freeman. San Francisco.

[4] O. Onicescu and G. Mihoc. Sur les chaines de variables statistique. Bull. Soc. Math. de France 59 (1935), 174-192.

[5] S. Karlin. Some Random Walks Arising in Learning Models. Pac. J. Math. 3 (1953), 725-756.

[6] L. Dubins and D. Freedman. Invariant Probabilities for Certain Markov Processes. Ann. Math. Stat. 37 (1966), 837-848.

[7] J. Bellissard, D. Bessis and P. Moussa. Chaotic States of Almost Periodic Schrödinger Operators. Phys. Rev. Lett. 49 (1982), 701-704.

[8] D. Bessis, J. S. Geronimo and P. Moussa. Function Weighted Measures and Orthogonal Polynomials on Julia Sets. Const. Approx. 4 (1988), 157-173.

[9] P. Diaconis and M. Shahshahani. Products of Random Matrices and Computer Image Generation in Random Matrices and Their Applications. 50 Conf. Math. A.M.S., Providence, RI (1986).

[10] S. Demko, L. Hodges and B. Naylor. Construction of Fractal Objects with Iterated Function Systems. SIGGRAPH '85 Proceedings (1985), 271-278.

[11] M. Berger and H. M. Soner. Random Walks Generated by Affine Mappings. J. Theor. Prob. 1 (1988), 239-254.

[12] M. F. Barnsley, S. G. Demko, J. H. Elton and J. S. Geronimo. Invariant Measures for Markov Processes Arising from Iterated Function Systems with Place-Dependent Probabilities. Ann. Inst. H. Poincaré Prob. et Stat. 24 (1988), 367-394.

[13] M. F. Barnsley and J. H. Elton. A New Class of Markov Processes for Image Encoding. Adv. Appl. Prob. 20 (1988), 14-32.

[14] H. Brolin. Invariant Sets Under Iteration of Rational Functions. Ark. Mat. 6 (1965), 103-144.

[15] M. F. Barnsley, J. S. Geronimo and A. N. Harrington. On the Invariant Sets of a Family of Quadratic Maps. Commun. Math. Phys. 88 (1983), 479-501.

[16] P. Moussa. Iteration des polynomes et proprietes d'orthogonalite. Ann. Inst. H. Poincaré 44 (1986), 315-325.

[17] M. F. Barnsley, J. S. Geronimo and A. N. Harrington. Almost Periodic Jacobi Matrices Associated with Julia Sets. Comm. Math. Phys. 99 (1985), 303-317.

Hannay Angles and Classical Perturbation Theory

S. Golin[1,2,*], *A. Knauf*[1], *and S. Marmi*[3]

[1]Fachbereich Mathematik, Technische Universität Berlin, Straße des 17. Juni 136, D-1000 Berlin 12, Germany
[2]Dipartimento di Fisica and INFN, Università di Bologna, Via Irnerio 46, I-40126 Bologna, Italy
[3]Dipartimento di Fisica, Università di Bologna, Via Irnerio 46, I-40126 Bologna, Italy

We report on previous and recent work in classical perturbation theory related to the Hannay angles. They are a means of measuring an anholonomy effect in classical mechanics closely corresponding to the Berry phase in quantum mechanics.

1 Introduction

In this note we report on the use of classical perturbation theory in the investigation of the Hannay angles. In particular, we will present work in progress.

The Hannay angles were introduced by Hannay [9] as a means of measuring an anholonomy effect in classical mechanics closely corresponding to the Berry phase [3,4] in quantum mechanics. Since the Hannay angles may be given a geometrical interpretation [7,11] they are called *geometrical angles.*

To be more explicit, we consider a classical Hamiltonian system depending on external parameters. Let this system, by fiat, be integrable for each fixed value of these parameters. In terms of action-angle variables the Hamiltonian can consequently be written as a function of the actions alone. If, however, we allow for a slow and periodic time-dependence on the parameters, the dynamics in terms of action-angle variables (J, φ) will be determined by the Hamiltonian

$$h = h_0(J, \tau) + \varepsilon h_1(J, \varphi, \tau),$$

where εh_1 is just the time-derivative of the generating function which gov-

*Supported by the Deutsche Forschungsgemeinschaft (DFG) and the Istituto di Fisica Nucleare (INFN), Sezione di Bologna

Springer Proceedings in Physics, Vol. 47 Number Theory and Physics
Editors: J.M. Luck · P. Moussa · M. Waldschmidt

erns the transformation to (J,φ). $\varepsilon > 0$ is a small quantity and $\tau := \varepsilon t$ is slow time. Because of the periodic time dependence of the parameters, h_0 and h_1 are T-periodic in τ for some positive T. Set $f(J,\varphi,\tau) := -\nabla_\varphi h_1(J,\varphi,\tau)$, $g(J,\varphi,\tau) := \nabla_J h_1(J,\varphi,\tau)$, and $\omega(J,\tau) := \nabla_J h_0(J,\varphi,\tau)$; the Hamiltonian equations of motion are

$$\begin{aligned} \dot{J} &= \varepsilon f(J,\varphi,\tau), \\ \dot{\varphi} &= \omega(J,\tau) + \varepsilon g(J,\varphi,\tau). \end{aligned} \tag{1}$$

The *adiabatic Hannay angles* are defined for small $\varepsilon > 0$ by

$$\begin{aligned} \Delta\varphi &:= \varphi\left(\frac{T}{\varepsilon}\right) - \varphi^0 - \int_0^{T/\varepsilon} \omega(J(t),\varepsilon t)\,dt \\ &= \varepsilon \int_0^{T/\varepsilon} g(J(t),\varphi(t),\varepsilon t)\,dt, \end{aligned} \tag{2}$$

where $\varphi^0 := \varphi(0)$.

In fact, we invoked averaging theory in [5,7] to explain under what mathematically precisely defined circumstances the adiabatic Hannay angles may be replaced by the *geometrical Hannay angles*

$$\int_0^T g^{av}(J^0,\tau)\,d\tau, \tag{3}$$

where $J^0 := J(0)$, $g^{av}(J,\tau) := \int_0^{2\pi} \ldots \int_0^{2\pi} g(J,\varphi,\tau)\,d^n\varphi/(2\pi)^n$ and n is the number of degrees of freedom. It should be noted that in order to determine the geometrical Hannay angles one need not to solve the canonical equations (1). It is precisely this fact that makes the averaged quantity (3) so useful, for the solution of the non-autonomous system (1) cannot, in general, be obtained analytically.

In [7] we gave a geometrical interpretation to the geometrical Hannay angles (cf. also [11]). They measure the holonomy of an Ehresmann connection which is constructed in a canonical way from the symmetry group of the given Hamiltonian system.

The adiabatic Hannay angles and the geometrical Hannay angles are related (cf. [5,7]) by the adiabatic theorem of classical mechanics and its generalization to an averaging theorem for phase space functions. For this purpose we are now going to state these two basic theorems.

Let us consider system (1) for any initial values $(J^0, \varphi^0) := (J(0), \varphi(0)) \in G \times \mathbf{T}^n$, where G is a bounded open subset of $\mathbf{R}^n$. It is crucial that (1) verifies a non-degeneracy condition, such as rank $\partial\omega(J,\tau)/\partial J = n$ for all τ. The adiabatic theorem [12] can be formulated as follows (for a more precise formulation we refer to [7]).

Theorem 1 *Let $\rho > 0$, then*

$$\text{meas}\ \{(J^0, \varphi^0) \in G \times \mathbf{T}^n \mid \max_{t\in[0,T/\varepsilon]} |J(t) - J^0| > \rho\} = \mathcal{O}\left(\frac{\sqrt{\varepsilon}}{\rho}\right). \tag{4}$$

While Theorem 1 provides a mathematical justification for the physical assertion of the adiabatic invariance of the actions, one still needs to generalize it to general phase space functions so as to relate adiabatic and geometric Hannay angles. For a phase space function $a(J, \varphi, \tau)$ define

$$A(t) := \frac{\varepsilon}{T} \int_0^t a(J(u), \varphi(u), \varepsilon u)\, du, \tag{5}$$

$$\bar{A}(t) := \frac{\varepsilon}{T} \int_0^t a^{av}(J^0, \varepsilon u)\, du, \tag{6}$$

where $a^{av}(J,\tau) := \int_0^{2\pi} .. \int_0^{2\pi} a(J,\varphi,\tau)\, d^n\varphi/(2\pi)^n$. When $t = T/\varepsilon$, $A(t)$ and $\bar{A}(t)$ denote respectively the time average of the phase space function $a(I, \varphi, \varepsilon)$ and of its average over the (fast) angular variables. Related to these functions one can prove a phase space averaging theorem [5,7]. We state it here for the n–dimensional case.

Theorem 2 *Let $\rho > 0$, then*

$$\text{meas}\ \{(J^0, \varphi^0) \in G \times \mathbf{T}^n \mid \max_{t\in[0,T/\varepsilon]} |A(t) - \bar{A}(t)| > \rho\} = \mathcal{O}\left(\frac{\sqrt{\varepsilon}}{\rho}\right). \tag{7}$$

Applying Theorem 2 to $a = \partial h_1/\partial J$ we thus get an adiabatic approximation to (2). In other words, we have proven that the adiabatic Hannay angles are given by

$$\Delta\varphi = \int_0^T g^{av}(J^0, \tau)\, d\tau + \mathcal{O}(\varepsilon^b), \tag{8}$$

for all $b \in [0, 1/2)$ and for all initial conditions which do not belong to a set of small measure; i.e., the difference between the geometrical and the adiabatic angles is typically small, as asserted by Hannay.

2 Can One Measure Hannay Angles?

So far it seems that Hann. y angles are a well-founded concept. However, there are serious experimental troubles involved in their determination. In [7] we argued that the experimental verification of the Hannay angles may be cumbersome because it involves, in general, high precision measurements as the angles are typically small as compared to the *dynamical angles* $\int_0^{T/\varepsilon} \omega(J(t), \varepsilon t)\, dt$. But there is a further difficulty. From (2) one reads off that a measurement of the Hannay angles involves an experimental determination of the angle $\varphi(T/\varepsilon)$ together with an *a priori* determination of the dynamical angles.

However, in general the dynamical angles cannot be easily estimated; neither can one solve (1) and determine $J(t)$ explicitly nor can one directly invoke averaging theory. Note that Theorem 2 cannot be used because there is no ε in front of the dynamical angles. Therefore the naive approximation $\int_0^{T/\varepsilon} \omega(J^0, \varepsilon t)\, dt$ to the dynamical angles is too crude (as the domain of integration has length T/ε). In fact, one can show for the simple example $h = J^2/2 + \varepsilon \cos\varphi$ that

$$\int_0^{T/\varepsilon} \left[\omega(J(t), \varepsilon t) - \omega(J^0, \varepsilon t)\right] dt = T\frac{\cos\varphi^0}{J^0} + \mathcal{O}(\varepsilon); \qquad (9)$$

and the secular term prevents the right hand side from vanishing as $\varepsilon \to 0$.

On the other hand this example also suggests a remedy to the problem, namely averaging over the initial angles φ^0. In other words, upon averaging on the initial torus J^0 with the Liouville measure one may expect $\int_0^{T/\varepsilon} \omega(J^0, \varepsilon t)\, dt$ to be a good approximation of the dynamical angles. Such an averaging procedure seems also to be acceptable from the experimental point of view.

The hypothesis that our system (1) is Hamiltonian has not yet been fully exploited, but it will be crucial for the following theorems. Denote by $< \cdot >$ averaging with respect to the initial angles φ^0, i.e. with respect to the initial torus.

For single-frequency systems one can make use of a method introduced in [5] to get the following result [6].

Theorem 3 *Let (1) describe a Hamiltonian system with one degree of freedom. Then*

$$\left\langle \int_0^{T/\varepsilon} \left[\omega(J(t),\varepsilon t) - \omega(J^0,\varepsilon t)\right] dt \right\rangle = \mathcal{O}(\varepsilon) . \tag{10}$$

Apart from some smoothness conditions the only essential assumption entering Theorem 3 is that the angular frequency ω is assumed to be nowhere zero. For the next theorem that deals with the case of several degrees of freedom one needs the same non–degeneracy condition as in Theorem 2 together with the assumption that *the unperturbed Hamiltonian* h_0 *do not depend explicitly on* τ, i.e. $h_0 = h_0(J)$ (cf. [8]).

Theorem 4 *Let (1) describe a Hamiltonian system with several degrees of freedom and assume that* h_0 *is* τ*-independent. Let* $a \in [0,1)$. *Then there exist constants* $c > 0$, $b_0 \in (0,1)$ *such that*

$$\text{meas} \left\{ (J^0,\varphi^0) \in G \times \mathbf{T}^n \left| \left\langle \int_0^{T/\varepsilon} \left[\omega(J(t)) - \omega(J^0)\right] dt \right\rangle > c\varepsilon^{1-a} \right. \right\} = \mathcal{O}(\varepsilon^b), \tag{11}$$

for all $b \in [0,b_0)$.

The condition of the time independence of h_0 in Theorem 4 is essential: Starting outside resonances there is no awkward time dependence that pushes the system into the resonant zones in action space and therefore perturbation theory yields for these systems better results than, for instance, Theorems 1 or 2.

This situation, although rather special, is not uncommon. Actually, it occurs in the only two non-trivial examples of systems with at least two degrees of freedom and non-vanishing Hannay angles which are known to us. The first is a satellite in the gravitational field of a slowly rotating oblate (or prolate) planet [7]. And, second, Kugler [10] investigated a vibrating string set at an angle with respect to the axis of a slowly rotating base (a Foucault pendulum in disguise, as he calls it). Common to both examples is that in terms of coordinates with respect to co-rotating frames of reference there appears no explicit time dependence in the unperturbed Hamiltonian h_0.

In fact, if there were a time dependence in the unperturbed Hamiltonian h_0 of a system with several degrees of freedom one has immediately an argument against the statement of Theorem 4. Consider the Taylor expansion of $\omega(J,\tau)$ up to second order with respect to J. Then the second order term involves the square of the deviation of the action from the initial action, i.e., is of the order $\mathcal{O}(\varepsilon^{2b})$, for all $b \in (0,1/2)$, and averaging over

φ^0 will not improve the rate of convergence to zero, due to positivity of the square. The oscillations in φ^0 are here of no help: one can only obtain an approximation to the dynamical angles with an error of order $\mathcal{O}(\varepsilon^{2b-1})$ whilst the Hannay angles are an $\mathcal{O}(1)$-effect. Since $b < 1/2$, in the adiabatic limit, the error term will exceed the Hannay angles.

To our knowledge there is no class of Hamiltonian systems in the literature where one can prove analytically that the evolution in the actions is really not less than $\mathcal{O}(\varepsilon^b)$, for all $b \in [0, 1/2)$, when h_0 is explicitly τ-dependent. However, here is a simple non-Hamiltonian system, where the conclusion of Theorem 4 cannot be drawn. Take

$$\begin{aligned} \dot{J}_1 &= 0, \\ \dot{J}_2 &= \varepsilon \cos \varphi_1, \\ \dot{\varphi}_1 &= \chi(\tau) J_1^2, \\ \dot{\varphi}_2 &= J_2^2. \end{aligned} \tag{12}$$

If one chooses χ to be affine in some subinterval, where it is also assumed to have a zero, then the it is easy to verify that

$$\left\langle \int_0^{T/\varepsilon} \left[\omega_2(J(t)) - \omega_2(J^0)\right] dt \right\rangle \neq o(1), \tag{13}$$

where ω_2 is the second component of ω. Essentially, the integration boils down to a Fresnel integral.

Acknowledgments One of us (S. G.) would like to thank his colleagues in Bologna for their hospitality.

References

[1] V.I. Arnol'd: *Geometrical Methods in the Theory of Ordinary Differential Equations*, Springer (New York, Heidelberg, Berlin 1983)

[2] V.I. Arnol'd, V.V. Kozlov, A.I. Neishtadt: *Mathematical Aspects of Classical and Celestial Mechanics, Encyclopaedia of Mathematical Sciences, Vol. 3. Dynamical Systems, III*, Springer (Berlin, Heidelberg, New York 1988)

[3] M.V. Berry: *Quantal phase factors accompanying adiabatic changes*, Proc. Roy. Soc. London **A392**, 45–57 (1984)

[4] M.V. Berry: *Classical adiabatic angles and quantal adiabatic phase*, J. Phys. A: Math. Gen. **18**, 15–27 (1985)

[5] S. Golin: *Existence of the Hannay angle for single-frequency systems*, J. Phys. A: Math. Gen. **21** 4535–4547 (1988)

[6] S. Golin: *Can One Measure Hannay Angles?*, Preprint (1989)

[7] S. Golin, A. Knauf, S. Marmi: *The Hannay Angles: Geometry, Adiabaticity, and an Example*, to appear in Commun. Math. Phys. **122** (1989)

[8] S. Golin, S. Marmi: *A Class of Systems with Measurable Hannay Angles*, in preparation

[9] J.H. Hannay: *Angle variable holonomy in adiabatic excursion of an integrable Hamiltonian*, J. Phys. A: Math. Gen. **18**, 221–230 (1985)

[10] M. Kugler: *Motion in Non Inertial Systems, Theory and Demonstrations*, Preprint (1987)

[11] R. Montgomery: *The Connection Whose Holonomy is the Classical Adiabatic Angles of Hannay and Berry and Its Generalization to the Non-Integrable Case*, Comm. Math. Phys. **120**, 269–294 (1988)

[12] A.I. Neishtadt: *Averaging in multifrequency systems. II.*, Dokl. Akad. Nauk SSSR **226**,1295–1298 (1976); English translatian: Sov. Phys. Dokl. **21**, 80–82 (1976)

Nekhoroshev Stability Estimates for Symplectic Maps and Physical Applications

G. Turchetti

Dipartimento di Fisica della Università di Bologna,
Via Irnerio 46, I-40126 Bologna, Italy and
INFN sezione di Bologna, Italy

Abstract. The Nekhoroshev stability estimates for symplectic maps of $\mathbf{R}^{2d}$ are presented. After an illustration of the basic ideas in a simple hamiltonian model, the dynamics of a charged particle in the magnetic lattice of an accelerator is considered and it is shown to be conveniently described by a symplectic map. After recalling the basic properties of the Birkhoff normal forms, we state a general theorem on the stability of the orbits and sketch the proof. The factorial divergence of the Birkhoff series appears to arise from nonlinearity as well as from the divisors, and a functional equation with interesting analytic properties is obtained from the majorant series.

1. Introduction

The stability analysis of motions in the neighborhood of an equilibrium point is a crucial problem in hamiltonian mechanics and is quite relevant for physical applications. After the proof of the celebrated KAM theorem [1], the problem is conceptually solved. Indeed for systems with two degrees of freedom, the invariant tori are topological barriers to motion and guarantee the stability of an elliptic equilibrium point, while for systems with more than two degrees of freedom the complement of tori is a connected domain and a slow diffusion can occur [2], preventing any stability property for $t \in [0, \infty[$.

However, since any physical phenomenon occurs in a finite time interval, it is reasonable to formulate stability conditions for finite time intervals. Given an integrable hamiltonian system with a small perturbation, the variation of the actions can be bounded for $0 \leq t \leq T_0 \exp[b(\epsilon_0/\epsilon)^{1/a}]$ where ϵ measures the intensity of the perturbation. This general result, first proved by Nekhoroshev [3], relies on a careful partitioning of phase space to deal with the resonances and a subtle analysis of the corresponding perurbation series.

A simpler, but equally interesting version of the theorem, was recently proved in the case of anharmonic oscillators, for which, according to the Birkhoff method, a single perturbative expansion can be written in phase space. The order N remainder of the functional equations, which conjugate the hamiltonians with their normal forms and define recursively the Birkhoff series, behaves as N^{aN} implying the exponential estimate [4]. The stability time proves to depend critically on the number theoretical properties of the frequencies. The result shows also that the Birkhoff series are asymptotic and raises the interesting but difficult problem of understanding the analytic structure of the functions to which they are asymptotic.

In order to explain the basic ideas, we first consider a pedagogical example which, in spite of its triviality, contains the basic ingredients of the general case. The technique, developed for flows, has been successfully applied to the restricted three body problem in celestial mechanics [5]. There are also physical systems, naturally described by symplectic maps, which require stability estimates. Relevant examples are the motion of charged heavy particles (protons) in high energy accelerators [6,7] and anomalous diffusion in the toroidal machines for thermonuclear fusion [8,9].

We shall describe the simplest model of magnetic lattice and recall the Birkhoff method [10] for the normal forms, which allow one to capture the basic properties of the motion [11].

We shall state a theorem, recently proved [12], which extends to the symplectic maps the stability results known for the flows. The technical differences are quite significant: the Cauchy method of majorant series allows one to distinguish the growth coming from the small divisors from the growth (which still turns out to be N^{aN}) due to the nonlinearity. Our majorant series has a generating function with a fractal natural boundary [13]. The same majorant series occurs in the conjugation problem for analytic maps in the neighborhood of a rational fixed point [14].

We can observe that our result is somewhat more general because there are maps which cannot be naturally associated with flows [15], and that some new light is shed on the divergence mechanism of the Birkhoff series.

Finally, even though in the present form our estimate is surely not optimal, it provides a scaling law which could be checked against numerical experiments for simple maps of $\mathbf{R}^4$.

2. A simple model.

In order to illustrate the Nekhoroshev method we consider first a hamiltonian $H = H_0 + \epsilon H_1$ defined in $\mathbf{T}^2 \times \mathbf{R}^2$

$$H = \omega \cdot \mathbf{j} + \epsilon\,[\mathbf{u} \cdot \mathbf{j} + V(\phi)] \qquad \mathbf{j} \in \mathbf{R}^2,\ \phi \in \mathbf{T}^2 \tag{2.1}$$

where $\omega, \mathbf{u} \in \mathbf{R}^2$ and V is assumed to have zero mean value and to be analytic in the strip $S_\alpha : |\mathrm{Im}\ \phi_k| \leq \alpha,\ k = 1,2$. The equation which formally conjugates H with an integrable hamiltonian $\hat{H}$ through a change of coordinate $(\mathbf{j}, \phi) \to (\mathbf{J}, \boldsymbol{\Phi})$ can be immediately written, using for instance the generating function $\mathcal{F}(\phi, \mathbf{J}; \epsilon)$ which satisfies the Hamilton-Jacobi equation

$$\omega \cdot \frac{\partial \mathcal{F}}{\partial \phi} + \epsilon \left[\mathbf{u} \cdot \frac{\partial \mathcal{F}}{\partial \phi} + V(\phi) \right] \;=\; \hat{H}(\mathbf{J}). \tag{2.2}$$

It can be directly verified that a solution is given by

$$\hat{H}(\mathbf{J}) = (\boldsymbol{\omega} + \epsilon \mathbf{u}) \cdot \mathbf{J} \qquad \mathcal{F}(\boldsymbol{\phi}, \mathbf{J}; \epsilon) = \boldsymbol{\phi} \cdot \mathbf{J} + F(\boldsymbol{\phi}; \epsilon) \tag{2.3}$$

where F satisfies the equation $(\boldsymbol{\omega} + \epsilon \mathbf{u}) \cdot \partial F/\partial \boldsymbol{\phi} + \epsilon V(\boldsymbol{\phi}) = 0$ whose explicit solution reads

$$F(\boldsymbol{\phi}; \epsilon) = i\epsilon \sum_{\mathbf{k} \in \mathbf{Z}^2/\{0\}} \frac{V_{\mathbf{k}}\, e^{i\mathbf{k}\cdot\boldsymbol{\phi}}}{\boldsymbol{\omega} \cdot \mathbf{k} + \epsilon \mathbf{u} \cdot \mathbf{k}}. \tag{2.4}$$

In equation (2.4) $V_{\mathbf{k}}$ denote the Fourier coefficients of V, which, by the analyticity hypothesis, decrease exponentially: $|V_{\mathbf{k}}| \leq A \exp(-\alpha|\mathbf{k}|)$ where $|\mathbf{k}| = |k_1| + |k_2|$. Equation (2.4) shows that F is analytic in the whole complex ϵ plane except for a dense set of poles at $\boldsymbol{\omega} \cdot \mathbf{k} + \epsilon \mathbf{u} \cdot \mathbf{k} = 0$, which make the real axis a natural boundary. If we choose, for instance, $\boldsymbol{\omega} = (1, -\omega)$ and $\mathbf{u} = (0, 1)$ then the poles appear at $\epsilon = \omega - p/q$ where $p, q \in \mathbf{Z}$ and q does not divide p.
When ϵ is real and $\boldsymbol{\omega}\cdot\mathbf{k} + \epsilon\mathbf{u}\cdot\mathbf{k} \neq 0$, $\forall \mathbf{k} \in \mathbf{Z}^2/\{0\}$, the motion is quasiperiodic, as can be seen by solving the Cauchy problem for the Hamilton equations and it is precisely the loss of quasiperiodicity of the motion which causes the appearance of the poles at $\boldsymbol{\omega} \cdot \mathbf{k} + \epsilon \mathbf{u} \cdot \mathbf{k} = 0$.
In order to investigate the perturbative expansion of F we let $F_{\leq N} = \sum_{n=1}^{N} F_n \epsilon^n$ be the Taylor expansion of F at order N and $\epsilon^{N+1} \mathcal{R}_N$ its remainder, namely $F = F_{\leq N} + \epsilon^{N+1} \mathcal{R}_N$

$$F_n = i(-1)^{n-1} \sum_{\mathbf{k}} \frac{(\mathbf{u} \cdot \mathbf{k})^{n-1}}{(\boldsymbol{\omega} \cdot \mathbf{k})^n} V_{\mathbf{k}} e^{i\mathbf{k}\cdot\boldsymbol{\phi}}, \tag{2.5a}$$

$$\mathcal{R}_N = i(-1)^N \sum_{\mathbf{k}} \left(\frac{\mathbf{u} \cdot \mathbf{k}}{\boldsymbol{\omega} \cdot \mathbf{k}} \right)^N \frac{V_{\mathbf{k}} e^{i\mathbf{k}\cdot\boldsymbol{\phi}}}{\boldsymbol{\omega} \cdot \mathbf{k} + \epsilon \mathbf{u} \cdot \mathbf{k}}. \tag{2.5b}$$

If we assume that the unperturbed frequency is Diophantine, namely that two real positive constants γ, η exist such that

$$|\boldsymbol{\omega} \cdot \mathbf{k}|^{-1} \leq \gamma\, |\mathbf{k}|^{\eta} \qquad \eta \geq 1, \tag{2.6}$$

then it is easy to see that $\mathcal{R}_N$ can be bounded provided that we are at a positive distance from the real ϵ axis, the bound behaving as $N^{N(\eta+1)}$. This bound is not of much use because we are interested in real orbits. The Nekhoroshev method consists in bounding not $\mathcal{R}_N$ but rather the remainder R_N of the functional equation (2.2) when F is replaced by its truncated Taylor series $F_{\leq N}$. Indeed (2.2) is then replaced by

$$\boldsymbol{\omega} \cdot \frac{\partial F_{\leq N}}{\partial \boldsymbol{\phi}} + \epsilon \left[\mathbf{u} \cdot \frac{\partial F_{\leq N}}{\partial \boldsymbol{\phi}} + V(\boldsymbol{\phi}) \right] = \epsilon^{N+1} R_N(\boldsymbol{\phi}, \mathbf{J}; \epsilon). \tag{2.7}$$

At orders $n \leq N$ in ϵ equation (2.7) defines the F_n recursively, while for $n > N$ it determines the remainder R_N. After the initializing equation $\omega \cdot \partial F_1/\partial \phi + V = 0$ we have

$$\omega \cdot \frac{\partial F_n}{\partial \phi} + \mathbf{u} \cdot \frac{\partial F_{n-1}}{\partial \phi} = 0 \qquad 2 \leq n \leq N, \tag{2.8}$$

which still gives (2.5a) while for $n > N$ we obtain

$$R_N = \mathbf{u} \cdot \frac{\partial F_N}{\partial \phi} = (-1)^N \sum_{\mathbf{k}} \left(\frac{\mathbf{u} \cdot \mathbf{k}}{\omega \cdot \mathbf{k}} \right)^N V_{\mathbf{k}} e^{i\mathbf{k}\cdot\phi}. \tag{2.9}$$

In order to bound $|R_N|$ we use the diophantine condition (2.7),the exponential decrease of $|V_{\mathbf{k}}|$ and replace the sum with an integral

$$\begin{aligned} |R_N| &\leq (\|\mathbf{u}\|\gamma)^N \int_{\mathbf{R}^3} d\mathbf{k}\; e^{-\alpha|\mathbf{k}|} \|\mathbf{k}\|^N |\mathbf{k}|^{N\eta} \leq \\ &\leq (\|\mathbf{u}\|\gamma\, 3^{\eta/2})^N \int_0^{2\pi} d\theta \int_0^\infty d\rho e^{-\alpha\rho} \rho^{N(\eta+1)+1} = A\, B^N\, \Gamma(N(\eta+1)+2) \end{aligned} \tag{2.10}$$

where $\rho = \|\mathbf{k}\|$ and we have used $\|\mathbf{k}\| \leq |\mathbf{k}| \leq \sqrt{3}\|\mathbf{k}\|$ letting $\|\cdot\|$ denote the euclidean norm; $A = 2\pi/\alpha^2$, $B = \gamma\, 3^{\eta/2}\|\mathbf{u}\|/\alpha^{\eta+1}$.
It can be seen directly that also $\partial R_N/\partial \phi$ can be bounded in the same way and we can write the result as $\|\partial R_n/\partial \phi\| \leq C\, N^{N(\eta+2)}$ where C is simply related to A and B. Recalling that (2.2) implies $\boldsymbol{\Phi} = \phi$, the equations of motion for the hamiltonian $H = \hat{H}(\mathbf{J}) + \epsilon^{N+1} R_N$ are given by

$$\begin{aligned} \frac{d\mathbf{J}}{dt} &= -\epsilon^{N+1} \frac{\partial R_N}{\partial \phi} \\ \frac{d\phi}{dt} &= \omega + \epsilon \mathbf{u} \end{aligned} \tag{2.11}$$

and consequently we can easily estimate the time T for which the variation of the action will not exceed ϵ. Integrating the first equation (2.11) from 0 to t we find

$$\|\mathbf{J}(t) - \mathbf{J}(0)\| \leq \epsilon^{N+1} \left\| \frac{\partial R_N}{\partial \phi} \right\| t \leq \epsilon C \left[\epsilon N^{\eta+2}\right]^N t \leq \epsilon \tag{2.12}$$

and consequently (2.12) will hold for $t \leq T = T_0 \left[\epsilon N^{\eta+2}\right]^{-N}$ with $T_0 = C^{-1}$.
It is possible to minimize the remainder with respect to N keeping ϵ fixed so that the stability time T becomes maximum. Since the function

$$f(N, \epsilon) = (\epsilon N^a)^N = \exp\left[N \log \epsilon + aN \log N\right] \tag{2.13}$$

has its minimum for

$$N = \bar{N} = \frac{1}{e\epsilon^{1/a}}, \qquad f(\epsilon, \bar{N}) = \exp(-a\bar{N}) = \exp\left(-\frac{a}{e}\, \epsilon^{-1/a}\right) \tag{2.14}$$

in our model the Nekhoroshev stability time turns out to be

$$T = T_0 \exp\left(\frac{\eta+2}{e}\,\epsilon^{-\frac{1}{\eta+2}}\right) \tag{2.15}$$

and its dependence on the number theoretical properties of the frequency is evident.

We conclude this section by noticing that a similar analysis can be carried out for a map of $\mathbf{R}\times\mathbf{T}$ defined by

$$\begin{cases} \phi' = \phi + \omega + \epsilon \\ j' = j + \epsilon V(\phi) \end{cases} \tag{2.16}$$

and that the extensions to higher dimensional phase spaces $\mathbf{R}^d\times\mathbf{T}^d$ are straightforward.

3. Symplectic maps and beam stability.

Among the physical systems which are more naturally described by symplectic maps, particle accelerators are probably the most relevant. The stability analysis they involve can be compared with the stability analysis of the solar system, which is described by Hamilton's equations. It is sufficient to keep in mind that in a hadronic accelerator of the next generation with a radius of 5 kilometers, the number of turns in a one-day run is typically 10^9, not very different from the number of revolutions of an intermediate planet in the estimated lifetime of the solar system. An accelerator is typically composed of a few hundreds of almost identical cells, whose action is described by a non-linear map, and the desired stability results involve $10^{11}\sim10^{12}$ iterations, far beyond the possibility of any numerical simulation.
In an accelerator a particle moves on a circular trajectory under the action of a constant *dipolar* field. The radius R of the orbit is related to the momentum p of the particle and to the intensity B_{dip} of the dipole magnetic field by

$$p = B_{\text{dip}}\, R. \tag{3.1}$$

The focusing of the beam in the x, y is plane, orthogonal to the orbit (the z axis being tangent to the reference circular orbit and y being the vertical axis) is obtained with *quadrupolar* fields which exert a linear attracting force $F_x = -kx$ along one axis and linearly repulsive $F_y = ky$ on the orthogonal one. The action of two consecutive quadrupoles rotated by $\pi/2$ is the same as the action of a convergent and divergent lens in each plane and the overall effect, with a judicious choice of the strength k and the length ℓ_{quad} of the quadrupole, is convergent. Mathematically this means that if we compute the transfer matrix in the phase space x, p_x, y, p_y for two rotated quadrupoles we obtain a block diagonal symplectic matrix with the trace of each block smaller than 2, so that the origin is an elliptic fixed point. For the basic *FODO* cell which is built with a *focusing* quadrupole, some dipoles and a *defocusing* quadrupole, it is very easy to write the transfer map $\mathbf{M}$ by solving the differential equations $\ddot{x} = 0, \quad \ddot{y} = 0, \ s_0 \leq s \leq s_0 + \ell_{\,\mathrm{dip}}$ for the dipoles and $\ddot{x} \pm kx = 0, \quad \ddot{y} \mp ky = 0, \ s_\pm \leq s \leq s_\pm + \ell_{\,\mathrm{quad}}$ for the quadrupoles (where s denotes the arc length). The transfer map is written as

$$\mathbf{M} = \mathbf{URU}^{-1}, \qquad \mathbf{U} = \begin{pmatrix} U_x & \mathbf{0} \\ \mathbf{0} & U_y \end{pmatrix}, \quad \mathbf{R} = \begin{pmatrix} R(\omega_x) & \mathbf{0} \\ \mathbf{0} & R(\omega_y) \end{pmatrix}$$

$$U_{x,y} = \begin{pmatrix} \beta_{x,y}^{1/2} & 0 \\ \gamma_{x,y} & \beta_{x,y}^{-1/2} \end{pmatrix} \qquad R(\alpha) = \begin{pmatrix} \cos\alpha & -\sin\alpha \\ \sin\alpha & \cos\alpha \end{pmatrix}. \tag{3.2}$$

The nonlinearity arises when the dipoles are superconducting, since multipolar fields due to the residual currents are unavoidable. The use of superconducting magnets is necessary at very high energies in order to keep the radius at reasonable values according to (3.1). Taking into account only the sextupole field in the dipoles, the equations of motion read

$$\begin{cases} \ddot{x} - \frac{1}{2}k_2(x^2 - y^2) = 0 \\ \ddot{y} + k_2 xy = 0 \end{cases} \qquad \text{for } s_0 \leq s \leq s_0 + \ell_{\,\mathrm{dip}} \tag{3.3}$$

and the transfer map can be obtained by writing the solution of (3.3) as a series in the initial conditions. A symplectic *(thin lens)* approximation to the transfer map is obtained by supposing all the nonlinearity to be concentrated in the midpoint of the dipole, that is, by including a factor $\ell_{\,\mathrm{dip}}\, \delta(s - s_0 - \ell_{\,\mathrm{dip}}/2)$ in the r.h.s. of (3.3). The resulting transfer matrix for the *FODO* cell is

$$\begin{pmatrix} x' \\ p_x' \\ y' \\ p_y' \end{pmatrix} = \mathbf{M} \begin{pmatrix} x \\ p_x + \frac{1}{2}k_2\ell_{\,\mathrm{dip}}\,(x^2 - y^2) \\ y \\ p_y - k_2\ell_{\,\mathrm{dip}}\; xy \end{pmatrix}. \tag{3.4}$$

4. Normal forms.

In the case of a linear map with an elliptic fixed point at the origin, we can introduce a set of coördinates such that the orbits are on $\mathbf{T}^2$ tori. Letting

$$\begin{pmatrix} \hat{x} \\ \hat{p}_x \\ \hat{y} \\ \hat{p}_y \end{pmatrix} = \mathbf{U}^{-1} \begin{pmatrix} x \\ p_x \\ y \\ p_y \end{pmatrix} \qquad z_1 = \hat{x} + i\hat{p}_x, \quad z_2 = \hat{y} + i\hat{p}_y \tag{4.1}$$

the new coordinates $\hat{x}, \hat{p}_x, \hat{y}, \hat{p}_y$ are called *normal coordinates* and $\hat{x}^2 + \hat{p}_x^2$, $\hat{y}^2 + \hat{p}_y^2$ are the invariants of the map (called Courant-Sneider invariants by accelerator theorists). Using complex coordinates the map reads

$$z' = e^{i\omega} z \qquad z = (z_1, z_2) \in \mathbf{C}^2, \quad e^{i\omega} = \mathrm{diag}(e^{i\omega_x}, e^{i\omega_y}) \tag{4.2}$$

together with its complex conjugate. As a consequence the normal form in the linear case is given by a direct product of rotations. The nonlinear map will read

$$z' = F(z, z^*) = e^{i\omega} z + [F]_2 + \ldots + [F]_n + \ldots \tag{4.3}$$

where $[\;]_n$ denotes the projection on the homogeneous polynomial in z, z^* of order n.
In the general case we say that a map $G : \mathbf{C}^2 \times \mathbf{C}^2 \to \mathbf{C}^2$ is in *normal form* with respect to $\omega \in \mathbf{R}^2$ if it commutes with the group of rotations $e^{i\omega}$ namely

$$\Delta G \equiv G(e^{i\omega} z, e^{-i\omega} z^*) - e^{i\omega} G(z, z^*) = 0. \tag{4.4}$$

If the map G is symplectic and ω non-resonant namely $\omega \cdot \mathbf{k} = 0,\ \mathbf{k} \in \mathbf{Z}^2 \Rightarrow \mathbf{k} = 0$ then

$$G(z, z^*) = e^{i\Omega(zz*)} z \qquad \Omega(zz^*) \equiv \Omega(z_1 z_1^*, z_2 z_2^*) \tag{4.5}$$

where $\Omega(\rho_1, \rho_2)$ is a real function such that $\partial \Omega_l / \partial \rho_j = \partial \Omega_j / \partial \rho_l$. As a consequence a symplectic normal form is a rotation with varying angle, the invariants being still $z_1 z_1^*$, $z_2 z_2^*$. We can observe that the map (4.5) naturally provides also the *interpolating flow* $\exp(i\Omega t) z$ of the orbits of the map. If ω is resonant then the invariants are no longer evident but are provided by the interpolating hamiltonian, whose flow interpolates the orbits of G; however, its construction as a formal series, whose existence is proved, is not straightforward [11].

If a given map F is not in normal form we look for a change of variables $z \to \varsigma$ which brings it to the normal form (4.5). Letting

$$z = \Phi(\varsigma, \varsigma^*) = \varsigma + \sum_{n \geq 2} [\Phi]_n(\varsigma, \varsigma^*) \tag{4.6}$$

be a symplectic diffeomorphism close to the identity the following functional equation must be satisfied:

$$F \circ \Phi - \Phi \circ e^{i\Omega} \varsigma = 0, \tag{4.7}$$

where we have used the notation $F \circ \Phi = F(\Phi(\varsigma, \varsigma^*), \Phi^*(\varsigma, \varsigma^*))$ and similarly for the second term. It is a well-known consequence of the KAM theorem that, even for a diophantine ω, equation (4.7) cannot have an analytic solution. As a consequence, denoting with $[\Phi]_{\leq N}$ and $[\Omega]_{\leq N}$ the Taylor series of Φ and Ω truncated at order N, we replace equation (4.7) with

$$F \circ [\Phi]_{\leq N} - [\Phi]_{\leq N} \circ e^{i[\Omega]_{\leq N}} \varsigma = R_N \tag{4.8}$$

where R_N is a remainder given by a series which begins at order $N+1$. Letting $[U]_n = [\varsigma \exp(i[\Omega]_{\leq N})]_n$ and letting $[S]_n$ be the homogeneous component of order n in the l.h.s. of equation (4.8) once we subtract the order n unknowns $[U]_n$ and $\Delta[\Phi]_n$, the recurrent equations are obtained by projecting (4.8) into the space of normal form polynomials and its complement. Letting Π be the projector into the space of normal form polynomials we have

$$\begin{aligned} [U]_n &= \Pi\,[S]_n \\ \Delta(1-\Pi)[\Phi]_n &= (1-\Pi)[S]_n. \end{aligned} \tag{4.9}$$

It has been proved [16], using a symplectic generator, that equations (4.9) admit solutions $[\Phi]_n$ and $[U]_n$ symplectic, up to order $n \leq N$. As a consequence a real frequency function $[\Omega]_{\leq N}$ is found. The solution $[\Phi]_{\leq N}$ is not unique but depends on a gauge $e^{i\Xi(\varsigma\varsigma^*)}$ where Ξ is an arbitrary function [16]. We stress that in this problem the natural expansion parameter is

$$\|z\| = \max_j |z_j| \tag{4.10}$$

and consequently we expand any function in a series of homogeneous polynomials in z and z^*. Letting $\Psi = [\Phi]_{\leq N}^{-1}$ define the inverse transformation we introduce the remainder $\hat{R}_n(t)$ corresponding to (4.8) iterated $t \in \mathbf{Z}$ times according to

$$\Psi \circ F^t \circ [\Phi]_{\leq N} = e^{it[\Omega]_{\leq N}} \varsigma + \hat{R}_N(t). \tag{4.11}$$

It is easy to show that the $\hat{R}_N(t)$ satisfy a recursion, which allows one to estimate them

$$\begin{aligned} \hat{R}_N(t+1) = &\left[\left(e^{it[\Omega]_{\leq N}}\varsigma\right) \circ \left(e^{i[\Omega]_{\leq N}}\varsigma + \hat{R}_N\right) - \left(e^{it[\Omega]_{\leq N}}\varsigma\right) \circ \left(e^{i[\Omega]_{\leq N}}\varsigma\right)\right] \\ &+ \left[\hat{R}_N(t) \circ \left(e^{i[\Omega]_{\leq N}}\varsigma + \hat{R}_N\right)\right], \end{aligned}$$

$$\hat{R}_N \equiv \hat{R}_N(1) = \Psi \circ \left[[\Phi]_{\leq N} \circ e^{i[\Omega]_{\leq N}}\varsigma + R_n\right] - \Psi \circ [\Phi]_{\leq N} \circ e^{i[\Omega]_{\leq N}}\varsigma. \tag{4.12}$$

5. The main theorem.

We can now state the main stability theorem which holds for maps $F : \mathbf{C}^{2d} \to \mathbf{C}^d$ holomorphic in a neighborhood of the origin that we assume to be the unit polydisc $\|z\| \leq 1$, with a linear part $e^{i\omega} z$, $z \in \mathbf{C}^d$ where ω satisfies the diophantine condition

$$|e^{i\mathbf{k}\cdot\omega} - 1|^{-1} \leq \gamma|\mathbf{k}|^{\eta} \quad \mathbf{k} \in \mathbf{Z}/\{0\}. \tag{5.1}$$

Letting $F = (F^1, \ldots, F^d)$ and $F^j = \sum_{\mathbf{k}} F^j_{\mathbf{k}} \, z_1^{k_1} z_1^{*k_2} \ldots z_d^{k_{2d-1}} z_d^{*k_{2d}}$ be the Taylor series, we define the following norms

$$\begin{aligned} \|[F^j]_n\|_1 &= \sum_{k_1+\ldots+k_{2d}=n} |F^j_{\mathbf{k}}|, \\ \|F^j\|_r &= \sum_n \|[F^j]_n\|_1 \, r^n, \\ \|F\|_r &= \max_{1\leq j\leq d} \|F^j\|_r. \end{aligned} \tag{5.2}$$

Theorem. *Let* $\|z\| \leq \rho/2$ *and* $z^{(t)} = F(z^{(t-1)}, z^{*(t-1)})$ *for* $1 \leq t$, $t \in \mathbf{Z}$ *, then* $\|z^{(t)}\| \leq (1 + \frac{1}{24})\rho$ *for all* $t \leq T$ *where*

$$T = \frac{7}{8}\sqrt{\frac{\rho}{6}} \exp\left[\frac{9+\eta}{2e}\left(\frac{\rho_*}{\rho}\right)^{\frac{1}{9+\eta}}\right] \tag{5.3}$$

where

$$\rho_* = \frac{2}{49\gamma} \qquad \rho \leq \rho_*(3e)^{-9-\eta}$$

and we have assumed that

$$\|[F]_2\|_1 \leq 3^{-\eta}\gamma^{-1}, \qquad \|[F]_n\|_1 \leq 1 \qquad n > 2.$$

Sketch of the proof. The first step consists in observing that equations (4.9) imply that

$$\|[U]_n\|_1 \leq \|[S]_n\|_1, \qquad \|(1 - \Pi)[\Phi]_n\|_1 \leq (\sigma_n/2)\|[S]_n\|_1 \tag{5.4}$$

if we define $\sigma_n = 2\gamma(n+1)^{\eta}$.
We recall that

$$[S]_n = [F \circ [\Phi]_{\leq N} - [\Phi]_{\leq N} \circ [U]_{\leq N}]_n - \Delta[\Phi]_n - [U]_n \tag{5.5}$$

where the last two terms just cancel two identical contributions in the large square bracket so that $[S]_n$ only depends on $[\Phi]_j$, $[U]_j$ with $j \leq n-1$, and that for any two holomorphic maps Q, P, which leave the origin invariant, the following estimates hold:

$$\|[Q \circ P]_n\|_1 \leq \sum_{s=1}^{n} \|[Q]_s\|_1 \sum_{\substack{m_1+\ldots+m_s=n \\ m_j \geq 1}} \|[P]_{m_1}\|_1 \ldots \|[P]_{m_s}\|_1. \tag{5.6}$$

Using (5.5) and (5.6) one first proves that

$$\sigma_n \|[S]_n\|_1 \leq (3\sigma_N)^{n-2} \mu_n \tag{5.7}$$

where μ_n satisfy the recursion

$$\mu_n = \sum_{s=2}^{n-1} \mu_s \sum_{\substack{m_1+\ldots+m_s=n \\ m_j \geq 1}} \mu_{m_1} \ldots \mu_{m_s}, \quad 3 \leq n, \tag{5.8}$$

initialized by $\mu_1 = \mu_2 = 1$. Using the symplecticity conditions one can indeed show that $\|\Pi[\Phi]_n\|_1 \leq (\mu_n/2)(3\sigma_N)^{n-2}$. The growth of the sequence μ_n is bounded by $\mu_n \leq (n!)^3$ so that from (5.4) and (5.7) it follows that for any $n \geq 2$

$$\|[\Phi]_n\|_1 \leq A^{n-2}, \qquad \|[U]_n\|_1 \leq A^{n-2}, \qquad A = 6\gamma(N+1)^{9+\eta} \tag{5.9}$$

while $\|[U]_1\|_1 = \|[\Phi]_1\|_1 = 1$. It is then easy to show that for any n one has $\|[\exp i[\Omega]_{\leq N}]_n\|_1 \leq (2A)^{n-2}$ which implies $\|[R_N]_n\|_1 \leq (4A)^{n-2}$.
It is then found that $\|R_N\|_r \leq (2A)^{-2}(1-4Ar)^{-1}(4Ar)^{N+1}$ for $r < 1/4A$ and estimating the inverse function Ψ one has $\|\hat{R}_N(1)\|_r \leq (4Ar)^{N+1}$ for $r < 1/18A$. Finally the remainder for the map iterated t times is estimated in a ball of radius r_t for $t \leq T$ where $r_0 = r < 1/18A, \ldots, r_t = r(1-t/2T)$ so that $r_T = r/2$ and one has

$$\|\hat{R}_N(t)\|_{r_{t-1}} \leq 4T^2(4Ar)^{N+1}, \qquad t \leq T. \tag{5.10}$$

The bound on the remainder $\hat{R}_N(1)$ given by $(4Ar)^{N+1} = [24\gamma(N+1)^{9+\eta}]^{N+1}$ reaches a minimum for fixed r when $N = \bar{N} \equiv e^{-1}(r_*/r)^{\frac{1}{9+\eta}} - 1$ where $r_* = 1/24\gamma$ and the minimum is $\exp[-(9+\eta)e^{-1}\ (r_*/r)^{1/9+\eta}]$. The only constraint is given by $\bar{N} \geq 2$ and reads $r \leq r_*(3e)^{-9-\eta}$.

Finally as a stability condition we require that if $\|\varsigma(0)\| \leq r/2$ then the image $\varsigma(t)$ of the orbit in normal coordinates, given by the r.h.s. of equation (4.11), stays in a polydisc of radius r: $\|\varsigma(t)\| \leq \|\varsigma(T)\| \leq \|\varsigma(0)\| + \|\hat{R}_N(T)\|_{r/2} \leq r$. As a consequence the time T, for which an orbit initially in a polydisc of radius

$r/2$ remains in a polydisc of radius r, is given by the condition $\|\hat{R}_N(t)\|_{r/2} \leq 4T^2(4Ar)^{N+1} \leq r/2$ and the optimal result obtained for $N = \bar{N}$ reads

$$T = \sqrt{\frac{r}{8}} \left[\frac{9+\eta}{2e} \left(\frac{r_*}{r} \right)^{\frac{1}{9+\eta}} \right]. \tag{5.11}$$

The final result is obtained by transforming these results into the ordinary coordinates. Letting $\|z(0)\| \leq \rho/2$ it is easy to check that letting $\rho = r\,48/49$ one obtains the translation of (5.11) into the ordinary coordinates expressed by (5.3). For a detailed proof see [12].

To conclude, we observe that the divergent character due to nonlinearity is captured by the sequence μ_n given by (5.8); it is interesting to observe that its generating function $g(x) = \sum_{n\geq 1} \mu_n x^n$ satisfies the functional equation

$$3g(x) = g(g(x)) + x^2 + 2x \qquad g(0) = 0,\ g'(0) = \frac{1}{2} g''(0) = 1 \qquad x \in \mathbf{C} \tag{5.12}$$

and that $g(x)$ is analytic in a neighborhood of $x = -1$ with a fractal natural boundary which includes the origin [13], a situation reminiscent of the analytic structure in ϵ of the function F considered in section 2.

REFERENCES

[1] A.N. Kolmogorov, Dokl. Akad. Nauk SSSR, **98**, 4, 527-530, (1954).
V.I. Arnol'd, Russ. Math. Surv., **18**, 5, 9-36, and **6**, 85-191 (1963).
J. Moser, Nachr. Akad. Wiss. Göttingen, Math.-Phys. Kl. **1**, 1, (1962).

[2] V.I. Arnol'd, Sov. Math. Dokl., **5**, 581-585,(1964).

[3] N.N. Nekhoroshev, Russ. Math. Surv., **32**, 6, 1-65, (1977).

[4] A. Giorgilli and L. Galgani, Cel. Mech., **37**, 95-112, (1985).
G. Benettin and G. Gallavotti, J. Stat. Phys., **44**, 293-338, (1986).
A. Giorgilli, *Recent developments of perturbation theory of classical hamiltonian systems*, in *Nonlinear Dynamics* Ed. by G. Turchetti, World Scientific, Singapore (1989)

[5] Giorgilli, A. Delshams, E. Fontich, L. Galgani and C. Simò, Journ. of Diff. Eq. **77**, 167-198 (1989).

[6] *Lecture Notes in Physics N° 247*, Springer Verlag (1986)

[7] A. Bazzani, P. Mazzanti, G. Servizi and G. Turchetti, Il Nuovo Cimento **102B**, 51-80, (1988).

[8] A.J. Liechtenberg, M.A. Liebermann *Regular and stochastic motions in dynamical systems*, Springer Verlag (1983)

[9] A. Bazzani, M. Malavasi, S. Siboni and G. Turchetti, *Perturbative methods in the analysis of magnetic fields structure*, Phys. Dep. Univ. of Bologna preprint (1989)

[10] G.D. Birkhoff, Acta Math., **43**, 1-119, (1920).

[11] G. Turchetti, *Perturbative Methods for Hamiltonian Maps*, in *Methods and Applications of Nonlinear Dynamics*, A.W. Sáenz editor, World Scientific, Singapore, 95-154, (1988).
G. Servizi, G. Turchetti, Il Nuovo Cimento **B95**, 121-154 (1986)

[12] A. Bazzani, S. Marmi, G. Turchetti *Nechoroshev estimates for isochronous symplectic maps* Phys. Dep. Univ. of Bologna preprint submitted to Celestial Mechanics.

[13] D. Bessis, S. Marmi, and G. Turchetti, work in progress.

[14] M. Malavasi, S. Marmi, G. Servizi and G. Turchetti in preparation

[15] J. Palis and J.C. Yoccoz, *Rigidity of Centralizers of Diffeomorphisms*, Ann. Ec. Norm. Sup., in press, (1988).

[16] A. Bazzani, *Normal Forms for symplectic maps in* $\mathbf{R}^{2n}$, Cel. Mech., **42**, 107-128, (1988).

p-adic Dynamical Systems

D. Verstegen

NIKHEF, P.O. Box 41882, NL-1009 DB Amsterdam, The Netherlands

We describe the conditions for the equivalence of a quadratic p-adic map to a quasiperiodic linear one near an indifferent periodic point. We also establish the existence of chaotic behaviour in p-adic iterations.

1 Introduction

Large classes of non-linear physical systems exhibit transitions to chaos which are universal. They can be modelled by simple non-linear maps and these maps are often one-dimensional[1,2]. It is certainly an intrinsically interesting question whether chaos and quasiperiodicity are properties of real or complex maps or whether they also occur in maps defined over other continuous number fields such as p-adics.

But there are also more direct motivations for exploring the properties of p-adic physical systems[3]. The p-adic fields $\mathbf{Q}_p$ are completions of the field Q of rational numbers with respect to the p-adic norm just as $\mathbf{R}$ is the completion of Q with respect to the absolute value norm. The $\mathbf{Q}_p$'s and $\mathbf{R}$ are natural partners in the theory of adeles [4]. This mathematical structure shows up in string theory, where the product of certain ordinary amplitudes times their p-adic analogues takes an extremely simple form [5,6].

The discovery of the ultrametric structure of the ground states in spin glasses[7] suggests that p-adic numbers might play a role there also.

This paper is largely based on work done in collaboration with E. Thiran and J. Weyers[8]. It is organized as follows: in section 2 we discuss some relevant properties of p-adic numbers. In section 3 we analyse the behaviour of the quadratic p-adic map near indifferent periodic points and we describe the conditions under which it is equivalent to a quasiperiodic linear map. In section 4 we analyse two examples of chaotic iterations.

These results are much easier to obtain than their real or complex analogues; this suggests that new concepts for dealing with non-linear dynamics might be tested first with p-adic numbers.

We do not aim at the most general analysis but rather at illustrating different characteristic patterns of such maps. We find many similarities with real or complex dynamics although we have no example of *transition* to chaos; this is not too surprising in view of the topological structure of Q_p.

2 The p-adic fields

We first briefly review some properties of p-adic numbers, with particular emphasis on the differences with real or complex numbers. More background material can be found in the mathematical literature [9].

The field Q_p of p-adic numbers is obtained by completing the field Q of rational numbers with respect to the p-adic norm. Any rational number x can be written as

$$x = p^{\alpha}\frac{a}{b} \tag{1}$$

where p is a fixed prime number, α, a and b are integers and p does not divide a nor b. The p-adic norm of x is defined as follows

$$|x|_p = p^{-\alpha}. \tag{2}$$

α is called the ordinal of x (at p) and is usually denoted $ord_p x$. Since no other norm than (2) will appear, we will use the simpler notations $|x|$ and ord x.

A convenient realization of Q_p is given by the set of power series

$$\{x = p^{ord\ x}(x_0 + x_1 p + x_2 p^2 + \cdots)\} \tag{3}$$

where the "digits" x_i are integers satisfying

$$0 \leq x_i \leq p-1; x_0 \neq 0. \tag{4}$$

The elements of the subfield Q are characterized by periodic sequences of x_i's just as they have a periodic decimal expansion in **R**.

The p-adic norm satisfies the "non archimedean" or "ultrametric" inequality

$$|x+y| \leq max\{|x|, |y|\}. \tag{5}$$

It follows that Q_p is a disconnected field: open subsets have no boundaries or more precisely they are both open and closed with respect to the topology induced by the norm. For example, the set of "p-adic integers" $\mathbf{Z}_p = \{x : |x| \leq 1\}$ can also be defined as $\{x : |x| < p\}$. Another consequence is that a large p-adic number cannot be constructed as the sum of many small ones or, in other words, there is no "path" in $\mathbf{Q}_p$ since the path concept relies on the possibility of covering a big distance in many small steps.No theorem of real analysis which relies on the connectedness of $\mathbf{R}$ will have a p-adic counterpart. In particular there is no intermediate value nor mean value theorem and hence continuous one-to-one functions from a compact to itself will not necessarily have a fixed point.

On the other hand, roundoff errors are much easier to control and the convergence condition for a series is simply that its main term goes to 0.

3 Attractive and indifferent periodic points

For simplicity we concentrate on quadratic maps although, clearly, our methods and results can be extended to other polynomial or rational maps as well. Note that because of the disconnectedness of $\mathbf{Q}_p$ there are many infinitely differentiable functions that have no simple "analytical" expression (an example is given at the beginning of next section).

Through a simple linear conjugacy, any quadratic map can be brought to the canonical form

$$f(x) = x^2 + c. \tag{6}$$

In this section we consider the region $|c| \leq 1$. The case $|c| > 1$ will be dealt with in the next section. We restrict ourselves to bounded orbits, which requires that the initial point satisfies $|x_{in}| \leq 1$.

Let's first look at this map modulo p, so that x and c can only take p values. Because of the finite number of possibilities every orbit eventually becomes periodic. Either it passes through 0 mod p or it does not. In the first case, the process approaches an attractive cycle because $|f'(x)| = |2x| < 1$ if $x = 0$ mod p and $|f'(x)| = 1$ otherwise. In the second case we have an approximate indifferent cycle:

$$f^n(x_0) - x_0 = 0 \pmod{p}. \tag{7}$$

If furthermore $(f^n)'(x_0) - 1 \neq 0$ mod p, then according to Hensel's lemma [9] there exists an exact solution to $f^n(x) = x$ with $x = x_0$ mod p (this is a sufficient but not a necessary condition for having an exact indifferent cycle).

We would now like to understand the dynamics near such an indifferent periodic point. If x^* is a periodic point of order n and if we note $x = x^* + y$, the map f^n is linearly conjugate to g:

$$g(y) = (f^n)'(x^*) \cdot y + \mu y^2 + \nu y^3 + \cdots + y^{2^n} \tag{8}$$

where $\mu, \nu, \ldots$ are p-adic integers; the coefficient of y will be noted ω. If x^* is a fixed point this reduces to

$$g(y) = \omega y + y^2. \tag{9}$$

It seems reasonable to ask if g is topologically conjugate, i.e. equivalent, to the linear map $L(y) = \omega y$ for small enough y[10]. L is particularly simple. In almost all cases, all points (except the fixed point 0) are "quasiperiodic":their period increases forever as the accuracy of the computation improves [11] and the orbit of any point covers densely some subinterval of $\mathbf{Z}_p$. If r is the smallest divisor of $p-1$ such that $\omega^r = 1$ mod p, the possible quasiperiods are rp^l $(l \geq 0)$. This is a simple consequence of the following: for $p \neq 2, |a^p - 1| = p^{-m-1}$ if $|a-1| = p^{-m}$. The linear map is not quasiperiodic in a finite number of cases, i.e. when $\omega^r = 1$ exactly.

To prove the topological conjugacy of g and L we must construct a homeomorphism U such that $U^{-1}LU = g$. We write

$$U(y) = q_1 y + q_2 y^2 + q_3 y^3 + \cdots \tag{10}$$

This p-adic problem is much simpler to solve than the analogous complex problem [2,12] because the convergence condition for U is simply

$$\lim_{m \to \infty} |q_m y^m| = 0. \tag{11}$$

We can choose $q_1 = 1$; the unknown coefficients, q_i, are then determined recursively from the identity $L.U = U.g$ or

$$\omega(q_1 y + q_2 y^2 + \cdots) = q_1(\omega y + y^2) + q_2(\omega y + y^2)^2 + \cdots \tag{12}$$

To avoid complicating the notations we have considered the case of an indifferent fixed point; the generalization to a cycle is easy. The recurrence relations are of the form

$$(\omega - \omega^{2m}) q_{2m} = \sum_{j=m}^{2m-1} \sigma_j q_j \tag{13}$$

with $|\sigma_j| \leq 1$, and similarly for q_{2m+1}. This implies

$$|q_m| \leq |(1-\omega)(1-\omega^2)\ldots(1-\omega^{m-1})|^{-1}. \tag{14}$$

Using

$$|1 - \omega^{rl}| = |l| \cdot |1 - \omega^r| \tag{15}$$

and the following bound for the factorial [4] :

$$|m!|^{-1} \leq p^{m(1-p^{-k})/(p-1)} \text{ if } p^k \leq m < p^{k+1}, \tag{16}$$

we find that the series for U converges iff r.ord y > ord$(\omega^r - 1)$. One can prove that U^{-1} converges under the same condition [8]. Thus if ω^r is very close to 1, g(y) and L(y) will be equivalent only for y very small. The investigation of a few examples suggests that the lack of convergence of the series for U when the above condition is not satisfied is related to the appearance of other cycles of g in the neighbourhood of the fixed point.

4 Chaos

Before addressing the case of the quadratic map with $|c| > 1$, we consider the following map from $\mathbf{Z}_p$ to $\mathbf{Z}_p$:

$$x = x_0 + x_1 p + x_2 p^2 + \cdots \to h(x) = x_1 + x_2 p + x_3 p^2 + \cdots \tag{17}$$

(here x_0 may be 0). This map is infinitely differentiable:

$$h'(x) = p^{-1}; h''(x) = 0. \tag{18}$$

Nevertheless, because of the disconnected topology of $\mathbf{Q}_p$, it is non-trivial:

$$h(x) \neq p^{-1} x \tag{19}$$

and it does not have a continuous real counterpart. h is the well-known "shift map": $(x_0, x_1, \ldots) \to (x_1, x_2, \ldots)$ and its dynamics is completely understood. One can distinguish two types of points: the elements of $\mathbf{Z}_p \cap Q$ are periodic or eventually periodic points; the elements of the complementary set are non periodic points of h.

The following properties of h are easily proved:

- periodic points are dense in $\mathbf{Z}_p$
- the orbit of any point is highly unstable; this is usually called "sensitive dependence on the initial conditions" and is due to the "expansiveness" of h: $|h'(x)| > 1$
- there are non periodic points whose orbit is dense, i.e. the orbit of one point comes arbitrarily close to all points in $\mathbf{Z}_p$; a simple example is provided by the sequence constructed by listing successively all possible blocks of length 1, 2, 3...

Following Devaney [2] such a map is chaotic. Its simplicity is quite remarkable !

We show next that a chaotic behaviour is also possible for polynomial maps. Iterating the map $f(x) = x^2+c$ ($|c| > 1$) with $|x_{in}^2| > |c|$ or $|x_{in}^2| < |c|$ leads to diverging sequences. Actually, bounded orbits are possible only if $c = -\gamma^2 (\gamma \in \mathbf{Q}_p)$. We will show that even in that case most points eventually end up at infinity. The remaining points belong to a Cantor set on which the dynamics is chaotic.

Which points will satisfy $|f(x)| = |\gamma|$ if $|x| = |\gamma|$? The answer is $x = \pm\gamma + y$ with $|y| = 1$. This suggests that, to find all points with bounded orbit, we perform an expansion in powers of γ^{-1}:

$$x = \lambda_{-1}\gamma + x_0 + x_1\gamma^{-1} + x_2\gamma^{-2} + \cdots \tag{20}$$

with $\lambda_{-1} = \pm 1$; the x_i's (which are not the p-adic digits) are to be determined.

Requiring f(x) to be of the same form as x (otherwise $|f^n(x)| \to \infty$ for $n \to \infty$) yields recursive relations for the x_i's. They can be solved by introducing for each i a new variable $\lambda_i = \pm 1$; x_i is then expressed in terms of $\lambda_{-1}, \ldots, \lambda_i$. The first step is

$$x = \lambda_{-1}\gamma + x_0 + \cdots \tag{21}$$

$$f(x) = 2x_0\lambda_{-1}\gamma + \cdots \tag{22}$$

We thus require $2x_0\lambda_{-1} = \lambda_0$ i.e. $x_0 = \frac{1}{2}\lambda_0\lambda_{-1}$. The next step is

$$x = \lambda_{-1}\gamma + \frac{1}{2}\lambda_{-1}\lambda_0 + x_1\gamma^{-1} + \cdots \tag{23}$$

$$f(x) = \lambda_0\gamma + \frac{1}{4} + 2\lambda_{-1}x_1 + \cdots \tag{24}$$

This fixes $x_1 : \frac{1}{4} + 2\lambda_{-1}x_1 = \frac{1}{2}\lambda_0\lambda_1$. One can easily show that only powers of 2 can appear in the denominator of the x_i's, hence $|x_i| \leq 1$ if $p \neq 2$. The series

$$x = \sum_{j=-1}^{\infty} x_j\gamma^{-j} = \lambda_{-1}\gamma + \frac{1}{2}\lambda_{-1}\lambda_0 + \frac{1}{4}\lambda_{-1}(\lambda_0\lambda_1 - \frac{1}{2})\gamma^{-1} + \cdots \tag{25}$$

(with $\lambda_i = \pm 1$) is thus convergent. We denote the set of such points by Λ.

Any point with a bounded orbit is in one-to-one correspondence with an infinite sequence of + or −'s: $(\lambda_{-1}, \lambda_0, \ldots)$. By construction f is equivalent to the shift map on these sequences. The dynamics is thus just the same as e.g. in the real case when $c < \frac{-5}{4} - \frac{\sqrt{5}}{2}$ [2]. f is chaotic on Λ which is a

Cantor set: it is a closed, non empty set which contains no intervals and in which every point is a accumulation point.

Λ can also be viewed as the closure of the repelling periodic points of f, so that Λ is the Julia set of f if we borrow the usual definition. A difficult problem in the complex case is the computation of the Hausdorff dimension of the Julia set [13]. The p-adic case is much simpler, the basic reason being that the expansion factor $|f'(x)|$ takes everywhere the same value $|\gamma|$ on Λ. This yields a Cantor set of the simplest type. The Hausdorff dimension D measures the growth of the number N of intervals needed to cover Λ when their measure δ goes to 0 : $N \propto \delta^{-D}$. Here, at level n, $\delta = |\gamma|^{-n+1}$ and $N = 2^n$, hence

$$D = \frac{ln2}{ln|\gamma|}. \tag{26}$$

In the previous example the Julia set was of dimension 1.

To conclude, we want to suggest two directions for further research. First it could be interesting to study iterations in the adelic context. Second one might get a better understanding of higher-dimensional maps (like the Hénon map) by studying them on p-adic fields.

References

[1] P. Cvitanovic, *Universality in Chaos*, a reprint selection (Adam Hilger Ltd, Bristol 1984).

[2] R.L.Devaney, *An Introduction to Chaotic Dynamical Systems* (The Benjamin/Cummings Publishing Compagny, Inc.1986).

[3] I.V. Volovitch, *Class. Quant. Gravity* **4**(1987)L83.

[4] I.M. Gel'fand, M.I. Graev and I.I. Pyatetskii-Shapiro, *Representation Theory and Automorphic Functions* (Saunders, London 1969).

[5] P.G.O. Freund and E. Witten, *Phys. Lett.* **B199** (1987) 191.

[6] Ph. Ruelle, E. Thiran, D. Verstegen and J. Weyers, UCL-IPT-89-01 preprint, to be published in *Mod. Phys. Lett. A*.

[7] R. Rammal, G. Toulouse and M.A. Virasoro, *Rev. Mod. Phys.* **58** (1986) 765.

[8] E. Thiran, D. Verstegen and J. Weyers, *J. Stat. Phys.* **54** (1989) 893.

[9] N. Koblitz, *p-adic Numbers, p-adic Analysis and Zeta Functions* (Springer Verlag, Berlin 1984).

[10] This problem has also been studied by: M. Herman and J. C. Yoccoz, in *Geometric Dynamics*, Proceedings, Rio de Janeiro 1981 (Springer-Verlag, Berlin Heidelberg 1983).

[11] S. Ben-Menahem, *p-adic Iterations*, TAUP 1627-88 preprint.

[12] C. L. Siegel and J. Moser, *Lectures on Celestial Mechanics* (Springer-Verlag, New-York 1971)

[13] H.O. Peitgen and P.H. Richter, *The Beauty of Fractals* (Springer-Verlag, Berlin Heidelberg 1986)

Part V

Further Arithmetical Problems, and Their Relationship to Physics

Dirichlet Series Associated with a Polynomial

P. Cassou-Noguès

Université de Bordeaux I, Unité associée au CNRS 226,
351 Cours de la Libération, F-33405 Talence Cedex, France

Abstract. This article is a survey recalling some properties of Dirichlet series associated with a polynomial. Most of the Dirichlet series we will study do not have a functional equation. We will be interested in their analytic continuation to the whole complex plane, their poles, their residues at simple poles and their values at negative integers.

1. Introduction

Let $P(X_1, ..., X_r)$ be a polynomial with r variables and positive real coefficients. The zeta function associated with P is defined by

$$Z(P)(s) = \sum_{n_1=1}^{\infty} ... \sum_{n_r=1}^{\infty} P(n_1, ..., n_r)^{-s}.$$

We will study some particular polynomials. We want to introduce some linear forms which will play an important role in the following [CN 1] . If $P(X_1, ..., X_r)$ is a polynomial it is well known that

$$\sum_{n_1 \geq 0} ... \sum_{n_r \geq 0} P(n_1, ..., n_r) T_1^{n_1} ... T_r^{n_r}$$

is a rational function. We write

$$\sum_{n_1 \geq 0} ... \sum_{n_r \geq 0} P(n_1, ..., n_r) T_1^{n_1} ... T_r^{n_r} = \sum_{j_1 \geq 1} ... \sum_{j_r \geq 1} \frac{\lambda_{j_1, ..., j_r}(P)}{(1 - T_1)^{j_1} ... (1 - T_r)^{j_r}}$$

2. The Riemann zeta function

We consider the polynomial $P(X) = X$. The associated Dirichlet series is the Riemann zeta function

$$\zeta(s) = \sum_{n=1}^{\infty} n^{-s},$$

Springer Proceedings in Physics, Vol. 47 Number Theory and Physics
Editors: J.M. Luck · P. Moussa · M. Waldschmidt

which is convergent for $Re(s) > 1$. One can prove that for $Re(s) > 1$,

(1) $$\Gamma(s)\zeta(s) = \int_{\gamma_\epsilon} \frac{(-z)^{s-1}e^{-z}}{1-e^{-z}} dz$$

where γ_ϵ is the path of integration starting at infinity on the real axis, encircling the origin counter-clockwise and returning to the starting point [W-W] . This defines the analytic continuation of $\zeta(s)$ on the whole complex plane. The function $\Gamma(s)\zeta(s)$ has possible poles for $s = 1, 0, -1, ..., -n, ...$ with rational residues. Then $\zeta(s)$ has a simple pole for $s = 1$ with residue 1 , is analytic at negative integers and

$$\zeta(-2k) = 0 \text{ if } n > 0$$

$$\zeta(1-2k) = (-1)^k B_k/2k \text{ if } k > 0$$

where B_k is the k-th Bernoulli number. It is also useful to write the values at negative integers in the following way [CN 1]:

$$\zeta(-k) = -\frac{1}{k+1} \sum_{j=1}^{k+2} \frac{\lambda_j(X^{k+1})}{j}$$

for all non-negative integers k .

3. The Hurwitz zeta function

Now we consider $P(X) = (a + vX)$. The associated Dirichlet series is the Hurwitz zeta function

$$\zeta_{(a,v)}(s) = \sum_{n=0}^{\infty} (a+nv)^{-s}.$$

It is convergent for $Re(s) > 1$ and can be continued to the whole complex plane using the formula

(2) $$\zeta_{(a,v)}(s) = \int_{\gamma_\epsilon} \frac{(-z)^{s-1}e^{-az}}{1-e^{-vz}} dz.$$

Then $\zeta_{(a,v)}(s)$ has a simple pole for $s = 1$ with residue $1/v$ and for all $k \geq 0$

(3) $$\zeta_{(a,v)}(-k) = \frac{-1}{v(k+1)} \sum_{j=1}^{k+2} \frac{\lambda_j((a+vX)^{k+1})}{j}.$$

One also uses the following formulas:

$$(4) \qquad \zeta_{(a,v)}(s) = \frac{1}{v}\left(\frac{1}{s-1} - \log v - \frac{\Gamma'(a/v)}{\Gamma(a/v)}\right) + (s-1)...$$

and

$$(5) \qquad \zeta'_{(a,v)}(0) = \log \frac{\Gamma(a/v)}{\sqrt{2\pi}} - \zeta_{(a,v)}(0)\log v.$$

The Hurwitz zeta function is useful in the study of Dirichlet L-series and zeta functions for abelian number fields. Let χ be a Dirichlet character mod f and

$$L(\chi, s) = \sum_{n=1}^{\infty} \chi(n) n^{-s}$$

the associated Dirichlet L-function. Then one can write

$$(6) \qquad L(\chi, s) = \sum_{a=1}^{f} \chi(a)\zeta_{(a,f)}(s)$$

and this formula gives the analytic continuation to the whole complex plane, the values at negative integers, the derivative at zero and a formula for $L(\chi,1)$ which is called the Lerch formula. Let p be a prime number. It is easy to define the p-adic Hurwitz zeta function using (3). Suppose that a and v are integers, $a \equiv 1 \bmod p$ and $p \mid v$. Then one has for any positive integer m [CN 1]

$$|\lambda_j((a+vX)^m)|_p \leq p^{-j+1}.$$

One defines

$$\zeta_{p,(a,v)}(-s) = \frac{-1}{v(s+1)} \sum_{j=1}^{\infty} \frac{\lambda_j((a+vX)^s)}{j},$$

which is a continuous function on the p-adic unit ball except at the point $s=1$ where there is a simple pole. For all $k \geq 0$

$$\zeta_{(a,v)}(-k) = \zeta_{p,(a,v)}(-k)$$

and we can prove formulas analogous to (4) and (5) using the p-adic gamma function defined by Morita [Mo 1], [F-G].

4. Barnes zeta function

Here we consider $P(X_1, ..., X_r) = a + v_1X_1 + ... + v_rX_r = L_{(a,V)}(X)$ and

$$Z_{(a,V)}(s) = \sum_{n_1=0}^{\infty} \cdots \sum_{n_r=0}^{\infty} L_{(a,V)}(m)^{-s}.$$

These Dirichlet series have been considered by Barnes [Ba 1], [Ba 2]. Using the same method as for the Hurwitz zeta function, one can prove that $Z_{(a,V)}(s)$ is convergent for $Re(s) > r$, can be continued to the whole complex plane as a meromorphic function with poles for $s = r, r-1, ..., 1$ and for all $k \geq 0$

$$Z_{(a,V)}(-k) =$$

$$\frac{(1)^r}{v_1...v_r(k+1)...(k+r)} \sum_{j_1 \geq 1} \cdots \sum_{j_r \geq 1} \frac{\lambda_{j_1,...,j_r}((a + v_1X_1 + ... + v_rX_r)^{k+r})}{j_1...j_r}.$$

Barnes used these Dirichlet series to define the multiple gamma function. Let

$$\log \rho_r(V) = -\lim_{a \to 0}(Z'_{(a,V)}(0) + \log a) \tag{7}$$

then the multiple gamma function is defined by

$$\log \frac{\Gamma_r(a,V)}{\rho_r(V)} = Z'_{(a,V)}(0). \tag{8}$$

One can also define p-adic zeta functions and p-adic multiple gamma functions which are related by formulas analogous to (6) and (7) [CN 2], [CN 3].

5. Shintani's zeta functions

The next example we want to study is the case where

$$P(X_1, ..., X_r) = \prod_{i=1}^{n} L_{(a_i,V_i)}(X).$$

The associated Dirichlet series is

$$Z(P)(s) = \sum_{n_1=0}^{\infty} \cdots \sum_{n_r=0}^{\infty} \prod_{i=1}^{n} L_{(a_i,V_i)}(n_1, ..., n_r)^{-s}.$$

These Dirichlet series have been used by Shintani [Sh 1] for computing the values at negative integers of the Dedekind zeta function for a totally real number field. $Z(P)(s)$ is convergent for $Re(s) > r/n$ and can be continued to the whole complex plane. To see this, one considers $\Gamma(s)^n\zeta(s)$, which can be expressed as a multiple integral generalizing the simple integral used for the Hurwitz zeta function. $Z(P)(s)$ has poles for $s = k/n$ where k is a relative integer less or equal to r and not divisible by n. Moreover there exists a polynomial Q_k, with coefficients in the field generated over the field of rational numbers by the coefficients of P, which is explicitly computable, such that

(9) $$\frac{\partial^r}{\partial x_1 ... \partial x_r} Q_k = P^k$$

and

(10) $$Z(P)(-k) = \sum_{j_1 \geq 1} ... \sum_{j_r \geq 1} \frac{\lambda_{j_1,...j_r}(Q_k)}{j_1 ... j_r}.$$

There are interesting formulas for the derivative $Z(P)'(0)$ in terms of the multiple gamma function [Sh 2], [CN 2].

6. Mellin's zeta functions

In 1900, Mellin [Me] studied

$$Z(P)(s) = \sum_{n_1=1}^{\infty} ... \sum_{n_r=1}^{\infty} P(n_1, ..., n_r)^{-s}$$

when P is a polynomial whose non-zero coefficients have a strictly positive real part. He showed that there exists a real number σ_0 such that $Z(P)(s)$ is convergent for $Re(s) > \sigma_0$ and that it can be continued to the whole complex plane as a meromorphic function whose poles are on the real line, and of order at most r . In 1927, Mahler [Ma 2] studied $Z(P)(s)$ with the following hypothesis. Let $P_m(X_1, ..., X_r)$ be the homogeneous part of P of highest degree. Then Mahler's hypothesis is

$$P(x_1, ..., x_r) > 0 \text{ for } x_1 > 0, ..., x_r > 0$$

and

$$P_m(x_1, ..., x_r) > 0 \text{ for } x_1 \geq 0, ..., x_r \geq 0 \text{ and } (x_1, ..., x_r) \neq 0.$$

He showed that $Z(P)(s)$ is convergent for $Re(s) > r/m$ and that

$$\operatorname*{Res}_{s=r/m} Z(P)(s) = \Gamma(r/m)^{-1} \int_0^{\infty} ... \int_0^{\infty} e^{-P_m(t_1,...,t_r)} dt_1 ... dt_r.$$

The results I will give now are proved for Mellin's hypothesis and $r = 2$ and for Mahler's hypothesis and any r [CN 4] [CN 5] . Let

$$P(X_1, ..., X_r) = \sum a_\alpha X^{\alpha_1} ... X^{\alpha_r}$$

and

$$SuppP = \{(\alpha_1, ..., \alpha_r) \text{ such that } a_\alpha \neq 0\}.$$

The Newton polyhedron of P is the union of the faces of the convex hull of the sets

$$\{(\alpha_1 - n_1, ..., \alpha_r - n_r) \text{ where } n_1, ..., n_r \text{ are positive real numbers.} \}$$

for $\alpha \in SuppP$. We write the equations of the faces of the Newton polyhedron of P

$$f_i(x_1, ..., x_r) = \frac{x_1}{a_{i,1}} + ... + \frac{x_r}{a_{i,r}} = 1$$

where the $a_{i,j}$ are rational numbers which can be infinite. Define

$$S(P) =$$

$$\cup\{f_i(1, ..., 1) - \sum_{\alpha \in SuppP} (1 - f_i(\alpha_1, ..., \alpha_r)\nu_\alpha), \nu_\alpha \text{ non-negative integer } \}$$

the union being taken over all the faces of the Newton polyhedron of P . Let us denote by $S(P)'$ the union of the set $S(P)$ and the set of natural numbers. One notices that $S(P)'$ is contained in the field of rational numbers. One can prove that all the poles of $\Gamma(s)Z(P)(s)$ are contained in the set $S(P)'$.

The set $S(P)'$ is a minimal set of possible poles, in the sense that for a given Newton polyhedron and for P generic with respect to this Newton polyhedron, $\Gamma(s)Z(P)(s)$ has poles at every element of $S(P)'$. Moreover if k is an integer, and if $-k$ is a simple pole of $\Gamma(s)Z(P)(s)$ there exists a polynomial Q_k such that (9) and (10) are true. If $-k \notin S(P)$ then $Q_k(0) = 0$ and $Z(P)(-k)$ belongs to the field K generated over the field of rational numbers by the coefficients of P. This is not always the case: an example is given by $P(X_1, ..., X_4) = X_1^3 + X_2^3 + X_3^3 + X_4^3$. We then have

$$Z(P)(0) = -\frac{8}{9}B_4\Gamma(\frac{1}{3})^3 + B_1^4.$$

Nevertheless, one can prove that the K-vector space generated by the residues at simple poles of $\Gamma(s)Z(P)(s)$ is of finite dimension, this dimension being bounded by the volume of the Newton polyhedron of P .

Some of these results have been extended by Sargos [Sa] and [Li] .

The methods used are refinements of Mellin's. They also can be used for studying the integrals

$$\int_1^\infty ... \int_1^\infty P(x_1, ..., x_r)^{-s} dx_1 ... dx_r$$

and

$$\int_0^1 ... \int_0^1 P(x_1, ..., x_r)^s dx_1 ... dx_r.$$

Using those last integrals, one gets interesting results on the singularities of P [CN 7].

Bibliography

[A] **Y. Amice** *Les Nombres p-adiques.* Presses Universitaires de France, 1975.

[A-F] **Y.Amice and J.Fresnel** *Fonctions zêta p-adiques des corps de nombres abéliens réels.* Acta Arith., 20, (1972), 353-384.

[Ba 1] **E. W. Barnes** *The theory of the double gamma function.* Philosophical Transactions of the Royal Society, 196, (1904), 373-425.

[Ba 2] **E. W. Barnes** *On the theory of the multiple gamma function.* Tran. Cambridge Philos. Soc., 19, (1904), 373-425.

[B-S] **Z. Borevich and I. Shafarevich** *Number Theory.* Academic Press: London and New York, 1966.

[CN 1] **P.Cassou-Noguès** *Valeurs aux entiers négatifs des fonctions zêta et fonctions zêta p-adiques.* Invent. Math., 51, (1979), 29-59.

[CN 2] **P.Cassou-Noguès** *Analogues p-adiques des fonctions gamma multiples.* Asterisque, 61, (1979), 43-55.

[CN 3] **P.Cassou-Noguès** *Applications arithmétiques de l'étude des valeurs aux entiers négatifs des séries de Dirichlet associées à un polynôme.* Ann. Inst. Fourier, 31, (1981), 1-35.

[CN 4] **P.Cassou-Noguès** *Intégrales et Séries de Dirichlet associées à un polynôme à deux indéterminées.* J. Number Theory, 23, (1986), 1-54.

[CN 5] **P.Cassou-Noguès** *Valeurs aux entiers négatifs des séries de Dirichlet associées à un polynôme II.* Amer. J. Math., 106, (1984), 255-299.

[CN 6] **P.Cassou-Noguès** *Valeurs aux entiers négatifs des séries de Dirichlet associées à un polynôme III.* Amer. J. Math., 109, (1987), 71-90.

[CN 7] **P. Cassou-Noguès** *Racines de polynômes de Bernstein.* Ann. Inst. Fourier, 36, (1986), 1-30.

[D] **J. Diamond** *On the values of p-adic L-functions at positive integers.* Acta arith., 35, (1979), 223-237.

[E] **H. Edwards** *Fermat's Last Theorem, a Genetic Introduction to Algebraic Number Theory.* Graduate Texts in Mathematics, Springer-Verlag, 1977.

[F] **J. Fresnel** *Nombres de Bernoulli et fonctions L p-adiques.* Ann. Inst. Fourier, 17, (1967), 281-333.

[F-G] **B. Ferrero and R. Greenberg** *On the behaviour of p-adic L-functions at $s = 0$.* Invent. Math., 50, (1978), 91-102.

[I-R] **K. Ireland and M. Rosen** *A Classical Introduction to Modern Number Theory.* Graduate Texts in Mathematics, Springer-Verlag. 1982.

[K] **N. Koblitz** *p-adic Numbers, p-adic Analysis, and Zeta-functions* Graduate Texts in Mathematics, Springer-Verlag, 1977.

[K-L] **T. Kubota and H.W. Leopoldt** *Eine p-adische Theorie der Zetawerte. I. Einführung in die p-adischen Dirichletschen L-Funktionen.* J. reine angew. Math., 214/215, (1964), 328-339.

[L 1] **S. Lang** *Cyclotomic Fields.* Graduate texts in Mathematics, Springer-Verlag, 1978.

[L 2] **S. Lang** *Cyclotomic Fields, II.* Graduate texts in Mathematics, Springer-Verlag, 1980.

[Li] **B. Lichtin** *Generalized Dirichlet Series and B-functions.* Compositio Math., 65, (1988), 81-120.

[Ma 1] **K. Mahler** *Introduction to the p-adic Numbers and their Functions.* Cambridge Tracts in Maths 64, Cambridge University Press 1973.

[Ma 2] **K.Mahler** *Über einen Satz von Mellin.* Math. Ann., 100, (1927), 384-395.

[Me] **H. Mellin** *Eine Formel für den Logarithmus transzendenter Funktionen von endlichem Geschlecht.* Acta Soc. Scient. Fennicae, 29, (1900), 3-49.

[Mo 1] **Y.Morita** *A p-adic analogue of the gamma function.* J. Fac. Sci. Univ. Tokyo, Sec IA, 22, (1975), 255-266.

[Mo 2] **Y. Morita** *On the Hurwirtz Lerch L-function* J. Fac. Sci. Univ. Tokyo, Sec.IA, 24, (1977), 29-43.

[Sa] **P. Sargos** *Séries de Dirichlet et polyèdres de Newton.* Préprint, (1987).

[Se] **J.-P. Serre** *Cours d'arithmétique.* Presses Universitaires de France, 1970.

[Sh 1] T.Shintani *On evaluation of zeta functions of totally real algebraic number field.* J. Fac. Sci. Univ. Tokyo, Sec IA, 23,(1978), 393-417.

[Sh 2] T. Shintani *On certain ray class invariants of real quadratic fields.* J. Fac. Sci. Univ. Tokyo, Sec. IA, 30, (1978), 139-167.

[Si] C.-L. Siegel *Lectures on Advanced Number Theory.* Tata Institute of Fundamental Research, Bombay, 1965.

[W-W] E. Whittaker and G. Watson *A C Course of Modern Analysis.* Cambridge University Press, 1958.

[Z] D. Zagier *Zetafunktionen und quadratische Körper.* Hochschultext, Springer-Verlag ,1981.

Some Remarks on Random Number Generators

M.L. Mehta

SPhT, CEN de Saclay, F-91191 Gif-sur-Yvette Cedex France

1 - In some applications one needs a sequence of random numbers. Actually no deterministic or reproducible law can ever produce truly random numbers. By random numbers one usually means a sequence of numbers generated by some fixed algorithm and having the properties of truly random numbers up to correlations extending to a finite distance. To generate them on a computer three methods are popular these days: (i) congruential, (ii) shift register, and (iii) lagged Fibonacci. They are popular because they are simple to use and reasonably good. However, each method has its defects. We describe them briefly below, following mostly Marsaglia[1] or Knuth[2], adding a remark or two from time to time. The difficulty is that most of our knowledge about them is empirical.

2 - A congruential generator uses the formula

$$x_n = ax_{n-1} + b \quad \mod \quad m \tag{1}$$

where m is a large number, usually a power of 2, say 2^{32}, which a computer can easily handle. As m is finite, the x_n eventually repeat, i.e. $x_{n+\nu} = x_n$ for some ν and every n. The smallest of such integers ν is known as the cycle length. The trick is to choose integers a and b, usually empirically, to have the multiplication simple and ν as large as possible. A good choice is $a = 2^\alpha \pm 1, b = 1$, if $m = 2^n$, since multiplication by such an a is simple to perform and α can be chosen to have the cycle length ν quite large.

The main defect of the congruential generator is that if we consider

$$(x_{j+1}, ..., x_{j+n})$$

as the coordinates of a vector $\mathbf{r}_j$ in n-space, then the points $\mathbf{r}_j$ lie on a lattice with a large unit cell, its volume being of the order of m^{n-1}. For a truly random sequence x_j, the points $\mathbf{r}_j$ should fill up the n-space uniformly, i.e. the unit cell volume should be of order 1.

For $n = 1$ or 2, especially 2, the lattice structure of the points $\mathbf{r}_j$ can be visualized by plotting them on squared graph paper. For $n \geq 3$ this is not so easy. In this case, one may test the randomness of the differences $x_{j+1} - x_j$ or of the ratios $(x_{j+3}-x_{j+2})/(x_{j+1}-x_j)$, or of the "$n$-volumes" $v_j = \det\ [x_{j+ni+k}]$, $i, k = 0, 1, ..., n-1$.

3 - A shift register generator uses the formula

$$x_n = x_{n-1}(I + L^a)(I + R^b) \tag{2}$$

where the x_n are written as binary vectors,

$$x_n = (b_{\beta-1}, b_{\beta-2}, ..., b_1, b_0), \quad b_j = 0 \ \text{ or } \ 1, \tag{3}$$

if $x_n = b_{\beta-1}2^{\beta-1} + ... + b_1.2 + b_0$, I is the $\beta \times \beta$ unit matrix, L is the $\beta \times \beta$ matrix with 1's in the slanting line just below the diagonal and zeros elsewhere, and R is the transpose of L. The effect of L is to shift the binary vector x to the left

$$(b_{\beta-1}, ..., b_1, b_0)L = (b_{\beta-2}, ..., b_1, b_0, 0) \tag{4}$$

and that of R is to shift it to the right. The integers a and b are properly chosen so as to have a large cycle length ν. For $m = 2^{31}$ good choices for (a, b) are (3,28), (6,25) and (13,18), the cycle length being the maximum possible, i.e. $2^{31} - 1$. For $m = 2^{32}$ a good choice is (15,17), the cycle length, though not $2^{32} - 1$, is still quite large.

As for the congruential generators the points $\mathbf{r}_j = (x_{j+1}, ..., x_{j+n})$ in n-space show regularities: they lie on the vertices of a distorted lattice.

4 - The lagged Fibonacci generator F(r,s,+) uses the formula

$$x_n = x_{n-r} + x_{n-s} \mod m. \tag{5}$$

Two theorems of Marsaglia[1] given below guide us in the choice of the integers r and s.

Theorem 1 - Let T be an $r \times r$ binary non-singular matrix and β any $1 \times r$ non-zero binary vector. Then the sequence β, βT, βT^2,...mod 2 (i.e. each component is taken modulo 2) has the maximum cycle length $k = 2^r - 1$ if and only if the order of T modulo 2 is k in the group of non-singular $r \times r$ binary matrices.

Theorem 2 - Let T be an $r \times r$ matrix of integers with an odd determinant and $\beta = (x_1, ..., x_r)$ a $1 \times r$ vector of integers, not all even. Then the sequence β, βT, βT^2,... mod 2^n (i.e. every matrix element is reduced modulo 2^n) has the period $(2^r - 1)2^{n-1}$ for every $n \geq 1$, if and only if T has order $j = 2^r - 1$ for mod 2, has order $2j$ for mod 4 and has order $4j$ for mod 8.

Remark 1 - If a non-singular $r \times r$ matrix T has order j mod 2, has order $2j$ mod 4 and has order $4j$ mod 8, where j is not necessarily equal to $2^r - 1$, then for any $1 \times r$ vector β with integer elements not all even, the sequence β, βT, βT^2, ... mod 2^n has the period $j.2^{n-1}$ for every $n \geq 1$.

Remark 2 - The number of vectors $\alpha = (x_1, ..., x_r)$ with integer components not all even and each of them less than 2^n is $2^{nr} - 2^{(n-1)r} = (2^r - 1)2^{(n-1)r}$. Out of these only $(2^r - 1)2^{n-1}$ are covered by β, βT, βT^2, ..., mod 2^n when β is any of the α. A large majority is unattained. Also the α with all components even integers are never attained. They are $2^{(n-1)r} - 1$ in number, excluding the zero vector. Thus the important question arises: How are the components of the $(2^r - 1)2^{n-1}$ vectors β, βT, ... distributed on $(0, 2^n - 1)$ when we start with any allowed β ? Does their distribution depend on the initial vector β ? Actually we should ask these questions not for the complete sequence of $(2^r - 1)2^{n-1}$ vectors, but for the partial sequence as we generate it.

Formula 5 can be written as

$$(x_{j+1}, ..., x_{j+r}) = (x_j, ..., x_{j+r-1})T \tag{6}$$

where the only non-zero elements of the $r \times r$ matrix T are

$$T_{i+1,i} = 1, \quad i = 1, 2, ..., r-1; \qquad T_{1,r} = T_{r-s+1,r} = 1. \tag{7}$$

To get a large cycle length one should choose r and s so that according to theorems 1 and 2 the order of T for mod 2 is $j = 2^r - 1$, for mod 4 is $2j$ and for mod 8 is $4j$. For certain values of r (i.e. r = 2, 3, 5, 7, 13, 19, 31, 61, 89, 107, 127, ...), $j = 2^r - 1$ is a prime number, and it is easier to ascertain the order of T for these values. This is so because one has only to use the squaring algorithm $r + 1$ times to check for $k = j + 1 = 2^r$ whether T^k is equal to T modulo 2 but not modulo 4 and T^{2k} is equal to T^2 modulo 4 but not modulo 8. For other values of r one has to check in addition that $T^\ell \neq 1$ mod 2 for ℓ any submultiple of $j = 2^r - 1$. Values of r and $s \leq 31$ for which T satisfies all the conditions of the two theorems are listed in table 1. For larger values of r, Marsaglia states that (55,24), (607,273) and (1279,418) are good choices.

Table 1. Some values of r, s for which the matrix T verifies the two theorems of Marsaglia. The matrix T is defined by $T_{i+1,i} = 1$, for $i = 1, 2, \ldots, r-1$; and $T_{1,r} = T_{s,r} = 1$, all other $T_{i,j} = 0$.

r	=	2	3	3	4	4	5	5	6	6	7	7	7	7	9
s	=	1	1	2	1	3	2	3	1	5	1	3	4	6	4
r	=	9	10	10	11	11	15	15	15	15	15	15	17	17	17
s	=	5	3	7	2	9	1	4	7	8	11	14	3	5	6
r	=	17	17	17	...	...	...	31	31	31	31	31	31	31	31
s	=	11	12	14	...	...	...	3	6	7	13	18	24	25	28

5 - As a side excursion in number theory, choose $r = 2$, $s = 1$; so that the sequence x_n consists of the ordinary Fibonacci numbers. The matrix $T = \begin{bmatrix} 0 & 1 \\ 1 & 1 \end{bmatrix}$ satisfies the requirements of theorems 1 and 2, namely the order of T mod 2 is 3, it is 6 mod 4 and 12 mod 8. The order of T mod 2^n is therefore $3 \times 2^{n-1}$. A slight generalization of theorem 2 is the following.

Theorem 3 - Let T be an $r \times r$ matrix with integer elements and its determinant not divisible by the prime number p, $p > 2$. Let β be a $1 \times r$ vector of integers, not all divisible by p. Then the sequence β, βT, βT^2,... mod p^n has the period $j.p^{n-1}$ for every n if and only if T has order j for mod p and has order $j.p$ for mod p^2.

Note that j need not be equal to $p^r - 1$; nor do we need, as for $p = 2$, that the order of T mod p^3 be $j.p^2$.

It is amusing to compute the order q of the 2×2 Fibonacci matrix T modulo any given integer m. In other words for any given m find the smallest integer $q = q(m)$ such that the ordinary Fibonacci numbers F_j satisfy the congruences $F_q = 0 \mod m$, $F_{q\pm 1} = 1 \mod m$. The above theorem together with the computation of some powers of T gives us the values $q(p^n) = q'(p).p^{n-1}$, with the following values of q'.

$$p = 2,\ 3,\ 5,\ 7,\ 11,\ 13,\ 17,\ 19,\ \ldots$$

$$q' = 3,\ 8,\ 20,\ 16,\ 10,\ 28,\ 36,\ 18,\ \ldots$$

Also if m and n are relatively prime, then $q(mn) = \ell(q(m), q(n))$, where $\ell(j,k)$ is the least common multiple of j and k.

6 - The lagged Fibonacci generator $F(r, s, +)$ has one serious defect. The numbers x_n, x_{n-r}, x_{n-s}, any n, have strong correlations. For example, $x_{n-r} < x_n < x_{n-s}$ or $x_{n-s} < x_n < x_{n-r}$ would never occur in an $F(r, s, +)$ sequence, whereas in a truly random sequence these inequalities have probabilities 1/6 each. The strong non-linearity introduced by the fact that the numbers x_j are always taken modulo m, does not modify this defect. To avoid this shortcoming it will be better to use an $F(r, s, t, +)$ formula

$$x_n = x_{n-r} + x_{n-s} + x_{n-t} \mod m. \tag{8}$$

The two theorems of Marsaglia mentioned above can certainly guide us in the choice of the integers r, s, t to have a very large cycle length. Unfortunately, for no values of r, s, $t \leq 40$, does T satisfy all the conditions of theorems 1 and 2. If we allow some of the coefficients to be 2, then there are many possibilities, e.g. there are 86 choices for $r = 17$ and 50 choices for $r = 31$. (Curiously, for $r = 13$ and $r = 19$ there are no such possibilities.) Some of them are listed in table 2.

Table 2. Some values of r, s, t, T_{sr} and T_{tr} for which the matrix T verifies the two theorems of Marsaglia. The matrix T is defined by $T_{i+1,i} = 1$, $i = 1, ..., r-1$; $T_{1r} = 1$, $T_{sr} = 1$ or 2, $T_{tr} = 1$ or 2, all other $T_{ij} = 0$.

r	=	17	17	...	17	17	17	...	31	31	...	31	31	...
s	=	2	2	...	13	13	13	...	2	2	...	4	4	...
t	=	4	6	...	14	15	15	...	4	7	...	19	19	...
T_{sr}	=	2	2	...	1	1	2	...	2	2	...	1	2	...
T_{tr}	=	1	1	...	2	2	1	...	1	1	...	2	1	...

For example, take the rule

$$x_n = 2x_{n-1} + x_{n-2} + x_{n-3} \mod m.$$

The corresponding 3×3 matrix T satisfies the conditions of theorem 2 and therefore the cycle length of the above sequence for $m = 2^n$ is $7 \times 2^{n-1}$. It will of course be better to choose a larger value of r to have a large cycle length.

An unsatisfactory feature of either $F(r,s,+)$ or $F(r,s,t,+)$ is that we do not even know, except empirically, that a finite sub-sequence of the x_n is equally distributed in the interval $(0, m-1)$; to say even less about higher correlations.

7 - Combination generators. To remedy the defects of individual generators, a combination of two or more independent generators is often recommended. Out of several possibilities the simplest one is the following. Let $(x) = x_1, x_2, \ldots$ and $(y) = y_1, y_2, \ldots$ be two sequences of random numbers produced possibly by different generators. Then the sequence $(z) = z_1, z_2, \ldots$, with $z_j = x_j + y_j \mod m$ should be preferred to either (x) or (y) for the following reason. The density of a single value, a pair of values, a triple of values, ... of (z) is more uniform than the one of either (x) or (y). In fact let, for any n, the probability of $x_n = j$ be p_j, that of $y_n = j$ be q_j and that of $z_n = j$ be r_j, so that $r_j = \sum_{k=0}^{m-1} p_{j-k} q_k$ (with the assumption that $p_{j-m} = p_j = p_{j+m}$.) If the (x) were uniformly distributed in $(0, m-1)$, then $p_j = 1/m$. So $d_1(x) = \sum_{j=0}^{m-1} (p_j - 1/m)^2$ is a measure of how badly the sequence (x) approaches uniform distribution. Now

$$d_1(z) = \sum_{j=0}^{m-1} \left(r_j - \frac{1}{m} \right)^2 = \sum_{j=0}^{m-1} \left(\sum_k p_{j-k} q_k - \frac{1}{m} \right)^2$$

$$\leq \sum_{j=0}^{m-1} \left(p_j - \frac{1}{m} \right)^2 = d_1(x).$$

This is so because $\sum_k q_k = \sum_k p_k = 1$, and

$$\left(\sum_k p_{j-k} q_k - \frac{1}{m} \right)^2 = \left(\sum_k \left(p_{j-k} - \frac{1}{m} \right) q_k \right)^2 \leq \sum_k \left(p_{j-k} - \frac{1}{m} \right)^2 q_k$$

since the average of the square is not less than the square of the average. Thus

$$d_1(z) \leq \sum_j \sum_k \left(p_{j-k} - \frac{1}{m} \right)^2 q_k = \sum_k q_k \sum_j \left(p_{j-k} - \frac{1}{m} \right)^2$$

$$= \sum_j \left(p_j - \frac{1}{m} \right)^2 = d_1(x).$$

Also let p_{ij} be the probability that $x_n = i$ and $x_{n'} = j$, with $\sum_i p_{ij} = \sum_i p_{ji} = p_j$. Denoting similar quantities for the sequence (y) by q_{ij}, q_j and for the sequence (z) by r_{ij}, r_j, one can show that

$$d_2(z) = \sum_{i,j} \left(r_{ij} - \frac{1}{m^2} \right)^2 \leq \sum_{i,j} \left(p_{ij} - \frac{1}{m^2} \right)^2 = d_2(x).$$

And so on.

So in a sense (z) is more random than either (x) or (y). Of course we would have liked to compare quantities of the form $\langle x_n x_{n+\alpha} \rangle - \langle x_n \rangle^2$, or better still, the same quantities over a limited range (the sum or average not running over the entire cycle but only over a portion of it) with the corresponding quantities for the sequence (z). It seems difficult even to compare $d_2'(x) = \sum_{i,j} (p_{ij} - p_i p_j)^2$ with $d_2'(z) = \sum_{i,j} (r_{ij} - r_i r_j)^2$.

8 - Until now we have considered a sequence of random integers with constant probability density between 1 and m, or on dividing by m, of real numbers with constant probability density in $(0, 1)$. In practice one often needs random numbers with a preassigned probability density function $p(x)$; for example, the Lorentz density with $p(x) = (2/\pi)(1+x^2)^{-1}$, or the step function with $p(x) = 1$ for $b < x < a$ and $p(x) = 0$ otherwise, or the normal distribution with $p(x) = (2\pi\sigma^2)^{-1/2}\exp(-x^2/2\sigma^2)$. In principle, if the differential equation $dy/dx = p(x)$ has the solution $x = F(y)$ and the sequence y_j has a constant probability density, then $x_j = F(y_j)$ will have the probability density $p(x)$. This method is practical in some cases and complicated in others. For example, it works nicely in the Lorentz case when $p(x) = (2/\pi)(1+x^2)^{-1}$, so that $x = tan(\pi y/2)$. On the other hand, for the normal distribution it is not at all easy and special methods have been developed for this case encountered frequently in applications. We will not go into details here.

References

[1] G. Marsaglia and L. H. Tsay, " Matrices and the structure of random number sequences", Lin. Alg. Appl., **67**, 147-185, (1985).

[2] D. Knuth, "The art of computer programming", vol. 2, Semi-numerical algorithms, Addison Wesley, Reading, Mass. (1981).

Bounds for Non-blocking Switch Networks

C.J. Smyth

University of Edinburgh, Department of Mathematics,
James Clerk Maxwell Building, King's Buildings, Mayfield Road,
Edinburgh, EH9 3JZ, UK

Abstract. We show that non-blocking $n \times n$ Beneš networks of 2×2 switches must have at least $n(n-1) - [n^2/4]$ switches. The proof uses a result concerning the maximum number of edges an n-node graph can have without containing any triangles.

With communications and computing in mind, much current effort is being directed to switching of optical signals. One common experimental device is a 2–inlet, 2–outlet (2×2) switch, which either switches two signals straight through (fig. 1.(a)) or crosses them over (fig 1.(b)). Networks of these switches can be built in integrated-optic form using waveguides made by Titanium diffusion on a substrate of Lithium Niobate ($LiNbO_3$) [2], [4].

Fig. 1 : 2–inlet 2–outlet switches.

We are interested here in such switch networks which can be built to form a non–blocking (see below) n–inlet n–outlet $n \times n$ switch. A small example is the 3×3 switch network below, formed from 2×2 switches (fig. 2). The main aim of this note is to show that a non-blocking Beneš network must have at

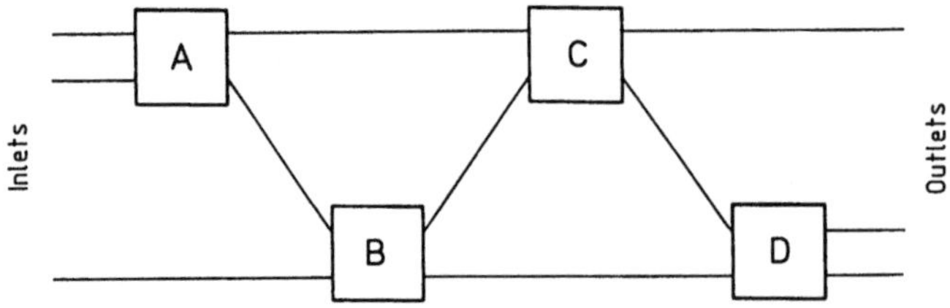

Fig. 2 : a 3×3 non-blocking Beneš network.

Springer Proceedings in Physics, Vol. 47 Number Theory and Physics
Editors: J.M. Luck · P. Moussa · M. Waldschmidt

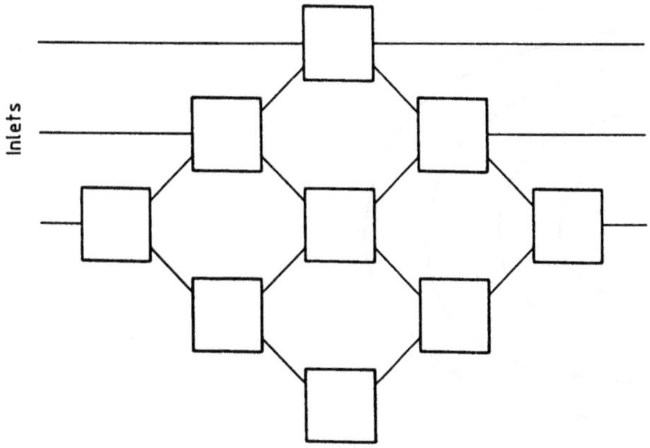

Fig. 3 : a 3×3 square network.

least $M_n \simeq (3/4)n^2$ switches (theorem 1). A Beneš network is a network of 2×2 switches where, in each network state, an inlet is connected to an outlet, and each switch has two inlet-to-outlet paths using that switch. Fig. 2 is an example of a Beneš network. Compare the square network of fig. 3, which is not a Beneš network.

An $n \times n$ Beneš network is *non blocking* if there is a non-empty set P of network states with the following property: for any state of P and any pair of inlet-to-outlet paths through the network in that state, we can change the state of some of the switches where the two paths cross, so that we are again in a state of P, and the two outlets are connected to the opposite inlets. That is, we have "swapped the outlets": if it was inlet 1 → outlet 1, inlet 2 → outlet 2, it is now inlet 1 → outlet2, inlet 2 → outlet 1.

This definition is easily (see [3]) shown to be equivalent, for Beneš networks, to the standard definition of non blocking: that a free inlet can be connected to a free outlet witout disturbing existing connections.

An algorithm for finding a set P ("preservable set") as above is described in [3]. Briefly, it is as follows: from the set of all states we remove the 0-forbidden states, which are these which have a pair of outlets which cannot be "swapped". Then, for $i = 1, 2, \ldots$, we remove the i-forbidden states. These are the states which can be "swapped" only by putting the network into an $(i-1)$-forbidden state. The state P is then the set of remaining states, i. e. those which are not i-forbidden for any i. The network is then non-blocking iff P is non-empty (see fig. 4).

For example, for the network of fig.2, the set P is the set of those states where the switches B and C are not both "straight through" (fig. 1. (a)).

We now state

Theorem A non blocking $n \times n$ Beneš network contains at least

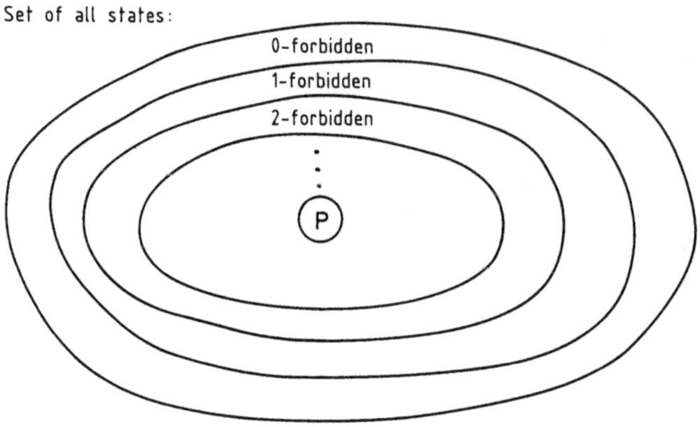

Fig. 4 : finding the set P of preservable states.

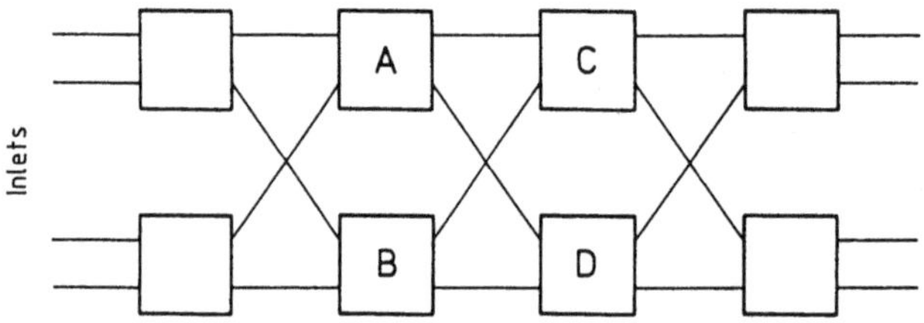

Fig. 5 : a 4×4 non-blocking Beneš network.

$$M_n = n(n-1) - [n^2/4]$$

switches.

This value is best possible for $n = 2$ (trivially) and for $n = 3$ (fig. 2), and $n = 4$ as shown by the non-blocking network of fig.5. The set P of preservable states for this network is the set of states where not both A = B and C = D. It is not known whether this result is best possible for $n \geq 5$. It seems unlikely ! In fact, I don't even know whether any non-blocking Beneš network has been found for $n \geq 5$!

The theorem shows that Beneš networks need 3/4 as many switches as square networks. This is not a great improvement, in view of the greater complexity of Beneš networks. However, Beneš networks have some advantages over square networks. For instance the path length (number of switches that an inlet-to outlet path passes through) is much more uniform, and the maximum path length is much smaller, than for square networks.

For the proof of the theorem, we label the outlets of our $n \times n$ Beneš network $1, 2, \ldots, n$, and, for a particular network state, label the switches by the pair i, j if that switch is connected to outlets i and j. Then we have the following

Lemma 1 Consider a preservable state of a Beneš network, and any triple (e. g. 1, 2, 3) of outlets. Then all three switch label pairs (1 2), (1 3), (2 3) occur and at least one of them is repeated.

Proof In any state where a label (e. g. 1 2) was not repeated, the paths ending in those outlets (e. g. 1, 2) would not cross, so those outlets could not be "swapped". Thus the state would be 0-forbidden. Therefore suppose that all pairs i, j occur, but suppose that *none* of the pairs (1 2), (1 3), (2 3) is repeated. Now write these labels in the order in which they appear in the networks, going from left to right, e. g. (1 2), (2 3), (1 3).

We now show that such a state is 1-forbidden. For if we swap the outlets of the center label (e. g. 2 3) the three labels in the resulting new network state become (1 3), (2 3), (1 3) with all others labels unchanged. Hence no label (1 2) occurs, so this new state is 0-forbidden, and the original state was 1-forbidden.

We also need

Lemma 2 The maximum number of edges in a graph with n nodes and containing no triangles is $[n^2/4]$.

The maximum occurs for the bipartite graphs $K_{\ell,\ell}$ (if $n = 2\ell$) or $K_{\ell,\ell+1}$ (if $n = 2\ell + 1$). For a proof see [1], p. 71–72.

We can now prove the theorem. For each preservable state of an $n \times n$ Beneš network, we define a graph with n nodes, with nodes i and j joined if the switch label (i j) is *not* repeated. Then by Lemma 1, the graph has no triangles. Hence by Lemma 2, it has at most $[n^2/4]$ edges, so there must be at least;

$$2\binom{n}{2} - [n^2/4] = M_n$$

label pairs, i. e. at least M_n switches.

References

[1] B. Bollobas, Graph Theory, Springer 1979.

[2] J.E. Midwinter, Photonic switching components: current status and future possibilities, Proc. 1st Photonic Switching meeting, vol. 25 in Springer Series in Electronics and Photonics, 1988, pp.15–23.

[3] C.J. Smyth , Non blocking photonic switch networks, I.E.E.E. SAC–6 (1988), pp. 1052–1062

[4] B.S. Wherett, Architectural aspects of optical computing, Proc. Workshop on Optical Bistability in Optical Computing, Huntsville U.S.A., Nov. 1986, SPIE Press.

The Ising Model and the Diophantine Moment Problem

P. Moussa

Service de Physique Théorique, CEN Saclay, F-91191 Gif-sur-Yvette Cedex, France

Abstract. The thermodynamical quantities which occur in the Ising model can be derived from the generating function of the moments of a measure with support on a finite interval. The moments belong either to a ring of polynomials with integer coefficients or to the ring of natural integers. Some general results are given on the nature of the possible generating functions once some additional assumptions are made about the support of the measure.

1. The Ising Model

Let Ω be a bounded subset of the lattice $\mathbf{Z}^d$, and say that two points in Ω are neighbours (in fact nearest neighbours) when they share all coordinates but one, which differs by ± 1. Points in Ω are sometimes called lattice sites and neighbouring sites are said to be linked on the lattice. With each site i in Ω we associate a variable σ_i, usually called a *spin* taking values ± 1. A configuration $\{\sigma_i\}$ is a choice of value for the spin at each site in Ω. The total number of possible configurations is 2^N, where $N = N_\Omega$ is the number of sites in Ω. With each configuration $\{\sigma_i\}$ is associated an energy $\mathcal{H}_\Omega\{\sigma_i\}$, given by

$$\mathcal{H}_\Omega\{\sigma_i\} \;=\; -J \sum_{n.n.\in\Omega} \sigma_i\sigma_j - h \sum_{k\in\Omega} \sigma_k \;. \tag{1}$$

The label $n.n. \in \Omega$ under the first sum means that the sum is taken over pairs (i,j) of nearest neighbours sites in Ω. We assume $J \geq 0$, which means that we consider the *ferromagnetic* case, and define h as the applied magnetic field. The partition function associated with the N sites in Ω is defined as follows:

$$Z_N^\Omega \;=\; \sum_{\{\sigma_i\},\ i\in\Omega} \exp\left(-\beta\mathcal{H}_\Omega\{\sigma_i\}\right) \;, \tag{2}$$

where β^{-1} is the absolute temperature and the sum is taken over all possible configurations, that is on the set of all possible values of all spins in Ω. The corresponding free energy per site F_N^Ω is customarily defined as

$$F_N^\Omega(\beta J, \beta h) \;=\; \left(\frac{-1}{\beta N}\right) \ln Z_N^\Omega \;. \tag{3}$$

Springer Proceedings in Physics, Vol. 47 Number Theory and Physics
Editors: J.M. Luck · P. Moussa · M. Waldschmidt

The limit $N \to \infty$, called the thermodynamic limit, has been shown to exist in the Ising model and in more general situations [1], provided that the size of the set Ω goes to ∞ together in each of the d possible directions, and then the limit

$$F(\beta J, \beta h) \;=\; \lim_{N\to\infty} F_N^{\Omega}(\beta J, \beta h) \tag{4}$$

is independent of the shape of Ω. Details can be found in the general references on the Ising Model [2,3,4]. It is customary to introduce the notation

$$x \;=\; e^{-2\beta J} \;,\; z \;=\; e^{-2\beta h} \;. \tag{5}$$

We reexpress implicitly the various functions as functions of x and z instead of the previous variables βJ and βh and we will use x as a temperature scale, x increasing from 0 to 1 when the temperature increases from 0 to ∞. The magnetization (per site) is defined as

$$I(x,z) \;=\; -\frac{\partial F}{\partial h} \;=\; 2\beta z \frac{\partial F(x,z)}{\partial z} \;. \tag{6}$$

The partition function Z_N^{Ω} can be expressed as the generating function of some graphical combinatoric factors. Setting N_L as the total number of links in Ω, we have [1]

$$Z_N^{\Omega}(x,z) \;=\; z^{-(N/2)} x^{-(N_L/2)} \sum_{n=0}^{N} P_n(x) z^n \;, \tag{7}$$

so that $Z_N^{\Omega}(x,z)$ is, up to some explicit multiplicative factor, a monic reciprocal polynomial of degree N in z, with coefficients themselves polynomials in x. In order to describe simply the combinatorial approach, it is convenient to consider a particular choice for Ω, which will lead to the same thermodynamic limit. We consider the so-called periodic boundary condition where $\Omega = \mathbf{T}_\ell^d = (\mathbf{Z}/\ell\mathbf{Z})^d$, so that $N = \ell^d$. This amounts to considering Ω as a size ℓ hypercube with opposite faces identified by the translation by ℓ in the appropriate direction. Each site in $\mathbf{T}_\ell^d$ has the same coordination number (i.e. number of nearest neighbours) $c = 2d$, and therefore we have $N_L = dN$ so that for

$$\Omega \;=\; \mathbf{T}_\ell^d \;, \tag{8}$$

we get

$$Z_N^{\Omega}(x,z) \;=\; z^{-(N/2)} x^{-(dN/2)} \sum_{n=0}^{N} \sum_{m=0}^{dN} w_N(n,m) z^n x^{2(dn-m)} \;, \tag{9}$$

where $w_N(n,m)$ is for $n \geq 1$ the number of possibilities of choosing in $\Omega = \mathbf{T}_\ell^d$, n points with m links. Obviously points and links can be visualized as a graph drawn on a subset of the d-dimensional lattice containing N sites and $w_N(n,m)$

interpreted as the number of different ways of drawing such a graph with n points (non-ordered) and m links. The exact evaluation of these occurrence numbers is very hard, but the existence of the thermodynamical limit allows one to describe their behaviour for large N. From (3), (4) and (9), one finds

$$F(x,z) \;=\; \frac{1}{2\beta}\ln z+\frac{d}{2\beta}\ln x-\frac{1}{\beta}\lim_{N\to\infty}\frac{1}{N}\ln\left(\sum_{m,n} w_N(n,m)z^n x^{2(dn-m)}\right)\;, \quad (10)$$

which can be rewritten as

$$F(x,z) \;=\; \frac{1}{2\beta}\ln z+\frac{d}{2\beta}\ln x-\frac{1}{\beta}\sum_{n=1}^{\infty}\sum_{m=0}^{\infty} p(n,m)z^n x^{2(dn-m)}\;. \quad (11)$$

For $n \geq 1$ the coefficients $p(n,m)$ can be obtained directly from the $w_N(n,m)$ through the following observation: $(1/N)w_N(n,m)$ is a polynomial in N and $p(n,m)$ is nothing but its value for $N = 0$. This very peculiar relation is a consequence of the existence of the thermodynamic limit [2,4].

The Ising model was first introduced to describe the existence of permanent magnetization, and the transition from the ferromagnetic to the paramagnetic state, but it also describes binary alloys [2] and the liquid-gas transition [5]. It can be generalised in many ways, for instance on other kinds of lattices, or by modifying the set of values taken by the σ_i, and finally the limit where the lattice spacing (fixed to 1 in our case) goes to zero is particularly interesting in view of the construction of quantum field theories.

2. The Lee-Yang representation

In a celebrated paper, Lee and Yang [5] gave the first proof of the existence of the thermodynamic limit and they showed [6] that when $0 \leq x \leq 1$ the partition function Z_N^Ω in (7) can vanish only when $|z| = 1$. A proof is given in [6,7]. Up to explicit factors in front, the partition function in (7) is a polynomial in z which can be factorised in elementary factors of degree two corresponding to pairs of complex conjugate roots. When N goes to ∞, the number of zeroes goes to ∞ but their density on the unit circle has a limit which allows us to write the free energy in (4) as

$$F(x,z) \;=\; -h-dJ-\frac{1}{\beta}\int_{\theta_0}^{\pi}\ln(z^2-2z\cos\theta+1)g(\theta,x)\,d\theta\;. \quad (12)$$

The integral is taken over the angle θ which parametrizes the unit circle (we need only the upper half thanks to complex conjugation). It can be shown [6,8] that the asymptotic density of zeroes $g(\theta,x)$—in fact its upper-half part—vanishes outside the interval $0 \leq \theta_0(x) \leq \pi$, as displayed by integration limits. In fact the density $g(x,\theta)$ is normalised as follows:

$$\int_{\theta_0}^{\pi} g(x,\theta)\, d\theta \;=\; \frac{1}{2}\,. \tag{13}$$

The corresponding representation for the magnetization (6) is

$$I(x,z) \;=\; 2(1-z^2)\int_{\theta_0}^{\pi} \frac{g(\theta,x)}{z^2 - 2z\cos\theta + 1}\, d\theta\,, \tag{14}$$

which has the following symmetry on reversing the magnetic field ($h \to -h$):

$$I(x,z) \;=\; -I(x,\frac{1}{z})\,. \tag{15}$$

At high temperature (x close to one), the Lee-Yang angle $\theta_0(x)$ is close to π, which means that a large part of the unit circle is free of zeroes. In this case, one sees from (14) that the magnetization $I(x,z)$ is analytic in z for $z=1$, and from (15) that $I(x,1)$ vanishes: there is no magnetization when the magnetic field h vanishes. On the other hand, at low temperatures, $\theta_0(x)$ vanishes, and the magnetization is no longer analytic at $z=1$, and the two different limits $I(x,1_+) = -I(x,1_-)$ may lead to opposite non-zero values: there may be a spontaneous magnetization in zero magnetic field. So we can define the critical value x_c such that $\theta_0(x)=0$ for $x \le x_c$ and $\theta_0(x) > 0$ for $x > x_c$. For temperatures higher than the critical temperature corresponding to x_c, there is no spontaneous magnetization, and for lower temperatures a spontaneous magnetization occurs. In this way, the Ising model provides an analytic description of the phase transition phenomenon.

3. Representation of the Ising model as a moment problem

From (12) we get the following expansion:

$$F(x,z) \;=\; -h - dJ - \frac{1}{\beta}\sum_{n=1}^{\infty} \mathcal{M}_n(x) z^n\,, \tag{16}$$

where

$$\mathcal{M}_0(x) \;=\; \frac{1}{2} \text{ and for } n \ge 1\,,\;\; \mathcal{M}_n(x) \;=\; \frac{-2}{n}\int_{\theta_0}^{\pi} \cos(n\theta)\, g(\theta,x)\, d\theta\,, \tag{17}$$

and from (6) we get

$$I(x,z) \;=\; 1 - 2\sum_{n=1}^{\infty} n\mathcal{M}_n(x) z^n\,. \tag{18}$$

Therefore the magnetization is nothing other than the generating function of a trigonometric moment problem with moments defined in (17). Comparison between (16) and (11) yields

$$\mathcal{M}_n(x) \;=\; \sum_{m=0}^{\infty} p(n,m) x^{2(dn-m)}\,. \tag{19}$$

In fact the sum over m in (19) is limited to $m \leq dn$ in view of the graphical interpretation of $p(n,m)$ and $w_N(n,m)$ given in section 2. As a consequence, $\mathcal{M}_n$ is an even polynomial of degree $2dn$ in x. The lowest degree term in $\mathcal{M}_n(x)$ has degree $2nd-(2n-2+L)$, where L is the maximum number of independent loops in a diagram with n points drawn on the lattice [4]. Extensive calculations up to high orders [4] show that $n\mathcal{M}_n(x)$ is a polynomial in x with *integer* coefficients. A complete proof is still lacking, except in particular cases. For example a proof exists for the non-trivial case of the so-called Bethe lattice [4,9], which is in some sense a lattice without loops. We shall in the following assume this integrity property for the coefficients of $\mathcal{M}_n(x)$.

A change of variable will allow us to reexpress the magnetization as the generating function of a moment problem on a real line. From (14) we get

$$I(x,z) = \frac{2(1-z^2)}{(1+z)^2}\int_{\theta_0}^{\pi}\left(1-\frac{4z}{(1+z)^2}\cos^2(\frac{\theta}{2})\right)^{-1} g(\theta,x)\, d\theta\,. \tag{20}$$

Therefore we define

$$v = \frac{4z}{(1+z)^2} = 1-\tanh^2(\beta h), \tag{21}$$

and get

$$\frac{I(x,z)}{2\tanh(\beta h)} = \int_{\theta_0}^{\pi}\frac{g(\theta,x)\, d\theta}{1-v\cos^2(\theta/2)} = \sum_{l=0}^{\infty} v^l \mu_l(x)\,, \tag{22}$$

with

$$\mu_0(x) = \frac{1}{2}\,,\quad \mu_l(x) = \int_{\theta_0}^{\pi}\left(\cos^2(\frac{\theta}{2})\right)^l g(\theta,x)\, d\theta\,. \tag{23}$$

Therefore the magnetization is the generating function of the moment problem using $\cos^2(\theta/2)$ as the integration variable. One easily gets [9] for $l \geq 1$ the relations

$$4^l\mu_l(x) = \binom{2l}{l}\mathcal{M}_0(x) - \sum_{p=1}^{l}\binom{2l}{l-p}p\mathcal{M}_p(x)\,, \tag{24}$$

$$\mathcal{M}_l(x) = (-1)^{l+1}\frac{2}{l}\mu_0(x) + \sum_{p=1}^{l}(-1)^{l-p+1}\frac{1}{p}\binom{l+p-1}{l-p}4^p\mu_p(x)\,. \tag{25}$$

From (24), and using the integrity property of the coefficients of $n\mathcal{M}_n(x)$, one easily deduces that $4^l\mu_l(x)$ is an even polynomial of degree $2dl$ in x with integer coefficients. Moreover, using bounds [7,9] on the Lee-Yang angle $\theta_0(x)$ one can prove [9] that $4^l\mu_l(x)$ is divisible by $(1-x^2)^l$. So we define $\mathcal{P}_l(x)$ by

$$\mathcal{P}_0(x) = \frac{1}{2}\,,\quad \text{and for } l \geq 1\,,\quad 4^l\mu_l(x) = (1-x^2)^l\mathcal{P}_l(x)\,, \tag{26}$$

which implies that $\mathcal{P}_l(x)$ is, for $l \geq 1$, an even polynomial with integer coefficients, of degree $2l(d-1)$ in x. In particular it is a constant when $d = 1$. Numerical computations up to high orders [9] show that all the coefficients of $\mathcal{P}_l(x)$ are *non-negative*, and we will now assume this positivity property. Through a rescaling of the integration variable one finally gets

$$G(w,x) = \sum_{l=0}^{\infty} w^l \mathcal{P}_l(x) = \int_0^{\Lambda(x)} \frac{\sigma(\lambda,x)d\lambda}{1-\lambda w} , \tag{27}$$

where

$$w = \frac{v(1-x^2)}{4} , \quad G(w,x) = \frac{I(z,x)}{2\tanh(\beta h)} \quad \text{and} \quad \Lambda(x) = \frac{4\cos^2(\theta_0(x)/2)}{(1-x^2)} , \tag{28}$$

the integration being performed on the variable

$$\lambda = \frac{4\cos^2(\theta/2)}{(1-x^2)} \quad \text{with weight} \quad \sigma(\lambda,x) = \frac{(1-x^2)g(\theta,x)}{2\sin\theta} . \tag{29}$$

We finally observe that $\Lambda(x)$ remains bounded by $4d^2$ near $x = 1$, due to the bounds on the Lee-Yang angle [4,9] at high temperatures. From the positivity assumption for the coefficients of $\mathcal{P}_l(x)$, we deduce that the radius of convergence, and $\Lambda(x)$ must be increasing functions of x so we have the bound

$$\Lambda(x) \leq 4d^2 . \tag{30}$$

Equation (27) implies

$$\mathcal{P}_0(x) = \frac{1}{2} , \text{ and for } l \geq 1 , \quad \mathcal{P}_l(x) = \int_0^{\Lambda(x)} \lambda^l \sigma(\lambda,x)d\lambda , \tag{31}$$

therefore the magnetization in the Ising model is related when $0 \leq x \leq 1$ to the generating function $G(w,x)$ of the moments of a measure with support on an interval $[0,\Lambda(x)] \subset [0,4d^2]$, the moments (of order greater than one) belonging to the ring of polynomials with integer coefficients. Furthermore, when x is the rational number p/q, we have $\mathcal{P}_l(p/q) = \mathcal{N}_l q^{2l(1-d)}$, where for $l \geq 1$,the $\mathcal{N}_l$ are integers. Through an additional rescaling of the variable λ, these integers $\mathcal{N}_l$ are easily seen to be moments of a measure with support in an interval $[0,\Lambda]$, where $\Lambda = 4d^2q^{2(d-1)}$. This approach to the Ising model as an integer moment problem (the "diophantine moment problem") was first introduced in [10]; additional details can be found in [11]. The question is: does the integrity condition on the moments restrict in a significant way the possible generating functions?

4. The diophantine moment problem: the discrete case

We shall consider a measure μ, implicitly assumed to be positive, with support $\mathcal{S}$ on the real line, and assume that the moments

$$\mu_n = \int_{\mathcal{S}} \lambda^n d\mu(\lambda) \tag{32}$$

are positive integers, and define its generating function $G(z)$ as

$$G(z) = \sum_{n=0}^{\infty} \mu_n z^n = \int_{\mathcal{S}} \frac{d\mu(\lambda)}{1-\lambda z} . \tag{33}$$

We have the following:

Theorem Suppose that there exists a polynomial $p(\lambda)$ with integer coefficients such that $|p(\lambda)| < 1$ for all $\lambda \in \mathcal{S}$, then every measure μ, with support in $\mathcal{S}$ and integer moments, is discrete, which means that it is a finite sum of Dirac measures, with positive weights, sitting on the zeroes of the polynomial $p(\lambda)$.

Proof Consider the sequence of numbers

$$K_n = \int_{\mathcal{S}} (p(\lambda))^{2n} d\mu(\lambda) . \tag{34}$$

Under the hypotheses we have

$$K_n - K_{n+1} = \int_{\mathcal{S}} \left(1 - (p(\lambda))^2\right) (p(\lambda))^{2n} d\mu(\lambda) \geq 0 . \tag{35}$$

Then the sequence K_n is a non-increasing sequence of positive integers, and therefore there exists an N such that K_n is constant for all $n \geq N$. The theorem results immediately from $K_N - K_{N+1} = 0$.

This theorem shows that there exist support sets $\mathcal{S}$ for which all generating functions of diophantine moment problems are indeed rational functions. The question asked at the end of the previous section was therefore meaningful. Let us give a simple example: $\mathcal{S} = [0,1]$, $p(\lambda) = \lambda(\lambda-1)$, μ has discrete masses at 0 and 1. In fact such a polynomial $p(\lambda)$ exists for any interval $\mathcal{S} = [0, \Lambda]$ when $\Lambda < 4$. For an explicit proof, see [11,16]. In this case, the discrete masses have to sit at values of the form $4\cos^2(\pi(p/q))$, where p and q are integers, as we shall see in the next section. On the other hand, we have

$$\frac{1}{\sqrt{1-4z}} = \frac{1}{\pi} \int_0^4 \frac{d\lambda}{\sqrt{\lambda(4-\lambda)}} \frac{1}{1-\lambda z} = \sum_{n=0}^{\infty} \binom{2n}{n} z^n , \tag{36}$$

which shows that such a polynomial $p(\lambda)$ does not exist on $[0,4]$. A complete treatment of the case $\Lambda = 4$ is given in the next section.

5. The diophantine moment problem on $[0,4]$

A complete answer to this question is too long to be reproduced here and for details see [10,11,12]. Nevertheless it is interesting to recall briefly (see section 3 above) the relation between the moment problem on $[0,4]$ and the trigonometric moment problem. Starting from

$$G(z) \;=\; \sum_{n=0}^{\infty} \mu_n z^n \;=\; \int_0^4 \frac{d\mu(\lambda)}{1-\lambda z} \;,\quad \mu_n \;=\; \int_0^4 \lambda^n d\mu(\lambda), \tag{37}$$

we introduce the function $I(u)$:

$$I(u) \;=\; G(0) + \frac{1-u}{1+u} G\left(\frac{u}{(1+u)^2}\right) \;=\; \sum_0^{\infty} C_n u^n \;. \tag{38}$$

We get, through the change of variable $\lambda \;=\; 4\cos^2(\theta/2)$,

$$\mu_n \;=\; \frac{1}{2}\int_{-\pi}^{+\pi} \left(4\cos^2(\frac{\theta}{2})\right)^n dg(\theta) \;,\; \text{and} \;\; I(u) \;=\; \int_{-\pi}^{+\pi} \frac{dg(\theta)}{1-u\exp(i\theta)} \;, \tag{39}$$

the measure $dg(\theta)$ being the image of the measure $d\mu(\lambda)$, and invariant under the change $\theta \to -\theta$. Similarly,

$$C_0 \;=\; I(0) \;=\; 2G(0) \;\;,\;\; \text{and for} \;\; n \geq 1 \;,\;\; C_n \;=\; \int_{-\pi}^{+\pi} \cos(n\theta) dg(\theta) \;. \tag{40}$$

We get the relations

$$C_0 \;=\; 2\mu_0 \;,\;\; \text{and for } n \geq 1 \;,\; C_n \;=\; (-1)^n 2\mu_0 + \sum_{p=1}^{n} (-1)^{n-p} \frac{n}{p} \binom{n+p-1}{n-p} \mu_p \;. \tag{41}$$

Conversely, we can recover $G(z)$ from $I(u)$:

$$G(z) \;=\; \frac{1}{\sqrt{1-4z}} \left(I\left(\frac{1-\sqrt{1-4z}}{1+\sqrt{1-4z}}\right) - \frac{I(0)}{2} \right) \tag{42}$$

and the μ_n from the C_n:

$$\mu_0 \;=\; \frac{1}{2} C_0 \;\;,\;\; \text{and for } n \geq 1 \;\;,\;\; \mu_n \;=\; \frac{1}{2}\binom{2n}{n} + \sum_{p=1}^{n} \binom{2n}{n-p} C_p \;. \tag{43}$$

The relations (38) and (42) show an equivalence between 1) functions $I(u)$ real analytic in the unit disc such that the boundary value of the real part of $I(u) - (1/2)I(0)$ when $|z| \to 1$ is a positive measure, and such that the expansion around zero has integer coefficients, the constant term being even, and 2) functions $G(z)$ analytic in a cut plane from $(1/4)$ to ∞, with a non-negative discontinuity across the cut, such that the expansion around zero has integer coefficients. We first observe that if $I(u)$ is a rational function of u, $G(z)$ is an algebraic function, with a quadratic singularity at $z = (1/4)$. In fact it is a non-trivial result that this is always the case as expressed by the following proposition.

Theorem Any function $I(u)$ satisfying (38) and (39), where the C_n take only a finite number of different values and $dg(\theta)$ is a positive measure, is a rational function with poles at roots of unity.

We only sketch the argument which goes through four steps: 1) Observe that due to the positivity of $dg(\theta)$, it results from (40) that we have $|C_n| \leq C_0$, so that each C_n can take at most $2C_0 + 1$ values. So the case where the C_n are integers fits the conditions of the theorem. 2) Use an adaptation of the Lagrange interpolation formula in the convolution algebra of measures on the unit circle to express the measure $dg(\theta)$ as a sum of a finite number of non-necessarily-positive measures having only trigonometric moments equal to 0 or 1. Such measures are called idempotent since the convolution product of the measure by itself reproduces the same measure. 3) Use a theorem by Helson [13] which essentially asserts that such an idempotent measure is expressible as a finite sum of discrete measures with masses on roots of unity, plus an absolutely continuous measure, the density of which is a polynomial in $\cos(\theta)$. One thus easily gets the above assertion that $I(u)$ is a rational function. 4) Thus the expansion (38) is periodic and its resummation shows that the only possible denominator is of the form $u^N - 1$, for some N, which completes the theorem.

Going back to the moment problem (37) on the real line, one sees that when the support of the measure $d\mu(\lambda)$ is on an interval $[0, \Lambda]$, with Λ strictly smaller than 4, there is no continuous part in the measure, and the discrete part has masses at points $4\cos^2(p/q)$, with p and q integers, corresponding to $\exp(i\theta)$ equal to roots of unity. It is interesting to notice that the magnetization in the Ising model for $d = 1$ corresponds here to the case $I(u)$ equal to a constant, which leads to the example (36) of the previous section [10].

6. The diophantine moment problem on $[0, \Lambda]$ for $\Lambda > 4$

Now we assume $\Lambda > 4$ and we shall say that $G(z) \in \mathcal{S}_\Lambda$ when

$$G(z) = \sum_{n=0}^{\infty} \mu_n z^n = \int_0^\Lambda \frac{d\mu(\lambda)}{1 - \lambda z} \; , \quad \mu_n = \int_0^\Lambda \lambda^n d\mu(\lambda), \tag{44}$$

with $d\mu(\lambda)$ a positive measure with support in $[0, \Lambda]$ and μ_n integers.

There is no general description of the set $\mathcal{S}_\Lambda$ but we shall describe a few properties.

a) Convolution Suppose $G(z) = \sum_{n=0}^{\infty} \mu_n z^n$ and $H(z) = \sum_{n=0}^{\infty} \nu_n z^n$ belong respectively to $\mathcal{S}_\Lambda$ and $\mathcal{S}_\Gamma$, then $G * H(z) = \sum_{n=0}^{\infty} \mu_n \nu_n z^n$ belongs to $\mathcal{S}_{\Lambda\Gamma}$. The function $G * H$ is the generating function of the moments of the convolution product of μ and ν. The union of all $\mathcal{S}_\Lambda$ for all $\Lambda > 4$ is a graded algebra, with degree equal to $\ln(\Lambda)$.

b) Polynomial transformation Suppose $G(z) \in \mathcal{S}_{q^2}$ with q integer bigger than 2. Then define the function

$$G_q(z) = \frac{1}{1-qz} G\left(\frac{z^2}{(1-qz)^2}\right) . \tag{45}$$

It can be shown that: $G_q(z) \in \mathcal{S}_{2q}$. For details see [10,11]. This property is somewhat surprising because since $q > 2$, the correspondence $G \to G_q$ maps $\mathcal{S}_{q^2}$ into one of its proper subsets $\mathcal{S}_{2q}$.

c) Maximum possible reduction By repeating transformations of the type defined in (45) several times, it is possible to map any $\mathcal{S}_\Lambda$ for $\Lambda > 4$ into $\mathcal{S}_9$. Then applying the transformation for $q = 3$ once leads into $\mathcal{S}_6$. The resulting function still belongs to $\mathcal{S}_9$, but the corresponding measure has vanishing density between 6 and 9. A new application of the transformation with $q = 3$ leads once more to a function in $\mathcal{S}_\Lambda$ with a value of Λ smaller than 6. By repeating the transformation with $q = 3$ several times one can get [10,11] a generating funtion of a measure with support in an interval $\left[(5-\sqrt{13})/2 - \epsilon\ ,\ (7+\sqrt{13})/2 + \epsilon\right]$ for any small ϵ. We get a maximum reduction to $[0.697\ldots, 5.302\ldots]$. Is it possible to get a further reduction? Note that each transformation is up to rational factors, a rational change of variable which will not introduce new kinds of singularities into the function. All possible kind of singularities are already present when Λ exceeds $5.302\ldots$.

d) Relation with the trigonometric moment problem Suppose Λ is an integer multiple of 4, which one can always realise by enlarging the integration interval with zero weight on the additional part. A rescaling of λ by a factor $(\Lambda/4)$ leads to a measure $\mu^{(r)}$ with support on $[0,4]$ and moments $\mu_n^{(r)}$ which are rationals, with denominator at most d^n, where $d = \Lambda/4$. The corresponding $C_n^{(r)}$ obtained from $\mu_n^{(r)}$ through (41) are also rational with denominator at most d^n. Conversely if $d^n C_n^{(r)}$ is integer, the $\mu_n^{(r)}$ obtained through (43) are such that $d^n \mu_n^{(r)}$ are all integers. As in section 5, the only exception is that $C_0^{(r)}$ must be even. As a result, one gets a one-to-one correpondence between $\mathcal{S}_{4d}$ and the set of generating functions of a trigonometric moment problem with moments such that $C_n d^n$ is integer and C_0 even.

e) Singular continuous measure The fixed points of the transformation (45) are themselves in $\mathcal{S}_{2q}$. The corresponding measure is itself singular and through a transformation $\lambda \to (1/\lambda)$ this measure is mapped [10,11] on the harmonic measure on the Julia set for the polynomial $(z-q)^2$. For $q > 2$ this Julia set is known to be a Cantor set, and the measure is singular continuous [14].

7. Conclusion

The short description of the general case given in the previous section shows that there is no chance of finding a general description of $\mathcal{S}_\Lambda$ for Λ large. Open problems are:

1) To find a description of $\mathcal{S}_\Lambda$ slightly larger than four, and to find whether there exist intermediate values of Λ, between 4 and 5.302..., for which only special kinds of singularities occur for $G(z)$.

2) To find suitable additional assumptions either on μ or on G which would lead us to a complete description of special subsets of $\mathcal{S}_\Lambda$.

3) To extend the description to the case where the moments belong to some ring of algebraic integers, such as $\mathbf{Z}(\sqrt{2})$ or $\mathbf{Z}(i)$.

4) To extend the description to the case where the moments are polynomial with integer coefficients in some extra variable.

The problem has some connections [11] with other analytic number theory problems such as:

1) The so-called arithmetic entire functions [15].

2) The problem of the existence of polynomials with integer coefficients taking bounded values on some sets $\mathcal{S}$ [16].

3) The localisation of algebraic integers [17,18].

4) The iteration of a polynomial with integer coefficients [11,18].

We hope that the connection with the Ising model and its combinatorial aspects which we have explained in the first part of this talk will raise new interest in both the Ising model and the number theory problems mentioned above. For the Ising model case, it would be helpful to complete the proof of the technical assumptions made in section 3, concerning the integrity property of the coefficients of $n\mathcal{M}_n$, and of the positivity properties of the coefficients of $\mathcal{P}_l$. We hope that our analysis may provide a frame for studying the Ising model in dimension $d \geq 2$, where almost no explicit form is known for the magnetization, except the case $d = 2$, for which the value in zero field ($z = 1$) has been obtained in the famous works by Onsager [19] and Yang [20].

References

[1] D. Ruelle, Statistical Mechanics : Rigorous Results, Benjamin Inc., Reading, Mass., 1969.

[2] C. Domb, Advances in Physics, **9**, p. 149–361, 1960.

[3] R. B. Griffiths, Rigorous results and theorems, in "Phase Transitions and Critical Phenomena", C. Domb and M. S. Green eds., Academic Press, vol. 1, p. 7–104, 1972.

[4] C. Domb, Ising models, in "Phase Transitions and Critical Phenomena", C. Domb and M. S. Green eds., Academic Press, vol. 3, p. 357–484, 1974.

[5] C. N. Yang and T. D. Lee , Phys. Rev. **87**, 404–409, 1952

[6] T. D. Lee and C. N. Yang **87**, 410–419, 1952
[7] D. Ruelle, Phys. Rev. Lett. **26**, 303-304, 1971
[8] D. Ruelle, Commun. Math. Phys. **31**, 265–277, 1973
[9] J. D. Bessis, J. M. Drouffe and P. Moussa, J. Phys. A.: Math. Gen. **9**, 2105–2124, 1976
[10] M. Barnsley, D. Bessis and P. Moussa, J. Math. Phys. **20**, 535–546, 1979
[11] P. Moussa, Ann. Inst. Henri Poincaré, **28 A**, 309–347, 1983
[12] V. S. Vladimirov and I. V. Volovitch, Teoreticheskaya i Mathematicheskaya Fizika, **53**, 3–15, 1982, English translation: Theor. Math. Phys., **53**, 943–951, 1982
[13] H. Helson, Proc. Am. Math. Soc. **4**, 686–691 , 1953, and: T.Itô and I. Amemiya, Bull. Amer. Math. Soc. **70**, 774–778, 1964
[14] H. Brolin, Arkiv För Mat. **6**, 103–144, 1965
[15] R. C. Buck, Duke Math. J. **15**, 879–891, 1948
[16] E. Hewitt and H. S. Zuckerman, Duke Math.J. **26**, 305–324, 1959
[17] D. H. Lehmer, Ann. Math. **34**, 461–479, 1933
M. Waldschmidt, L'Enseignement Mathématique, **26** Série **II**, 201–209, 1980
[18] P. Moussa, J. S. Geronimo, D. Bessis, C. R. Acad. Sc. Paris **299**, Série **I**, 281—284, 1984
[19] L. Onsager, Phys. Rev. **65**, 117–149, 1944
[20] C. N. Yang, Phys. Rev. **85**, 808–816, 1952

Statistical Theory of Numbers

B. Julia

Laboratoire de Physique Théorique de l'Ecole Supérieure,
24, Rue Lhomond, F-75231 Paris Cedex 05, France

Abstract. This article is three-sided: it is a pedestrian introduction to the Riemann hypothesis in the case of the classical zeta function, it provides also an elementary presentation of Quantum Statistical Mechanics and finally it is an attempt to present the reader with a precise dictionary between the fields of number theory -additive or multiplicative- on the one hand and of thermodynamics of "solved quantum systems" on the other, the physical terms mentioned in the mathematical part will be defined in the Physics section.

1. Introduction

In his famous memoir of 1859 [1] Riemann started the field of analytical number theory and stimulated the development of a good fraction of modern Mathematics. At that time Boltzmann was only 15 years old, he was to derive together with Gibbs the results of thermodynamics from the simpler principles of Statistical Mechanics. Let us take note that 1859 is also the date of the first communication by Maxwell on the dynamical theory of gases. After Planck Quantum Statistical Mechanics was developed in parallel with Quantum Mechanics, the Bose-Einstein and Fermi-Dirac statistics date back respectively to 1924 and 1926.

In 1989 we are accustomed to the occurrence of theta functions and modular forms in the quantum theory of strings and in the partition functions of statistical mechanics on a two-dimensional torus. These expressions are related to generating functions of additive number theory. Here we would like to argue that multiplicative number theory which is best encoded through Dirichlet series corresponds also to some physical systems; we do not know yet their Hamiltonian form but we know their spectrum and we can discuss their thermodynamics.

As one may expect the primes play the role of partons and their quantum "statistics" is at our disposal, different choices lead to distinct classical series. This is one meaning of the word statistics. Another one arises: the "statistical" distribution of the primes on the real line is also connected with the Riemann hypothesis. The title of this article has been chosen despite this ambiguity. A most interesting connection seems however the fact that one can reinterpret the pole of the zeta function at $s = 1$ as a breakdown of the grand-canonical ensemble

Springer Proceedings in Physics, Vol. 47 **Number Theory and Physics**
Editors: J.M. Luck · P. Moussa · M. Waldschmidt

at the Hagedorn [2] temperature because of the exponential increase of the density of energy levels for large energies. Lee-Yang type theorems restrict the locus of the zeros of partition functions that may be related to phase transitions. The Riemann hypothesis is similar. Here we shall meet zeros and also poles in solved systems. The thermodynamic formalism leads us to generalize various generating functions of number theory and may help in their study.

2. Some generating functionals of number theory

We shall begin this introduction to number theory with the Riemann hypothesis. Firstly we shall define the famous zeta function and a few generalizations. Secondly we shall review a number of equivalent forms of the Riemann hypothesis in particular the description of the density of large prime numbers that corresponds to it. Then there will be a general presentation of the generating functionals of additive and multiplicative number theory. Finally we shall briefly discuss the transformations of Mellin and Moebius which have nice physical interpretations.

2.1 The Zeta function of Riemann

The *definition* of the Riemann zeta function is as a Dirichlet series that converges for $Re(s) > 1$:

$$\zeta_R(s) \;=\; \sum_{n=1}^{\infty} \frac{1}{n^s}.$$

A property that distinguishes Dirichlet series from power series is the fact that c : "the abscissa of convergence" may differ from a : the abscissa of absolute convergence. Series of the type:

$$D(s = \sigma + it) \;=\; \sum_{n=1}^{\infty} a_n e^{-\lambda_n s}$$

converge for all values of s such that $\sigma > c$ (and never for $\sigma < c$), the $\lambda_n' s$ are an infinitely increasing sequence of positive numbers. The case $\lambda_n = n$ corresponds to the familiar power series whereas the case $\lambda_n = Log n$ will be called a Dirichlet series below. In this case one can show that

$$c \;\leq\; a \;\leq\; c+1.$$

Examples.

For the Riemann zeta function $a = c = 1$, there is a pole at $s = 1$ and this analytic function behaves as $1/(s-1) + \gamma_E$ (Euler's constant) as s tends to 1.

The alternate Dirichlet series with coefficients $a_n = (-1)^n$ has $c = 0$ and $a = 1$, in fact it is equal to the product of the zeta function by the expression $-(1-2^{1-s})$.

We shall have a lot to say about the inverse of the zeta function:

$$Z_M = \frac{1}{\zeta_R},$$

it is also a Dirichlet series with $a = 1$ and $c \leq 1$.

Let us now review three *properties* of the analytic continuation of ζ_R to the complex plane minus the point s = 1. It can be defined there by an integral representation:

$$\zeta_R(s) = \frac{\Gamma(1-s)}{2\pi i} \int_C \frac{e^{sLog(-x)}}{e^x - 1} \frac{dx}{x} \qquad (1)$$

where the contour C encircles the positive real half-axis in the positive direction. We note the following special values at integer points:

$$\zeta_R(-n) = \frac{(-1)^n B_{n+1}}{n+1}$$

where the odd Bernoulli numbers B_{2p+1} vanish identically. Euler showed the identities:

$$\zeta_R(2p) = -\frac{(2i\pi)^{2p} B_{2p}}{2(2p)!}$$

but it was only in 1978 that Apéry proved that $\zeta_R(3)$ is irrational!

The second crucial property is the functional equation. If one calls $\theta(x)$ the theta function of the one dimensional lattice one has:

$$\omega(x) = \frac{\theta(x) - 1}{2} = \sum_{n=1}^{\infty} e^{-n^2 x\pi}$$

and one can show for $Re(s) > 1$ the remarkable formula:

$$\Gamma(\frac{s}{2}) \pi^{-s/2} \zeta_R(s) = \int_0^{\infty} \omega(x) x^{s/2} \frac{dx}{x}. \qquad (2)$$

The Poisson summation formula implies (partly) the modularity of the theta function and leads to:

$$x^{1/4}\theta(x) = x^{-1/4}\theta(\frac{1}{x})$$

This translates through eq. (2) and the definition

$$\xi(s) = \Gamma(\frac{s}{2} + 1)\pi^{-s/2}(s-1)\zeta_R(s)$$

into the functional equation (a physicist would say the "duality") of Riemann:

$$\xi(s) = \xi(1-s).$$

Let us note however that the definition of ξ cancelled unnaturally and by brute force the pole at s = 1, the multiplication of eq. (2) by s(s-1)/2 however preserved the symmetry about the axis $Re(s) = 1/2$.

Finally the zeta function admits an "Euler product" decomposition over the primes (in the convergence domain of the Dirichlet series). This is a general result of *multiplicative* number theory: let us consider a so-called multiplicative arithmetic function f(n) (i.e. a function that is multiplicative for the product of two relatively prime integer arguments, $f(ab) = f(a)f(b)$ if $GCD(a,b) = 1$). Then

$$\sum_{n=1}^{\infty} f(n) = \prod_{p} (1 + f(p) + f(p^2)...).$$

Clearly n^s is multiplicative so the following decomposition holds:

$$\zeta_R(s) = \prod_p \frac{1}{(1-p^{-s})}.$$

As a consequence we discover another multiplicative arithmetic function the so-called Moebius function $\mu(n)$:

$$\frac{1}{\zeta_R} = Z_M(s) = \prod_p (1-p^{-s}) = \sum_{n=1}^{\infty} \frac{\mu(n)}{n^s}.$$

μ counts the parity of the number of distinct prime factors of n and vanishes whenever n is not "squarefree" (i.e. when n is divisible by a square).

Before turning our attention to various formulations of the Riemann hypothesis let us define briefly some *generalizations* of the Riemann zeta function.

First of all Dirichlet considered the (multiplicative) characters χ of the group $(Z/mZ)^*$ extended to take the value zero for arguments non prime to m. The order of this group is $\phi(m)$ (phi is called Euler's totient function) namely the number of nonnegative integers less than m and prime to it. Let us remark that $\phi(m)$ is also a multiplicative function. The corresponding Dirichlet series are called Dirichlet L functions:

$$L_\chi(s) = \sum_{n=1}^{\infty} \frac{\chi(n)}{n^s}$$

they generalize ζ_R. In fact the trivial character reproduces the zeta function times

$$\prod_{p|m} (1-p^{-s}).$$

For non trivial characters however the L functions converge for $Re(s) > 0$. The product of all L functions relative to the same integer m is related to another generalized zeta function, namely the zeta function of the (algebraic number) field generated over the rationals by the m-th roots of unity [3].

One may associate to an arbitrary number field K its Dedekind zeta function ζ_K defined by:

$$\zeta_K(s) = \sum_{n=1}^{\infty} \frac{N_n}{n^s}$$

where N_n is the number of "integral ideals" with "norm" equal to n. To fix quickly the ideas let us consider $K = Q[\sqrt{-1}]$. The integral ideals are the ideals in the ring of Gaussian integers $(Z + iZ)$ and their norms are the numbers of residue classes modulo these ideals. There is the notion of prime ideals and a corresponding product formula for the Dedekind zeta function. The convergence domain is $Re(s) > 1$ and there is a pole at $s = 1$ with residue proportional to the class number of the field. The class number h measures the non-unicity of the factorization of numbers into primes; for K the field of Gaussian fractions $h = 1$.

By analogy with the case $K = Q$ we can define generalized L_K functions. The generalized Riemann hypothesis says that the complex zeros of all these functions lie on the axis of abscissa 1/2. Analogs of the functional equation hold also.

Finally let us terminate this incomplete list with another famous instance of zeta function: Weil [4] considered the field of functions on an algebraic variety and defined a corresponding zeta function when all coefficients take their values in a finite field. One can prove the analog of the Riemann hypothesis in that case.

2.2 Equivalent forms of the (classical) Riemann hypothesis

The most elementary formulation of the Riemann hypothesis is in terms of the complex zeros of the zeta function. We have seen that zeta vanishes for even negative integers but its product with $(s-1)\Gamma(s/2+1)$ appears in ξ which does not vanish on the real axis.

The first form of the hypothesis is that ξ vanishes only on the axis $Re(s) = 1/2$.

The second form involves the inverse of the Riemann zeta function: Z_M. We have seen that the convergence abscissa is at most equal to 1, the Riemann hypothesis is equivalent to saying that it is equal to 1/2.

Thirdly Mertens defined a function M(x) in 1897 and conjectured that it should be less than or equal to $\sqrt{x}$. It is defined as the sum up to x of $\mu(n)$ $(= (-1)^f$ where f is the number of distinct prime factors of the squarefree number n or = 0 when n is divisible by a square). This is a special instance of the discrete version of integration by parts. It is called Abel summation. Take any C^1 function g(x) and define

$$A(x) = \sum_{n \leq x} a_n$$

then:

$$\sum_{n \leq x} a_n g(n) = A(x)g(x) - \int_1^x A(t)g'(t)\,dt \tag{3}$$

A(x) will be interpreted below as a (once integrated) density of states. Let us give *a few examples* of eq. (3) with $x = \infty$ and $g(x) = x^{-s}$:
the cases $a_n = 1$, $\mu(n)$, $|\mu(n)|$, $l(n)$, $\Lambda(n)$
correspond to (3) equal to respectively:
ζ_R, Z_M, $\zeta_R(s)/\zeta_R(2s)$, $\log \zeta_R$ and $-\zeta'_R/\zeta_R$.
The corresponding expressions A(x) are called:
E(x), M(x), Q(x), $\Pi(x)$ and $\psi(x)$.
E(x) is the integral part of x, Q(x) counts the squarefree numbers and

$$\Pi(x) = \pi(x) + 1/2\,\pi(\sqrt{x}) + 1/3\,\pi(\sqrt[3]{x}) + ...$$

where $\pi(x)$ counts the number of primes up to x; $l(n) = 1/k$ when $n = p^k$ and vanishes otherwise. Finally $\Lambda(n)$ (the von Mangoldt function) vanishes unless n is a power of a prime p in which case it takes the value $\log p$.

The third formulation of the Riemann hypothesis says that M(x) does not grow faster than $x^{1/2+\epsilon}$ for any positive ϵ. It has been shown recently that the Mertens conjecture is actually false. Let us remark that E(x) grows like x and Q(x) like $6x/\pi^2$ see for example [5]. So there must be an efficient compensation between squarefree numbers with odd and even numbers of factors.

The zeros with "small" imaginary part have been successively determined by numerical methods. The imaginary part is known exactly because all zeros so far are simple and the functional equation implies pairs of zeros unless they are on the symmetry axis. One knows there are no zeros off the axis with imaginary part less than 10^8. On the other hand one also knows that there is no zero in the following region:
$(s = \sigma + it)$ t big enough and

$$\sigma \geq 1 - \frac{C}{(\log t)^{2/3}(\log\log t)^{1/3}}.$$

Note that as t tends to infinity the right hand side tends to 1, in other words the best upper bound for the abscissa of convergence of Z_M has not decreased since Riemann. This is the sign of a nasty phenomenon at $Re(s) = 1$!

The fourth and our last form of the Riemann hypothesis is a precise description of the *distribution of large prime numbers.* More precisely one would like to know the asymptotic behaviour of $\pi(x)$ (or equivalently of $\Pi(x)$ or $\psi(x)$) for x tending to infinity. The "prime number theorem" envisioned by Legendre and Gauss around 1800 was proved in 1896 independently by Hadamard and de la Vallée Poussin. It says that $\pi(x)$ and $\Pi(x)$ behave as Li(x) (i.e. as $x/\log x$) for large x.

$$Li(x) = \int_2^x \frac{dt}{\log t}.$$

Another 50 years later an elementary proof of this fact was found! This theorem amounts to say that the zeta function of Riemann does not vanish on the axis $Re(s) = 1$.

The Riemann hypothesis is equivalent to

$$\psi(x) \sim x + O(x^{1/2}\log^2 x)$$

and also to

$$\frac{\pi(x) - Li(x)}{Li(x)} < x^{\epsilon - 1/2}$$

for any positive ϵ and any sufficiently large x.

One can see the relation between *the error term* in the prime number theorem and the zero-free region of the zeta function by recalling the Hadamard product formula over the complex zeros of zeta:

$$\xi(s) = \xi(0)\prod_\rho (1 - \frac{s}{\rho})$$

and by looking at the formula of Riemann:

$$\frac{d\Pi(x)}{dx}\log x = 1 - \frac{1}{x(x^2-1)} - 2\sum_{Re(t)>0} \frac{\cos(t\log x)}{\sqrt{x}}$$

where $\rho = 1/2 + it$.

2.3 Generating functionals

It is useful to distinguish right away the *additive* theory of numbers from the *multiplicative* one. The former treats typically problems of partitions of integers into "things": integers, squares, a square plus a multiple of a given number... The latter deals with factorization properties and with multiplicative functions in the sense defined above.

So let us begin with the *additive* case more familiar (to physicists at least) of partition numbers. And firstly let us review the classical case of sums of integers. One can generate the sequence of the numbers of partitions p(n) of n into integers by the "generating function":

$$(qZ_V)^{1/24} = 1 + \sum_{n=1}^{\infty} p(n)q^n.$$

Euler showed it to be equal to the inverse of an infinite product over all the positive integers m: $\prod(1-q^m)$. If one defines $q = e^{2\pi i\tau}$, one recognizes the partition function for the internal modes of the Veneziano open string model: Z_V with periodic time τ. The factor q that we have extracted corresponds to the tachyon mode in Physics and is crucial for the *modular invariance* of the partition function which is nothing but the -24th power of η_D the Dedekind eta function. The positivity of the coefficients of Z_V has a simple physical interpretation, namely they are dimensions of vector spaces. The coefficients of the q-series $\Delta = 1/Z_V$ are the famous $\tau(n)$'s of Ramanujan, they have peculiar multiplicative properties as we shall see!

Remarks

One may distinguish the unordered partitions with a fixed number k of summands $P_1(n,k)$ and compute their generating function (u is called fugacity):

$$1 + \sum_{k>0}\sum_{n>0} P_1(n,k)q^n u^k = \prod_{n>0} (1-uq^n)^{-1}.$$

Let us note also the remarkable "bosonization" formula (we shall discover another one below, eq. (5), and we shall discuss their physical interpretations in part 3.):

$$\prod_1^{\infty} (1-q^{2n+1})^{-1} = \prod_1^{\infty} (1+q^m). \quad (4)$$

It says that there are as many partitions of an integer into different integers as into odd integers (different or not).

Two more additive examples

Let us now discuss the partitions into squares: they are closely related to the theory of theta functions. It turns out that the combinatorics is simpler if we count ordered partitions with signs (i.e. factors of 2 for non zero terms). Then the first case ($k=1$) of partition into one square is either possible in two ways or not at all (n is non-zero): it is usually denoted $r_1(n)$ but let us rename it $P_2(n,1)$. The generating function for $k=1$ is nothing but $2\omega(x)$ where now $q = e^{-\pi x}$.

The next case $k=2$ leads to interesting number theory:

$$P_2(n,2) = 4(d_1(n) - d_3(n))$$

where $d_m(n)$ is the number of prime divisors of n congruent to m modulo 4. The generating function for $k = 2$ is obviously the square of the theta function (if one defines $P_2(0,k) = 1$). More generally the full generating function is given by the simple formula:

$$\frac{1}{1-u\theta(x)} = 1 + \sum_{k>0}\sum_{n>0} P_2(n,k)q^n u^k.$$

More generally to an integral lattice Λ is associated a theta function: Θ_Λ in which one has replaced $q^{n.n}$ by $q^{<n,n>/2}$, self-dual lattices correspond to modular forms. Θ_Λ is the generating function of the number of representations of an integer by the quadratic form $<,>$. The modular invariance follows from the obvious periodicity of the exponential map $q = e^{2i\pi z}$ and from the Poisson summation formula already encountered in the one dimensional case above.

In the theory of graded Lie algebras the degree is the additive quantity, one has the grading condition on the subspaces G_n of degree n: $[G_n, G_m] \subset G_{n+m}$. The Frenkel-Kac construction of the basic representation of affine Kac-Moody Lie algebras gives an identity for the generating function of the dimensions of the subspaces of various degrees. For example in the case of $\widehat{E_8}$ this generating function reads:

$$Z_{\widehat{E_8}} = \Theta_{\Lambda_{E_8}} . \eta_D{}^{-8}$$

where Λ_{E_8} is the root lattice of the ordinary Lie group E_8, it is the cube root of the famous modular invariant j.

The power series are the convenient tool for constructing these generating functions because their factorization corresponds precisely to the addition of exponents of the parameter q, and these are the additive quantities of the problem.

Let us now turn to the *multiplicative* case and let us start with some completely multiplicative examples; by definition we shall consider coefficients a_n such that $a_{nm} = a_n a_m$ for all m's and n's. The Riemann zeta function corresponds to a generating function for the trivial sequence $(1_n) = (1, 1, ...)$ because now we must use Dirichlet series $\sum a_n n^{-s}$. We already noticed that n^{-s} is multiplicative when we derived the existence of an Euler product decomposition over the primes in multiplicative situations. Again we are looking for product decompositions and the multiplicativity of the coefficients is preserved by the product operation. This operation is defined as follows: one multiplies formally two Dirichlet series and one looks at the coefficients of the product. Clearly complete multiplicativity is not preserved. For example one can easily show that the generating function for d(n), the number of divisors of n including

1 and n, is the square of $\zeta_R(s)$. Dirichlet's L_χ functions are also the generating functions of completely multiplicative characters.

More examples come to mind if one requires only multiplicativity of the coefficients. The inverse of ζ_R: Z_M is the generating function for the Moebius function $\mu(n)$. The generating function for the absolute value of the Moebius function has been given above, it is the quotient: $\zeta_F(s) = \zeta_R(s)/\zeta_R(2s)$ the proof through factorization involves the trivial identity:

$$(1+y) = (1-y^2)/(1-y). \tag{5}$$

We shall reinterpret that again as a bosonization formula in part 3.2.

Yet another example of multiplicativity is furnished by Euler's totient function whose generating function is $\zeta_R(s-1)/\zeta_R(s)$. To summarize: multiplicativity is preserved under product of generating functions and implies factorization over the primes. It is preserved by Moebius transforms as we shall see in the next section.

2.4 Moebius and Mellin transforms

The *Moebius transform* acts on sequences a_n and preserves multiplicativity. It is defined by the hatted sequence:

$$\widehat{a}_n = \sum_{d|n} a_d.$$

The first example is $\widehat{1}(n) = d(n)$. Another example is provided by the elementary formula:

$$n = \widehat{\phi}(n) .$$

We are led to examine the action of the Moebius transform on the generating functions. It amounts to a simple multiplication by ζ_R! Naturally the inverse operation is the division by the same function and involves the Moebius function $\mu(n)$ as follows:

$$a_n = \sum_{d|n} \mu(d)\widehat{a}(n/d)$$

A sign of the cancellations at work in the slow growth of the Mertens function is the equality

$$\widehat{\mu}(n) = \delta_{n0} :$$

the sum of the values of μ for all divisors of any integer n vanishes. So the compensation involves a subtle ordering of squarefree numbers.

Let us now return to the relations between the additive world of the power series and the multiplicative one of the Dirichlet series. The fundamental formula is the formula for the *Mellin transform.* Let us define $D(s) = \sum a_n n^{-s}$ and $P(q) = \sum a_n q^n$ and let us recall the definition of Euler's Γ function:

$$\Gamma(s) \;=\; "(s-1)!" \;=\; \int_0^\infty x^s e^{-x}\,\frac{dx}{x}.$$

Then we have:

$$D(s) \;=\; \frac{1}{\Gamma(s)} \int_0^\infty x^s P(e^{-x})\,\frac{dx}{x}. \tag{6}$$

Equation (6) has been encountered before (modulo a suitable rescaling) namely in the special case of eq. (2). The 1/2 in s/2 introduces squares of integers and is responsible for the apparition of the theta function. The naive application of eq. (6) leads to an integral representation of the zeta function that is valid for $Re(s) > 1$. The related representation given by eq. (1) allows another proof of the functional equation via the Cauchy theorem and the localization of the poles of the inverse Mellin transform $1/(e^x - 1)$ on the imaginary axis, see for example [6].

Let us give two more *examples.* The Mellin transform of the modular form Δ of weight 12 is useful because as we have noted the coefficients of this form are multiplicative. This is a special case of a general theory: we have already seen that the functional equation on the left hand side of eq. (6) corresponds to some modular covariance of the right hand side. If the right hand side P(q) is an eigenfunction of operators introduced by Hecke, then the sequence of coefficients is multiplicative and the Dirichlet series factorizes. Finally the Mellin transform of $(\theta^2 - 1)$ is the so-called Epstein zeta function for the lattice of roots of $A_1 \times A_1$. It is also equal to 4 times the product of the Riemann zeta function by the Dirichlet L function corresponding to the nontrivial character of $(Z/4Z)^*$. This concludes our mathematical introduction and allows the alert reader to anticipate the main tricks of the physical section to follow.

3. Quantum Statistical Mechanics

We have mentioned in the introduction and in part 2. a collection of Physics sounding words: duality (one of several!), bosonization, density of states or of energy levels, quantum statistics of partons, partition functions, fugacity, grand-canonical ensemble. The last five terms will be defined in section 3.1 on the basic tools of Quantum Statistical Mechanics (QSM).

In section 3.2 (QSM without Hamiltonian) we shall introduce number theoretical systems whose generating functions can be interpreted physically. We shall in passing discuss Kramers-Wannier duality and its similarity with the functional equation. Our main result is to interpret multiplicativity in the sense of the last section as the absence of interactions. Finally we shall define what we meant by bosonization in part 2.3.

The problem of phase transitions will be reviewed in 3.3, in particular the analogy between theorems restricting the location of Lee-Yang or Fisher zeros of partition functions with the Riemann hypothesis. This will also be the place for the presentation of Hagedorn catastrophes: poles of the partition functions due to exponential increase of the density of states.

In the conclusion we shall present natural models and functions that may help in the study of the Riemann hypothesis, in particular the introduction of a chemical potential (or of a "fugacity").

3.1 Basic tools of QSM

Let us begin with a traditional presentation of Quantum Statistical Mechanics. So let us assume that we are given a *Hamiltonian* system with a finite number of degrees of freedom. We shall use the language of the theory of gases. The optimistic approach is to try and solve for the eigenvalues of the Hamiltonian operator and to compute the *partition function.* For example one may try to find the *canonical ensemble partition function* defined by : $Z = Tr(e^{-sH})$ where H is the Hamiltonian. In practice this is not possible and one must make approximations and try to evaluate Z directly (see below).

In order to recover thermodynamics starting from the postulates of Statistical Mechanics one uses three different approximations. Firstly ergodicity (mixing) assumptions allow us to identify time average and the so-called micro-canonical average. The *micro-canonical ensemble* is a probability distribution that attributes equal weight to all quantum states allowed by the conservation equations of an isolated system (in particular the conservations of energy and of the number of particles).

Secondly in the limit of large (homogeneous) systems at equilibrium one has to prove that the statistical average over several different *ensembles* leads to the same result (the thermodynamic limit). So one may choose the easiest ensemble to compute with. For example the average over an ensemble may be equal to the most probable value in another one. But in the limit of large systems one may neglect the fluctuations and identify the averages.

Finally the thermodynamicist who is interested only in the large scale properties retrieves them from a single function: the thermodynamic potential. For example the partition function Z corresponds to the *canonical ensemble*, its logarithm $\log Z = \Phi$ is a thermodynamic potential. A nice introduction to Statistical Mechanics is contained in [7]. But let us start with a simpler instance of thermodynamic potential: the entropy of an isolated (micro-canonical) system.

The entropy is a function of the total energy E and of the number N of particles, all the other physical quantities can be derived from it. If the system can occupy a set of $z\,(E,N)$ equiprobable states its entropy is $\Sigma(E,N) = \log z$ in natural units. The temperature T is a derived concept, its inverse "s" (β according to the physicist's tradition) is equal to the derivative $\partial\Sigma/\partial E$. The derivative $-\partial\Sigma/\partial N$ times the temperature is called the "chemical" potential. In the case of an electron gas it is nothing but the electric potential (times the electronic charge), it is usually noted μ. The *fugacity* u is by definition $u = e^{s\mu}$. This leads us to the canonical (resp. *grand-canonical*) ensembles. They correspond to the situations where one fixes the temperature (resp. the temperature and the chemical potential) of the system, or more precisely s (resp. s and $r = -s\mu$) instead of their conjugate variable(s) E (resp. E and N). The corresponding thermodynamic potentials are Φ (the Helmholtz free energy times -s) resp. Ξ (the "pressure" times the volume times s).

$$\Phi(s,N) \;=\; \Sigma \;-\; sE \;=\; \log Z \tag{7}$$

resp.

$$\Xi(s,r) \;=\; \Sigma \;-\; sE \;-\; rN \;=\; \log\zeta \;=\; PV\,s. \tag{8}$$

ζ is the grand canonical partition function it is defined by:

$$\zeta \;=\; Tr(e^{-sH-rN}) \tag{9}$$

where N is the "number of particles" operator which we denote with the same symbol as its eigenvalues. Equations (7) and (8) are Legendre transforms with respect to the quoted variables of $\Sigma = \log z(E,N)$. z, Z and ζ are called *partition functions.*

Let us conclude this introduction with the example of a gas of free particles. A particle physicist calls partons elementary constituents of other observable particles, for example the quarks which are "free" in the limit of high energies. The *statistics of the partons* is usually bosonic or fermionic; in the absence of interactions this means that if the possible (*discrete* for simplicity) energy levels of the partons are distributed with density $p(E)$ the free Hamiltonian reads:

$$H \;=\; \int_{E_0}^{\infty} p(E)\,dE\;E\,a^+(E)a(E).$$

The operators a and a^+ obey canonical commutation (Bose) (resp. anticommutation (Fermi)) relations:

$$a(E)a^+(E') - (resp.+)a^+(E')a(E) \;=\; \delta_{E,E'}.$$

We omitted a possible degeneracy index on a and a^+ which arises if p(E) is greater than 1. The "number of states operator" is given by:

$$N \;=\; \int p(E)\,dE\;a^+(E)a(E).$$

In this paper we shall reserve the name *density of states* to the distribution $\rho(E)$ corresponding to the spectral decomposition of the Hamiltonian H. The once integrated density of states is the number of (discrete) states of energy less than E.

3.2 QSM without Hamiltonian

In this section we would like to do two things. Firstly we shall briefly analyze what can be done in the situation where one knows only the density of states and no other dynamical detail. And secondly we shall reconstruct a number of free theories from their partition functions. The study of integrable systems allows one to check approximation schemes but is neither economical nor generic. On the other hand the knowledge of a microscopic Hamiltonian helps in the study and exploitation of symmetries, it allows a concrete discussion of the "renormalization group" and it is indispensable for answering local questions. So the question arises to try and study directly a solved system, that is a system whose energy spectrum is completely known. Can we discover its phase diagram or find a "simple" Hamiltonian realization. We shall not attempt here to find a realistic realization in terms of experimental systems. But to a number of theoretical generating functions we shall associate simple Hamiltonians whose partition functions are precisely equal to them.

The important thermodynamical object is the partition function: for example the canonical one: $Z = Tr(e^{-sH})$. It is equal to:

$$Z = \int \rho_0(E)\, dE e^{-sE}. \qquad (10).$$

Note that the density corresponds to a fixed number of particles N_0 in this case. One recognizes in (10) a Laplace transform in the variable E or what is the same thing a Mellin transform in the variable $x = e^{-E}$. If the range of integration of E is $[0, \infty]$ it corresponds to $[0, 1]$ for x. As an illustration let us recall the defining formula for Euler's Γ function given in section 2.4, it corresponds to

$$\rho_0(E = -\log x) = e^{-e^{-E}}.$$

The spectrum is not bounded below but the density of states decreases extremely fast there, it tends to 1 for large positive energies.

Analogously the microcanonical entropy is readily computable once one knows the degeneracy of the energy levels corresponding to the given energy (and particle number). The macro-canonical pressure is also computable with the help of the generalized density of states $\rho(E, N)$. We have with obvious notations:

$$\zeta = \int \rho(E, N)\, dEdN e^{-sE-rN}.$$

Putting together the concepts of the last section and those of part 2. we are led to look for equalities between partition functions and generating functions. The inspiration may come from the Moebius function which strongly suggests a fermionic character for the prime numbers as does the concept of squarefree numbers. The key remark is now that energy is the additive quantity of thermodynamics, in multiplicative number theory the additive quantities are the logarithms of the integers, so let us identify the two theories.

A particle physicist naturally wants to break numbers into partons: they will have energies $\log p$ for p the prime numbers. The energy levels will depend on the *statistics* of these primes. In the fermionic case they will be the logarithms of squarefree numbers but if we insist on treating them as bosons they will correspond to the (multiplicative) nonzero integers. The unique factorization of integers ensures that each nonzero integer appears exactly once. Now the number of factors may or may not be fixed. It turns out that mathematicians have encountered the second (i.e. grand-canonical) situation which is usually easier to handle in Physics also. Let us call the bosonic case the free Riemann gas because its grand-canonical partition function at zero μ is nothing but the Riemann ζ_R function. Its inverse is not quite a partition function because of the minus signs. It is the supertrace of the operator e^{-sH}. The proper treatment of the fermionic case leads to the generating series of $|\mu(n)|$:

$$\zeta_F(s) \;=\; \zeta_R(s) \;/\; \zeta_R(2s) \tag{11}$$

Let us remark that our choice involves some arbitrariness but it seems the most natural. The same idea allows us to reinterpret several classical objects of Mathematics. The absence of n! factor implies that one should look at "Quantum" Statistical Mechanics and not at its classical limit. Furthermore the absence of chemical potential begs for the reintroduction of a fugacity in the classical formulas, this will be discussed in the conclusion and in a forthcoming paper to appear in the Comptes-Rendus of the French Academy of Sciences. The above equalities contain an inverse temperature s but μ has been set equal to zero ($u = 1$). Z_M can be interpreted as a partition function of particles with indefinite metric, the primes are then "ghost" states, let us call this system the Moebius gas.

It is obvious now to interpret the Moebius transform as the tensor product of a model by the Riemann gas, and its inverse as the tensoring by the Moebius gas. Equations derived from (3) for $x = \infty$ are of the form

$$\zeta(s) \;=\; \int_0^\infty s\; A(e^E)\; e^{-sE}\, dE$$

they are quite similar to eq. (10).

Let us now explain the term *bosonization.* The partition function of a bosonic oscillator as we have seen is of the form $\sum q^n, n \geq 0$ this sum is equal to $(1-q)^{-1}$. A fermionic sum terminates after two terms (the occupation number is 0 or 1), its value is $(1+q)$. In some instances partition functions of bosons may be equal to partition functions of fermions, there are famous examples in the theory of supersymmetric strings. Eq.(4) is obviously of that character,but also eq. (5): let us rewrite eq. (11) as: $\zeta_R(s) = \zeta_F(s)\zeta_R(2s)$ Now a system with energies E_i and temperature $(2s)^{-1}$ is equivalent to a system with energies $2E_i$ and temperature s^{-1}. So let us iterate, we get:

$$\zeta_R(s) = \zeta_F(s)\zeta_F(2s)\zeta_F(4s)...$$

the minimal energies of the successive systems tend to infinity so at finite temperature it should not be difficult to make sense of this infinite product. The upshot of this is a true equivalence between bosons and a ladder of fermion models.

The multiplicative property did correspond to a factorization over prime numbers in Arithmetics; in Physics the same fact is the sign of a decoupling of the degrees of freedom corresponding to the different primes. The above gases have even less interaction because each prime corresponds to a harmonic oscillator without self couplings (= non linearities).

It is tantalizing to try to identify the functional equation of Riemann with a duality à la Kramers-Wannier. In the example of the two dimensional (canonical) Ising model we have a (grand-canonical) lattice gas model where the chemical potential is the exterior magnetic field and the nearest neighbour coupling a short range interaction, the spin is essentially the occupation number. In two dimensions and at zero chemical potential the duality relates values of ζ near $s = \infty$ and near $s = 0$ (see for example [8]). The critical temperature is left invariant by the duality!

3.3 Zeros and poles of partition functions

The phase transitions are associated to a loss of analyticity of the thermodynamic potential. Some derivative thereof blows up at the phase transition and the simplest reason for such an event is the occurrence of a zero of the partition function. In the thermodynamic limit such zeros can be real, but one studies them first in the case of a finite number of degrees of freedom in the complex plane. A famous theorem of Lee and Yang, see [9] for example, asserts that in the complex plane of chemical potential there is no zero of the partition function of the two-dimensional Ising lattice gas unless the potential is purely imaginary. M. Fisher [9] has studied analogous properties in the temperature (or s) complex plane. We would like now to stress the importance and the generality of poles of partition functions.

Clearly when the density of energy eigenstates increases exponentially in the ultraviolet integrals of the type of eq.(10) diverge. Let us use our solved models from number theory to make this more precise. The Riemann gas levels are the logarithms of the integers they coalesce to build a level density e^E and clearly the grand-canonical partition function cannot be defined above the Hagedorn temperature T=1 [2]. This is our familiar pole at s=1 for the Riemann ζ function. This is a general property of highly degenerate systems like string models for example. The choice of thermodynamic ensemble is crucial in such cases [10].

4. Conclusion

We are left with a dictionary and a program. One of the natural questions to ask is whether the introduction of a chemical potential that is the replacement of the zeta function by a function of two variables s and u helps to avoid the pole at s=1. The classical example of Bose condensation shows that one must allow non-vanishing chemical potentials above the critical temperature of decondensation to find the equilibrium distribution, but in the case of fugacity $u = -1$ the Dirichlet series is well known and it is not known to converge better than ζ_R. We shall return in the announced mathematical article to the formulation of a generalized Riemann hypothesis for the function:

$$\zeta_G(s,u) = \sum u^{\Omega(n)}/n^s$$

where $\Omega(n)$ is the number of prime factors (distinct or not) of n. The case of Dirichlet L functions is also interesting because of the absence of pole.

Let us apologize to the many people who have made one or all of these trivial remarks before and are not quoted here. We keep discovering precursors of what we once thought was original but rather formal work. Some dates are given in [11] Mackey discussed the bosonic case but did not really introduce fermions... Several other approaches ought to be related to the present one: the work of M. Berry and others on dynamical systems with closed orbits of length $\log p$ and the Selberg ζ function, the detailed study of zeros of partition functions and their concentration on sets of low dimensionality etc...

References

[1] B. Riemann, Monatsberichte der Berliner Akad. (1859) November.
[2] R. Hagedorn, Suppl. Nuovo Cimento 3 (1965) 147. See also Yu. B. Rumer, J. Exp. Theor. Phys. 38 (1960) 1899.

[3] J. P. Serre, Cours d'Arithmétique PUF, Paris 1977.
[4] A. Weil, Collected papers Vol.I, 399-410.
[5] G.H. Hardy and E.M. Wright, An introduction to the theory of numbers Clarendon Press, Oxford 1945.
[6] H.M. Edwards, Riemann's zeta function Academic Press, New York 1974.
[7] K. Huang, Statistical Mechanics Wiley, New York 1963.
[8] R.J. Baxter, Exactly solved models in Statistical Mechanics Academic Press, New York 1982.
[9] M. E. Fisher in W.E. Brittin, Boulder Lectures in theoretical Physics 1964 Un. of Colorado press, Boulder 1965.
[10] B.L. Julia, in preparation,
[11] B.L. Julia, 1985 unpublished.
B.L. Julia, Journal de Physique 50 (1989) 1371.
D. Spector, Utrecht preprint THU '88/41.
J. Hannay, unpublished according to M. Berry.
G.W. Mackey, Unitary group representation in Physics, Probability and number theory Benjamin, Reading 1978.
Unknown...

Algebraic Number Theory and Hamiltonian Chaos

F. Vivaldi

School of Mathematical Sciences, Queen Mary College,
University of London, Mile End Road, London E1 4NS, UK

1. Introduction

Three great themes were developed in algebraic number theory during the nineteenth century, namely quadratic forms, algebraic numbers and ideals, which call to mind the names of Gauss, Kummer, and Dedekind, respectively. They represent three ways of dealing with one and the same problem, the failure of the fundamental theorem of arithmetic (unique factorization into primes) in algebraic number fields. Had the fundamental theorem been valid, Gauss would have probably answered any conceivable question about binary quadratic forms while in his teens, and Kummer would have proved Fermat's last theorem. As to ideal theory, it wouldn't have been developed for quite a while.

It is worth noting that the more modern ideal theory (1871) has not superseded the older developments, and in many instances one has to return to quadratic forms and algebraic numbers in order to obtain quantitative results.

During this century, new connections were discovered between algebraic numbers and integral matrices. These ideas were largely developed in the context of representation theory, and resulted in a process of cross fertilization from which both algebraic number theory and matrix theory drew great benefit.

The link between algebraic number theory and chaotic dynamics is the result of the discovery that an important class of fully chaotic hamiltonian systems can be represented by means of integral matrices. This result has opened the study of chaos to the methods of algebraic number theory, which have helped unveil the intimate structure of the laws governing chaotic motions. At the same time, number-theoretical concepts have been rephrased in the language of dynamics, thereby adding one extra dimension to the many-sided theory of numbers.

We intend to illustrate some aspects of the connections between quadratic forms, algebraic numbers, ideals, integral matrices and certain chaotic dynamical systems. All the results presented here have been published elsewhere [1–3].

2. Some typical problems

1. Quadratic forms. The basic problem in the theory of binary quadratic forms is representation. One wishes to characterize the integers m for which the dio-

Springer Proceedings in Physics, Vol. 47 — Number Theory and Physics
Editors: J.M. Luck · P. Moussa · M. Waldschmidt

phantine equation

$$m = ax^2 + bxy + cy^2 = Q(x, y),$$

has integral solutions x and y, and construct all such solutions for a given m. Here a, b and c are also integers, and the discriminant $d = b^2 - 4ac$ of Q is assumed not to be a perfect square. The natural starting point is the case $m = p$, a prime, upon which the general solution for m composite is built. A beautiful partial solution to this problem was first given by Gauss [4]. A complete solution is still somewhat elusive [5].

2. Algebraic numbers. Let ω be a root of the monic quadratic polynomial $f(x) = x^2 + bx + c$, that is, $\omega^2 = -b\omega - c$, where b and c are integers, and ω is not rational. By repeated applications of this relation, all integral polynomials in ω can be reduced to linear polynomials. The set of such linear expressions has the structure of a ring, which we denote by $\mathbf{Z}[\omega]$. Its elements are called *quadratic integers* . Geometrically, $\mathbf{Z}[\omega]$ is a two-dimensional lattice, which we represent as $\mathbf{Z}[\omega] = [1, \omega]$ (integral basis), meaning that each element in $\mathbf{Z}[\omega]$ can be written uniquely as an integral linear combination of 1 and ω.

The main problem is factorization. Let $\xi \in \mathbf{Z}[\omega]$. Do elements ξ_1 and ξ_2 in $\mathbf{Z}[\omega]$ exist such that the factorization $\xi = \xi_1\xi_2$ is possible, and if so, can this process be continued until one arrives at a unique decomposition of ξ into 'primes'? This problem is solved once the solution is known for the case $\xi = p$, a rational prime (note that $p \in \mathbf{Z}[\omega]$). The uniqueness problem is complicated by the existence of nontrivial *units* (the divisors of 1), which play the same role as the integers ± 1 in $\mathbf{Z}$. If ω is real, there are infinitely many units in $\mathbf{Z}[\omega]$, but factorizations which differ by units should be regarded as identical.

3. Ideals. An ideal J in $\mathbf{Z}[\omega]$ is a lattice with the additional property of being invariant under multiplication by all elements of $\mathbf{Z}[\omega]$. An ideal admits an integral basis $J = [\omega_1, \omega_2]$, and it can be visualized geometrically as a sublattice of $\mathbf{Z}[\omega]$. The basic problem of ideal theory is again factorization, and we wish to know if J can be expressed uniquely as the product of prime ideals. We recall that if $J_1 = [\alpha_1, \beta_1]$ and $J_2 = [\alpha_2, \beta_2]$ are ideals in $\mathbf{Z}[\omega]$, then $J_1 J_2 = [\alpha_1\alpha_2, \alpha_1\beta_2, \alpha_2\beta_1, \beta_1\beta_2] = [\gamma, \delta]$, where the reduction to a two-elements basis is possible because any lattice in $\mathbf{Z}[\omega]$ is (at most) two-dimensional. Again, the solution is based on the decomposition of the rational primes ideals $J = (p) = [p, p\omega]$, for any prime ideal in $\mathbf{Z}[\omega]$ must divide a unique (p).

Ideals are then regrouped into classes by means of the equivalence relation that places the ideals J and I in the same class if $\alpha J = \beta I$ for some $\alpha, \beta \in \mathbf{Z}[\omega]$. The number of ideal classes in $\mathbf{Z}[\omega]$ is called the *class number*.

4. Integral matrices. Common problems involving integral matrices include conjugacy and factorization of characteristic polynomials. Let

$$M = \begin{pmatrix} a & b \\ c & d \end{pmatrix}; \quad a, b, c, d \in \mathbf{Z},$$

and let $f(x)$ be its characteristic polynomial. Then $f(x)$ is integral and monic, and if its roots are not rational, they are quadratic integers.

While factoring $f(x)$ in $\mathbf{Q}$ or $\mathbf{C}$ is trivial, the factorization of $f(x)$ in integers modulo p (p a prime) is much more interesting. Such a factorization takes the form $f(x) \equiv (x-a)(x-b) \pmod{p}$, for some integers a and b. If a and b are distinct, then the matrix M acting on the module $\mathbf{Z}^2/p\mathbf{Z}^2$ admits two distinct invariant eigenspaces.

Two integral matrices M and M' are said to be conjugate if there exists an integral matrix X, with unit determinant, such that $MX = XM'$. Conjugate matrices have the same eigenvalues, but the converse is not true, and the problem is to count and classify conjugacy classes.

5. Chaotic dynamics. Let ϕ be a smooth map of the 2-dimensional torus onto itself. A periodic orbit of ϕ is a set of points $\{x_1, \ldots, x_n\}$ on the torus such that $\phi(x_k) = x_{k+1}$ and $\phi(x_n) = x_1$. Periodic orbits are very important because they contain a finite amount of information, no matter how complicated is the dynamics. For this reason the 'modern' approach to dynamical systems (after Poincaré) places great emphasis on the study of periodic orbits.

A mapping ϕ is chaotic if nearby points separate from each other at an exponential rate. This makes predictions about the future of the system very difficult to obtain, and one often turns to periodic orbits as approximants for non-periodic behaviour. Because in some chaotic systems the periodic orbits are dense, they become particularly valuable. One basic problem in dynamics is the classification of chaotic maps on certain manifolds, and the construction and classification of their periodic orbits.

3. Chaotic dynamics and integral matrices

Chaos takes on many different forms. For instance, chaos and order may coexist in the same system, so that exponential instability of motion is not necessarily a global property. Generally speaking, systems which are globally chaotic are simpler, and this explains why so much attention has been devoted to the most chaotic of all hamiltonian systems, the Anosov systems [6]. They are closely related to integral matrices, the connection being based on the following result [7]

Theorem (Manning, 1974). *Every Anosov diffeomorphism of the n-dimensional torus is topologically conjugate to a linear Anosov diffeomorphism.*

In essence, this theorem says that when a map on the torus is completely chaotic, what really matters is its action on the generators, and this leads to linear relations. Thus, for instance, in two dimensions one need only consider 2×2 integral matrices, with trace greater than 2 (in absolute value), and unit determinant. These are hyperbolic toral automorphisms, whose eigenvalues are *units* in quadratic fields (they are quadratic integers whose product is 1, whence they divide 1).

Next we note that the periodic orbits of a hyperbolic toral automorphism coincide with points having rational coordinates. All rational points of the two dimensional torus, sharing the same denominator f, form a lattice. By clearing denominators, we obtain the lattice $\mathbf{Z}^2/f\mathbf{Z}^2$, whose points are integers pairs (x,y), and which is invariant under the action of unimodular matrices. We are interested in classifying orbits on it, but this problem is made difficult by the lack of a multiplicative structure in $\mathbf{Z}^2/f\mathbf{Z}^2$.

4. Matrices and algebraic numbers

Let M is a 2×2 integral matrix, and λ one of its eigenvalues. We can always solve the eigenvalue equation for M in such a way that the eigenvectors have components in $\mathbf{Z}[\lambda]$

$$\begin{pmatrix} a & b \\ c & d \end{pmatrix} \begin{pmatrix} \omega_1 \\ \omega_2 \end{pmatrix} = \lambda \begin{pmatrix} \omega_1 \\ \omega_2 \end{pmatrix}; \qquad \omega_1, \omega_2 \in \mathbf{Z}[\lambda]. \tag{1}$$

Consider now the module $J = [\omega_1, \omega_2]$. It can be shown that J is actually an *ideal* in $\mathbf{Z}[\lambda]$ (see [5], appendix 2). Because a change of basis for J corresponds to conjugacy for M, and because any multiple of J in $\mathbf{Z}[\lambda]$ also leads to an eigenvector of M, we conclude that this procedure extablishes a one-to-one correspondence between ideal classes in $\mathbf{Z}[\lambda]$ and matrix classes. So we get at once a complete classification of Anosov diffeomorphisms in terms of a basic quantity of algebraic number fields. In particular, the number of conjugacy classes of Anosov diffeomorphisms is equal to the class number of $\mathbf{Z}[\lambda]$.

We can go further. If $\xi = m\omega_1 + n\omega_2 \in J$, from (1) we obtain $\lambda\xi = (ma+nc)\omega_1 + (mb+nd)\omega_2 = m'\omega_1 + n'\omega_2$. Denote the transpose of M by M^T. Then $M^T(m,n) = (m',n')$, which means that the action of M^T on the lattice $\mathbf{Z}^2/f\mathbf{Z}^2$ can be replaced by multiplication by λ in J/fJ

$$\xi_{t+1} \equiv \lambda \xi_t \,(\mathrm{mod}\ fJ).$$

This is a nice result, because it provides the discrete phase space with a natural arithmetical structure which turns a problem of dynamics into one of arithmetic.

5. An example

We illustrate the connection between quadratic forms, algebraic numbers, ideals, integral matrices and toral automorphisms by means of an example, which describes an important algebraic phenomenon called *splitting*. For the sake of simplicity, we have chosen a case in which the fundamental theorem of arithmetic does hold, but we shall briefly comment on the effects of non unique factorization.

1. Quadratic forms. The prime $p = 13$ is representable by the quadratic form $Q(x,y) = x^2 - 3y^2$.

$$13 = Q(5,2) = 5^2 - 3.2^2.$$

The discriminant of Q is equal to 12. One should not be distracted by the non-uniqueness of this representation, e.g. $Q(5,2) = Q(4,1)$, which merely corresponds to a change of coordinates.

2. Algebraic numbers. The prime $p = 13$ is the product of two distinct primes in $\mathbf{Z}[\sqrt{3}]$

$$13 = (5+2\sqrt{3})(5-2\sqrt{3}).$$

The discriminant of $\mathbf{Z}[\sqrt{3}]$ is 12. As above, we note $(5+2\sqrt{3})(5-2\sqrt{3}) = (4+\sqrt{3})(4-\sqrt{3})$, but $5+2\sqrt{3} = (4-\sqrt{3})(2+\sqrt{3})$, and $(2+\sqrt{3})$ is a *unit* because $(2+\sqrt{3})(2-\sqrt{3}) = 1$, so this does not violate unique factorization.

3. Integral matrices. Let

$$M = \begin{pmatrix} 2 & 3 \\ 1 & 2 \end{pmatrix}; \qquad \lambda^2 = 4\lambda - 1, \quad \lambda = 2 \mp \sqrt{3}.$$

The characteristic polynomial of M, of discriminant 12, has two distinct roots modulo 13

$$f(x) = x^2 - 4x + 1 \equiv (x-6)(x-11) \pmod{13}.$$

Then the matrix M acting on the module $\mathbf{Z}^2/13\mathbf{Z}^2$ has two distinct eigenvectors, namely $(5,2)$ and $(5,-2)$.

4. Ideals. The ideal (13) in $\mathbf{Z}[\sqrt{3}]$ splits into the product of two distinct prime ideals

$$(13) = [13, 13\sqrt{3}] = [5+2\sqrt{3}, 1+3\sqrt{3}][5-2\sqrt{3}, 1-3\sqrt{3}].$$

Note that $\mathbf{Z}[\sqrt{3}] = \mathbf{Z}[\lambda]$.

5. Chaotic dynamics. Let $x = (m/13, n/13)$, where m and n are integers, be a rational point on the two dimensional torus. Then the orbit of M through x is periodic and its period T does not depend on m and n. T divides $p-1 = 12$, and there are two sublattices which are dynamical invariants. The orbits belonging to these lattices are called *ideal orbits*.

The process of splitting described in the above examples does not occur for all primes, and indeed we have

1. The diophantine equation $\pm 7 = x^2 - 3y^2$ has no solutions in integers.

2. The prime $p = 7$ remains prime in $\mathbf{Z}[\sqrt{3}]$.
3. The characteristic polynomial of M is irreducible modulo 7.
4. The ideal $(7) = [7, 7\sqrt{3}] \in \mathbf{Z}[\sqrt{3}]$ is prime.
5. Let $x = (m/7, n/7)$, be a rational point on the torus. Then the orbit of M through x is periodic and its period T does not depend on m and n. T divides $p + 1 = 8$, and there are no invariant sublattices.

It turns out that, with a finite number of exceptions, all primes can be regrouped into two infinite families, whose elements behave like 13 or 7, respectively. The construction of these families is done by means of *arithmetic progressions* modulo the discriminant $d = 12$ of $\mathbf{Z}[\lambda]$. Specifically, all primes $p \equiv \pm 1 \pmod{12}$ belong to the former family, while the primes $p \equiv \pm 5 \pmod{12}$ belong to the latter. The exceptions are 2 and 3, which are the discriminant divisors.

From the dynamical viewpoint, the characterization of splitting in terms of ideals (or of characteristic polynomials) is the most informative, because ideals have a geometrical representation in the phase space. Because ideals are quite insensitive to non unique factorization, so are the dynamical properties of periodic orbits, and the only precaution to be taken is the exclusion of a finite number of primes (this is when $\mathbf{Z}[\lambda]$ is not a maximal order [5]).

By contrast, the arithmetic of quadratic forms and algebraic numbers becomes considerably more complicated when unique factorization fails, and the lack of a direct dynamical interpretation makes this aspect of the theory less attractive for our purposes. Quadratic forms are too fine a tool for dealing with the dynamical manifestations of splitting. Some aspects of the theory of quadratic forms do however appear in the study of toral automorphisms, for instance when classifying periodic orbits.

6. Chaos in discrete exponentials and logarithms

We shall interpret in dynamical language the irregular behaviour characterizing discrete exponential functions and their inverses. The key is the study of ideal orbits, which hold a unique place within the inexaustible variety of chaotic motions.

When splitting occurs, the factorization $f(x) = x^2 - tx + 1 = (x - \lambda)(x - \lambda^{-1})$ of the characteristic polynomial of M is represented modulo p by the factorization $f(x) \equiv (x - g)(x - g^{-1}) \pmod{p}$. The map $\lambda \to g$ is a ring homomorphism, whence $g + g^{-1} \equiv t \pmod{p}$ and $gg^{-1} \equiv 1 \pmod{p}$.

Because ideals are also discrete eigenspaces, motions of ideal orbits decompose coordinatewise. So if $M(x_n, y_n) = (x_{n+1}, y_{n+1})$ and $(p) = P_1 P_2$ we have

$$x_{n+1} \equiv g x_n \pmod{p}, \quad y_{n+1} \equiv g y_n \pmod{p}, \tag{2}$$

for orbits on P_1, and the same expression with g^{-1} in place of g for orbits on P_2.

Equation (2) says that for an ideal orbits, the dynamical evolution is given by multiplication by an integer g modulo a prime p. Its solution is given by the discrete exponential function $x_n \equiv g^n \pmod p$ (by shifting n, if necessary). This could hardly be made simpler, and it should be regarded as the arithmetical essence of hamiltonian chaos. Note that similar devices form the basis for the construction of random numbers generators (see [8], chapter 3).

Because discrete exponentiation g^n describes motions in chaotic system, its inverse, the discrete logarithm (also called the *index*), will likewise exibit complicated behaviour. The index $\mathrm{Ind}_g(n)$ of n in base g is defined as the unique residue class modulo $p-1$ satisfying the congruence $g^{\mathrm{Ind}(n)} \equiv n \pmod p$. Here g must be a primitive root modulo p, which means that we are dealing with an ideal orbit of period exactly equal to $p-1$ (cf. section 5). Important functions built by means of indices are the so called *Dirichlet's characters* modulo p, defined as [9]

$$\chi_m(n) = \exp\left(\frac{2\pi i m\,\mathrm{Ind}(n)}{p-1}\right). \tag{3}$$

Thus $\chi_m(n)$ is a root of unity, which depends irregularly on n. Characters appear in chaotic dynamics in the spectral fluctuations of ideal orbits. Specifically, the discrete Fourier transform f_m of the motion $x_n = g^n \pmod p$ is found to be [3]

$$f_m = U(m)W(m), \tag{4}$$

where

$$U(m) = \sum_{n=1}^{q} \chi_{-m}(n); \qquad q = (p-1)/2, \tag{5}$$

and

$$W(m) = \frac{p}{p-1}\frac{\chi_{-m}(2)}{1-2\chi_{-m}(2)}.$$

In expressions (4) the function W determines the gross features of the spectrum, while the irregular function U is responsible for the fluctuations. The latter satisfies the statistical property

$$\langle |U|^2 \rangle_m = q \qquad (m \text{ odd}),$$

by means of which some average properties of the spectrum can be characterized.

It is useful to visualize U as a 'random walk' in the complex plane, consisting of q steps of unit length, and direction determined by the discrete logarithm, according to (3). If the walk is then continued so as to include a total of $p-1$ steps, the irregular path followed in the first part is retraced exactly, and the walk terminates at the origin, by virtue of the known identity

$$\sum_{n=1}^{p-1} \chi_m(n) = 0; \qquad m \neq p-1,$$

which is miraculously hiding the great irregularity of the character functions.

Acknowledgements

I thank S.E.R.C. for supporting this research.

References

[1] I. C. Percival and F. Vivaldi, Arithmetical properties of strongly chaotic motions, *Physica 25*D (1987) 105–130.
[2] F. Vivaldi, Arithmetical theory of Anosov diffeomorphisms, *Proc. R. Soc. Lond. A* **413** (1987) 97–107.
[3] M. Bartuccelli and F. Vivaldi, Ideal orbits of toral automorphisms, preprint QMC DYN 88-3, Queen Mary College, London (1988). (Submitted to *Physica D.*)
[4] C. F. Gauss, *Disquisitiones Arithmeticae*, Leipzig (1801). (English transl. Yale, New Haven and London (1966).)
[5] H. Cohn, *A classical invitation to algebraic numbers and class fields*, Springer-Verlag, New York (1978).
[6] V. I. Arnold, and A. Avez, *Ergodic problems of classical mechanics*, Benjamin, New York (1968).
[7] A. Manning, Classification of Anosov diffeomorphisms on tori, *Lecture Notes in Mathematics*, No. 468, Springer-Verlag, New York (1975).
[8] D. E. Knuth, *The art of computer programming*, Addison-Wesley, Reading, Mass. (1981).
[9] T. A. Apostol, *Modular functions and Dirichlet series in number theory*, Springer-Verlag, New York (1976).

Semiclassical Properties of the Cat Maps

J.P. Keating

H.H. Wills Physics Laboratory, Tyndall Avenue, Bristol BS8 1TL, UK

1 Introduction

The cat maps are a family of completely chaotic Hamiltonian dynamical systems whose classical and quantum mechanics are especially well understood.

The classical mechanics was the subject of an illuminating paper by Percival and Vivaldi [1], who gave a complete classification of the periodic orbits in terms of the theory of quadratic number fields. The quantum mechanics was developed by Hannay and Berry [2], who gave an exact and explicit quantisation scheme. Consequently the cat maps provide a particularly useful set of systems with which to investigate the influence of chaotic classical motion on quantum phenomena.

The aim of this article is to describe some asymptotic properties of the long periodic orbits of the classical motion and then to show how these properties may be used to explain the behaviour of certain quantum (i.e. non-classical) phenomena in the semiclassical limit (i.e. as Planck's constant tends to zero). The motivation is that it appears that typical quantum systems exhibit universal behaviour in this limit [3]. For example the statistics of the eigenvalues take one of a small number of universal forms. It is thus of interest to investigate these statistics for the quantised cat maps.

2 The Cat Maps

The system to be considered has for its phase space the 2-torus. The dynamics are then generated by a hyperbolic automorphism of the torus

$$\begin{pmatrix} q_{n+1} \\ p_{n+1} \end{pmatrix} \equiv \begin{pmatrix} t_{11} & t_{12} \\ t_{21} & t_{22} \end{pmatrix} \begin{pmatrix} q_n \\ p_n \end{pmatrix} \quad (\text{mod } 1) , \tag{1}$$

where the matrix has integer entries (for continuity), unit determinant (for area preservation) and has $|t_{11}+t_{22}|>2$ for hyperbolicity. Such maps are known as *cat maps* [4].

All points on the torus with rational coordinates are fixed points of some power of the map and these represent the only fixed points. Furthermore the denominator of a rational point is preserved under the action of the map. Consequently, when studying periodic orbits, it is natural to consider the lattice of points on the torus with a given denominator, g. Under the action of the map a point $(Q_n/g, P_n/g)$ remains on the $g \times g$ lattice. The numerators map according to

Springer Proceedings in Physics, Vol. 47 Number Theory and Physics
Editors: J.M. Luck · P. Moussa · M. Waldschmidt

$$\begin{pmatrix} Q_{n+1} \\ P_{n+1} \end{pmatrix} \equiv \begin{pmatrix} t_{11} & t_{12} \\ t_{21} & t_{22} \end{pmatrix} \begin{pmatrix} Q_n \\ P_n \end{pmatrix} \pmod{g} . \tag{2}$$

All orbits on the lattice are periodic but not, in general, with the same period. The period of the lattice, $n(g)$, may be defined to be the least common multiple of the periods of all of the orbits which lie on it:

$$\begin{pmatrix} t_{11} & t_{12} \\ t_{21} & t_{22} \end{pmatrix}^{n(g)} \equiv \begin{pmatrix} 1 & 0 \\ 0 & 1 \end{pmatrix} \pmod{g} . \tag{3}$$

It is useful at this point to consider the equation

$$b^{\alpha(g)} \equiv 1 \pmod{g} , \tag{4}$$

where b is an integer and $(b, g)=1$. Then, if g is a prime p, it follows from Fermat's theorem that the smallest solution of (4) is given by

$$\alpha(p) = \frac{p-1}{m} , \tag{5}$$

where m is an integer divisor of p-1 which must be determined by trial and error for a given b and p. For example, if b=10 then (5) gives the period of the decimal expansion of $1/p$.

For the cat maps the form of the smallest solution of (3), with g a prime p, may be obtained from the work of Percival and Vivaldi [1]. If d is the discriminant of the quadratic field which contains the eigenvalues of the map (i.e. d is an integer determined by the map), then for $p \nmid d$

$$n(p) = \frac{p-(d/p)}{m} , \tag{6}$$

where (d/p) is the Kronecker symbol [1] (which may take values ± 1) and m is an integer divisor of $(p-(d/p))$ which must be determined by trial and error for the particular map and prime p. Similarly the form of the period function for primes which divide d (of which there are only a finite number) and of all powers of primes may be obtained.

Once the values of the period function for primes and powers of primes are known the values for composite arguments may easily be calculated. If the prime factorisation of g is given by

$$g = \prod_p p^{\alpha_p(g)} , \tag{7}$$

then the period function, defined by (3), will be the least common multiple of the periods of the prime factors,

$$n(g) = [\{ n(p^{\alpha_p(g)}) \}] . \tag{8}$$

The period function, $n(g)$, is a very erratic function of the denominator g. It is natural therefore to study the distribution of values which it takes and its behaviour when averaged over a small range of values of g (i.e. a range $R(g)$ centered at g, where, as $g \to \infty$, $R(g) \to \infty$ but $R(g)/g \to 0$). These properties are best explored using the methods of probablistic number theory (for simple introductions see Schroeder [5] and Kac [6]; for a more detailed review see Elliott [7]). Of particular interest is their behaviour as $g \to \infty$, that is as the lattice of points on the torus with denominator g probes the phase space on the finest scales.

There are well known difficulties involved in the direct application of probablistic methods to number theory [6]; nevertheless the results obtained are invariably the same as those which follow from a rigorous analysis. In the following study we content ourselves with a 'naive' application of the methods of probability theory.

Consider the distribution of values which $n(g)$ takes for a given map as g runs through a series of primes. From (6) this involves finding the distribution of values which the divisor, m, takes.

Let $C_t(k,i)$ be the number of primes less than or equal to the ith prime which, for a particular map (matrix) t, give a divisor $m = k$ in the solution (6) of the equation (3). Computer tests suggest that the limit

$$D_t(k) = \lim_{i \to \infty} \frac{C_t(k, i)}{i} \tag{9}$$

exists and so we may consider the probability that the divisor, defined by (6), takes the value k.

The methods used in calculating this probability distribution are best explained by considering the analogous distribution for the divisors defined by the solution (5) of equation (4). That is we seek the probability, $P(k)$, that, for a particular integer b (which is not a perfect square), the divisor which appears in (5) takes the value k.

Consider first the distribution of values of the divisor for a range of values of b with p fixed. Obviously we need only consider $1 \le b \le p-1$. It was shown by Gauss [8] that the number of integers b in this range which give a divisor k in (5) is $\phi[(p-1)/k]$, where $\phi(n)$ is Euler's totient function [9]. It appears that these particular b are randomly distributed among the p-1 values in the range. Furthermore there appears to be no correlation between the divisors associated with a given integer b for different primes p. Thus we assume that, if $Q(k)$ is the probability that p-1 has k as a divisor,

$$\begin{aligned} P(k) &= Q(k) \left\langle \frac{\phi\left(\frac{p-1}{k}\right)}{p-1} \right\rangle_{p \equiv 1 \pmod{k}} \\ &= \frac{Q(k)}{k} \left\langle \prod_{q \mid \frac{p-1}{k}} \left(1 - \frac{1}{q}\right) \right\rangle_{p \equiv 1 \pmod{k}}, \end{aligned} \tag{10}$$

where q represents a prime and the average is over primes, p, which are congruent to 1 (mod k).

Dirichlet's theorem states that there are infinitly many primes in each of the $\phi(n)$ residue classes which are coprime to a given integer n. Moreover, asymptotically the density of primes in each such class is the same [10]. Thus

$$Q(k) = \frac{1}{\phi(k)} \tag{11}$$

and the probability that an integer of the form $(p-1)/k$ is divisible by a prime q (i.e. that $p\equiv 1 \pmod{qk}$ given $p\equiv 1 \pmod{k}$) is

$$\begin{aligned} W_k(q) &= \frac{1}{\phi(q)} = \frac{1}{q-1} \quad \text{if } q \nmid k \\ &= \frac{\phi(q)}{\phi(q^2)} = \frac{1}{q} \quad \text{if } q \mid k . \end{aligned} \tag{12}$$

Now if

$$\begin{aligned} \delta_{q \mid \frac{p-1}{k}} &= 1 \qquad \text{if } q \mid \frac{p-1}{k} \\ &= 0 \qquad \text{if } q \nmid \frac{p-1}{k} \end{aligned} \tag{13}$$

then

$$P(k) = \frac{1}{k\phi(k)} \left\langle \prod_q \left(1 - \frac{1}{q}\delta_{q \mid \frac{p-1}{k}}\right) \right\rangle_{p\equiv 1 \pmod{k}} . \tag{14}$$

The divisibility of an integer by a prime is independent of its divisibility by other primes and so the product and averaging operations in (14) may be interchanged. Then using (12) gives

$$P(k) = \frac{A}{k^2} \prod_{q \mid k} \left[\frac{1 + \frac{1}{q}}{1 - \frac{1}{q(q-1)}} \right], \tag{15}$$

where

$$A = \prod_p^{\infty} \left(1 - \frac{1}{p(p-1)}\right) . \tag{16}$$

The probability (15) may easily be shown to be normalised. It may also be noted that the distribution is independent of the integer b.

The value of $P(1)$ is just the primitive form of Artin's conjecture, which is described in an article by M.Ram Murty [11]. Lower order corrections to the conjecture are described in the same article and these will also apply to the full distribution (15).

Using the formalism of Percival and Vivaldi [1] the probability distribution for the cat maps may be derived in an analogous way [12]. It is found that the distribution takes exactly the same form as in the above case, i.e. it is given by (15). The density of primes with $(d/p)=1$ and which give a divisor $m=1$ has been independently studied by Bartucccelli and Vivaldi [13], who obtained results consistent with the distribution (15).

Next we consider the behaviour of $n(g)$ as $g\to\infty$ through the integers. From (8) we have

$$n(g) \leq \prod_{p|g} n\left(p^{\alpha_p(g)}\right) \quad . \tag{17}$$

It is known that the values of $\omega(g)$, the number of distinct prime factors of g, are Gaussian distributed about $\log\log(g)$ [6]. Combining this fact with the form of the period function for powers of primes, it may be shown, using probablistic arguments [12], that, for a given map,

$$\left\langle \frac{n(g)}{g} \right\rangle_g \leq C(d)(\log(g))^{\langle 1/m\rangle -1} \quad , \tag{18}$$

where $\langle\ \rangle_g$ denotes an average over values of g and m is the divisor defined by (6).

The value of $\langle 1/m\rangle$ may be obtained from (15):

$$\left\langle \frac{1}{m} \right\rangle = \prod_p^{\infty}\left(1 - \frac{p}{p^3-1}\right) \tag{19}$$

which converges to a value 0.5759.... Hence, as $g\to\infty$

$$\left\langle \frac{n(g)}{g} \right\rangle_g \to 0 \quad , \tag{20}$$

which contradicts a conjecture of Hannay and Berry [2].

The average of the probability, (12), over values of k,

$$W(q) = \frac{1}{q}\sum_{k=1}^{\infty} P(kq) + \frac{1}{q-1}\left(1 - \sum_{k=1}^{\infty} P(kq)\right), \tag{21}$$

may be calculated using (15) to be

$$W(q) = \frac{1}{q-1}\left(1 - \frac{1}{q^2(q-1)}\right). \tag{22}$$

Thus, for primes $q>>1$, $W(q)\approx 1/q$ and so integers of the form (6) may be treated as 'typical' with respect to division by these primes. Combining this with some further simplifying approximations it is possible to estimate the way in which the limit (20) is approached.For example it is found that

$$\langle \log(n(g)) \rangle_g \sim \log(g) - \eta \log\log(g) \times \log\log\log(g) \tag{23}$$

as $g \rightarrow \infty$, where η is a positive constant.

Under these same approximations certain features of the distribution of values which the function $n(g)$ takes may also be explained. For example, the probability that $n(g)$ takes the value y may be shown to be a very slowly varying function of y, for $y << g$.

3 The Quantised Cat Maps

The quantum mechanical description of the cat maps was developed by Hannay and Berry [2]. They showed that for quantisation of *any system*, whose phase space is the unit 2-torus, to be consistent:

a) Planck's constant, h, may only take inverse integer values,

b) if h=1/N then the wavefunctions, in both the position and momentum representations, may be non-zero only at rational points with denominator N.

For a given value of Planck's constant the wavefunctions at points with denominator N map according to

$$\Psi_{Q_2}^{(n+1)} = \sum_{Q_1=1}^{N} U_{Q_1,Q_2} \Psi_{Q_1}^{(n)}, \tag{24}$$

where U is the quantum propagator (a unitary operator). Hannay and Berry gave a unique prescription for associating a quantum matrix with a given cat map, t:

$$\left[U_N^{(t)}\right]_{Q_1,Q_2} = \sqrt{\frac{t_{12}}{iN}} \left\langle e^{\frac{i\pi}{Nt_{12}}\left[t_{11}(Q_1+mN)^2 - 2(Q_1+mN)Q_2 + t_{22}Q_2^2\right]} \right\rangle_m, \tag{25}$$

where the average is over all integer values of m.

It may be shown that, for a given N, the propagator associated with a cat map t, when raised to some power, $n_q(N)$, is equal to the identity matrix (up to an unimportant overall phase factor). If N is odd then the quantum period function, $n_q(N)$, is equal to the classical period function, $n(N)$, for the map t. If N is even then the two functions differ by a possible factor of 2. Consequently $n_q(N)$ has the same asymptotic and distributional properties as $n(N)$.

The eigenstates of the propagator are defined by

$$U^m | k \rangle = \exp(im\theta_k) | k \rangle, \tag{26}$$

where the θ_k are the *eigenangles*. The periodicity of the propagator forces each of the N

eigenangles to lie on one of $n_q(N)$ evenly spaced points around the unit circle. Thus, from (20), as $N \to \infty$ (i.e. in the semiclassical limit) the eigenangles tend, on average, to exhibit infinite degeneracy. This behavior is not typical: quantum systems with chaotic classical limits generically exhibit eigenvalue repulsion [3].

The density of states for the eigenvalues is defined by

$$\rho(\theta) = \sum_{k=1}^{N} \sum_{j=-\infty}^{\infty} \delta(\theta - \theta_k + 2\pi j) \tag{27}$$

Applying the Poisson summation formula gives

$$\rho(\theta) = \frac{N}{2\pi} + \frac{1}{\pi} \mathrm{Re} \sum_{m=1}^{\infty} \mathrm{Tr}(U_N^m) \exp(-im\theta) \tag{28}$$

and, since the propagator is periodic, all information about the spectrum may be obtained from the terms with $1 \le m \le n_q(N)$.

The connection between the eigenstates and the classical motion may be made by expressing the trace of the mth power of the propagator (25) as a sum over the fixed points of t^m [12]: if

$$t^m = \begin{pmatrix} T_{11}^{(m)} & T_{12}^{(m)} \\ T_{21}^{(m)} & T_{22}^{(m)} \end{pmatrix} \tag{29}$$

then

$$\mathrm{Tr}\left[U_N^{(t)}\right]^m = \frac{1}{\sqrt{-R^{(m)}}} \sum_{k,n \in \square} \mathrm{e}^{-\frac{i\pi N}{R^{(m)}}\left[T_{21}^{(m)} k^2 - T_{12}^{(m)} n^2 + 2\left(T_{22}^{(m)} - 1\right) kn\right]}, \tag{30}$$

where

$$R^{(m)} = 2 - \mathrm{Tr}(t^m) \tag{31}$$

and the sum is over integers k,n which are the coordinates of points which lie in the parallelogram formed by acting on the unit square with $(t^m - \mathrm{I})$. Each such pair of integers is uniquely associated with a fixed point of t^m through

$$t^m \begin{pmatrix} q \\ p \end{pmatrix} = \begin{pmatrix} q \\ p \end{pmatrix} + \begin{pmatrix} k \\ n \end{pmatrix}. \tag{32}$$

The sum (30) is of exactly the same form as the general semiclassical formulae which connect the trace of a quantum propagator to a sum over periodic orbits of the associated classical motion [14,15]. For the quantised cat maps these sums are exact (rather than asymtotic approximations) and computable (since all of the fixed points of t^m may be generated explicitly).Moreover, since the propagator is periodic, the spectrum is completely determined by the periodic orbits with period $\le n_q(N)$ (of which there are only a finite number). It is found that the sum (30) is more convenient to evaluate than directly taking the trace of (25).

4 The statistics of the Eigenangles

The quadratic long range statistics (e.g. the rigidity or the variance [16]) of the eigenangles of the quantised cat maps may be expressed as transforms of the spectral form factor $K_N(m)$:

$$K_N(m) = \left| \mathrm{Tr}\left[U_N^{(t)}\right]^m \right|^2 . \tag{33}$$

Using (24) the trace of the propagator may be evaluated for a certain family of cat maps to give the following results: if

$$N = \prod_p p^{\alpha_p(N)} \tag{34}$$

then

$$K_N(m) = \prod_p K_{p^{\alpha_p(N)}}(m) , \tag{35}$$

where, for $p \nmid d$

$$\begin{aligned} K_p(m) &= 1 \qquad \text{if} \quad n(p) \nmid m \\ &= p^2 \qquad \text{if} \quad n(p) \mid m \end{aligned} \tag{36}$$

with similar results for primes which divide the discriminant and powers of primes.

It may be seen, from (35), that the quadratic long range statistics of the eigenangles depend upon the prime factorisation of N. This dependence on Planck's constant is not found for other quantum systems (even those with a toral phase space) and so it is natural to investigate the form of the statistics when averaged over a range of values of N.

It may be seen, from (36), that the statistics are related to the period function $n(g)$. Thus the average of the statistics is connected to the distribution of values which the period function can take. Using the results obtained about this distribution an approximation for the averaged form factor may be derived [12]: for a given map

$$\begin{aligned} \langle K_N(m) \rangle_N &\approx \alpha(N)\log(m) + \beta(N), \quad m << N \\ &\approx \gamma(N), \qquad\qquad m >> N . \end{aligned} \tag{37}$$

This is not the form factor of a 'typical' quantum system with a chaotic classical limit,which has been shown to increase *linearly* with m before tending to a constant [17]. Consequently the quadratic long range statistics, when averaged over N, do not take the universal forms found for most quantum systems.

Acknowledgements

I am grateful to M.V.Berry and J.H.Hannay for a number of stimulating discussions. During the course of this work I received financial support from the Science and Engineering Research Council.

References

[1] I.Percival and F.Vivaldi, Physica 25D, 105 (1987)
[2] J.H.Hannay and M.V.Berry, Physica 1D, 267 (1980)
[3] M.V.Berry, Proc.Roy.Soc.Lond. A413, 183 (1987)
[4] V.I.Arnol'd and A.Avez, Ergodic Problems of Classical Mechanics(Benjamin, Reading, MA) (1968)
[5] M.R.Schroeder, Number Theory in Science and Communications (Springer-Verlag) (1984)
[6] M.Kac, Probability and Related Topics in the Physical Sciences (Interscience, London) (1959)
[7] P.D.T.A. Elliott, Probablistic Number Theory Vols. 1 and 2 (New York : Springer-Verlag) (1979)
[8] C.F.Gauss, Disquisitiones Arithmeticae (Fleisher-Liepzig) (1801)
[9] A.Baker, A Concise Introduction to the Theory of Numbers (C.U.P.) (1984)
[10] W.Scharlau and H.Opolka, From Fermat to Minkowski : Lectures on the Theory of Numbers and its Historical Development(New York : Springer-Verlag) (1985)
[11] M.Ram Murty, The Math.Intell 10, No 4, 59 (1988)
[12] J.P.Keating, Ph.D. Thesis, University of Bristol (1989).This work will be published in more detail elsewhere.
[13] M.Bartuccelli and F.Vivaldi, Preprint QMC DYN 88-3, (1988)
[14] M.C.Gutzwiller, J.Math.Phys. 12, 343 (1971)
[15] R.Balian and C.Bloch, Ann.Phys.(NY) 69, 76 (1972)
[16] F.J.Dyson and M.L.Mehta, J.Math.Phys, 4, 701 (1963)
[17] M.V.Berry, Proc.Roy.Soc.Lond. A400, 229 (1985)

Index of Contributors

MIX
Papier aus verantwortungsvollen Quellen
Paper from responsible sources
FSC® C105338

If you have any concerns about our products,
you can contact us on
ProductSafety@springernature.com

In case Publisher is established outside the EU,
the EU authorized representative is:
Springer Nature Customer Service Center GmbH
Europaplatz 3, 69115 Heidelberg, Germany

Printed by Libri Plureos GmbH
in Hamburg, Germany